# MANAGING HUMAN CAPITAL IN TODAY'S GLOBALIZATION

## A Management Information System Perspective

# MANAGING HUMAN CAPITAL IN TODAY'S GLOBALIZATION

## A Management Information System Perspective

Heru Susanto, PhD
Fang-Yie Leu, PhD
Chin Kang Chen, PhD
Fadzliwati Mohiddin, PhD

APPLE
ACADEMIC
PRESS

Apple Academic Press Inc.
3333 Mistwell Crescent
Oakville, ON L6L 0A2, Canada

Apple Academic Press Inc.
1265 Goldenrod Circle NE
Palm Bay, Florida 32905, USA

*Exclusive worldwide distribution by CRC Press, a member of Taylor & Francis Group*
No claim to original U.S. Government works

International Standard Book Number-13: 978-1-77188-738-0 (Hardcover)
International Standard Book Number-13: 978-0-42945-789-0 (eBook)

### Library and Archives Canada Cataloguing in Publication

Title: Managing human capital in today's globalization : a management information system perspective /
Heru Susanto, PhD, Leu Fang-Yie, PhD, Chin Kang Chen, PhD., Fadzliwati Mohiddin, PhD

Names: Susanto, Heru, 1965- author. | Fang-Yie, Leu, author. | Chen, Chin Kang, author. |
Mohiddin, Fadzliwati, author.

Description: Includes bibliographical references and index.

Identifiers: Canadiana (print) 20190072407 | Canadiana (ebook) 2019007244X | ISBN 9781771887380
(hardcover) | ISBN 9780429457890 (eBook)

Subjects: LCSH: Personnel management—Technological innovations. | LCSH: Management information
systems.

Classification: LCC HF5549.5.T33 S87 2019 | DDC 658.3—dc23

### Library of Congress Cataloging-in-Publication Data

Names: Susanto, Heru, 1965- author. | Fang-Yie, Leu, author. | Chen, Chin Kang, author. |
Mohiddin, Fadzliwati, author.

Title: Managing human capital in today's globalization : a management information system perspective /
Heru Susanto, PhD, Leu Fang-Yie, PhD, Chin Kang Chen, PhD., Fadzliwati Mohiddin, PhD

Description: Toronto ; New Jersey : Apple Academic Press, 2019. | Includes bibliographical references and
index.

Identifiers: LCCN 2019007156 (print) | LCCN 2019008748 (ebook) | ISBN 9780429457890 (ebook) |
ISBN 9781771887380 (hardcover : alk. paper)

Subjects: LCSH: Personnel management--Technological innovations. | Management information systems.

Classification: LCC HF5549.5.T33 (ebook) | LCC HF5549.5.T33 S87 2019 (print) | DDC 658.3--dc23

LC record available at https://lccn.loc.gov/2019007156

Apple Academic Press also publishes its books in a variety of electronic formats. Some content that appears in print may not be available in electronic format. For information about Apple Academic Press products, visit our website at **www.appleacademicpress.com** and the CRC Press website at **www.crcpress.com**

# ABOUT THE AUTHORS

**Heru Susanto, PhD**, is a researcher of the Data Security Group at the Indonesian Institute of Sciences. He is also an honorary professor at the Department of Information Management, College of Management, Tunghai University, Taichung, Taiwan. Also, Dr. Heru worked as an assistant professor in Technology Management, School of Business, University Technology of Brunei. Dr. Heru has experience as an IT professional and as web division head at IT Strategic Management at Indomobil Group Corporation. He has worked as the Prince Muqrin Chair for Information Security Technologies at King Saud University, Riyadh, Saudi Arabia. His research interests are in the areas of information system, computer security, information security management system, and information system and technology as an enabler of organizations' business process re-engineering and e-marketing. He has over 15 years of professional teaching and research experience at the university level and over 20 years of working on research for various studies in the field, with more than 27 books, seven full authored books, 20 book chapters, and more than 35 international publication in peer-review journals included in: Scopus, Science Direct, Springer, ISI, and Global Indexed.

**Fang-Yie Leu, PhD**, is currently a Professor in the Computer Science Department at Tunghai University, Taiwan. He is on the editorial boards of several journals. Professor Leu organizes international workshops on mobile commerce, cloud computing, network and communication security (MCNCS), and compact wireless extravehicular activity (EVA) communications system (CWECS). He is an IEEE member and now serves as a technical program committee (TPC) member of many international conferences. He was also a visiting scholar at the University of Pittsburgh, Pennsylvania. His research interests include wireless communication, network security, grid applications, and security. Dr. Leu received his bachelor's, master's, and PhD degrees all from the National Taiwan University of Science and Technology, Taiwan.

**Chin Kang Chen, PhD**, is a Lecturer at the Universiti Brunei Darussalam, Begawan, Brunei. He also has industry experience from working on projects with local telecommunications companies and government ministries and agencies. His research interests include information security management system, IT governance, mobile devices, cloud computing, and e-government. Dr. Chen received his bachelor's of engineering from the Royal Melbourne Institute of Technology, Australia; his master's in digital communication from Monash University, Melbourne, Victoria; his master's in information technology from Queensland University of Technology, Australia; and his PhD from the University of Brunei.

**Fadzliwati Mohiddin, PhD**, is a program leader and principal lecturer at the School of Business, Universiti Teknologi Brunei (UTB). Prior to joining UTB, she was the Chief Information Officer, the Director of ICTC, and the Dean of the Faculty of Business and Management Sciences, Sultan Sharif Ali Islamic University (UNISSA). She also held the post of Deputy Dean at the Faculty of Business, Economics and Policy Studies, Universiti Brunei Darussalam from 2009 to 2010. She lectures in business information systems and general management. She was involved with several ICT projects that include knowledge management systems and e-learning systems for the Ministry of Education. In addition, she has been appointed as a judge for several business and ICT competitions, such as the Asia Pacific ICT Award (APICTA) and the Brunei ICT Award (BICTA) since 2010. She holds a BA, Management Studies (Universiti Brunei Darussalam); an MBA (Lancaster University, UK); and a PhD, Information Systems (Curtin University of Technology, Western Australia). Her current research interest includes information systems success, knowledge management, e-government, and general management.

# CONTENTS

# ABBREVIATIONS

| | |
|---|---|
| 3D | three-dimensional |
| AI | artificial intelligence |
| APTK | Agensi Pekerjaan Tempatan Dan Pembangunan Tenaga Kerja |
| BPR | business process reengineering |
| BSC | balanced scorecard framework |
| CAD | computer-aided design |
| CAP | Computer Accommodations Program |
| CI | continuous improvement |
| CIP | continuous improvement program |
| CRM | customer relationship management |
| CSI | Civil Service Institute |
| CSLA | Student Learning Collaboration Analysis |
| DGH | District General Hospital |
| DSS | decision support system |
| e-compensation | electronic compensation |
| e-education | electronic education |
| e-HRM | electronic human resource management |
| e-learning | electronic learning |
| e-library | electronic library |
| e-mail | electronic mail |
| e-performance management | electronic performance management |
| e-PM | electronic performance management |
| e-recruitment | electronic recruitment |
| e-training | electronic training |
| EDI | electronic data interchange |
| EDP | electronic data processing |
| EEO | equal employment opportunity |
| EFT | electronic fund transfer |
| EGNC | E-Government National Centre |
| EHIS | electronic hospital information system |
| EHS | Environmental, Health and Safety |
| EPA | Environmental Protection Agency |
| EPPs | Educator Preparation Programs |

| | |
|---|---|
| ERP | enterprise resource planning |
| ES | expert systems |
| FAO | Food and Agriculture Organization |
| FMCG | fast moving consumer goods |
| H&S | health and safety |
| HCM | human capital management |
| HIS | health information system |
| HMIS | health management information system |
| HR | human resource |
| HRD | human resource department |
| HRIS | human resource information system |
| HRM | human resource management |
| HRMIS | human resource management information system |
| HRMS | human resources management system |
| i-CRM | interactive CRM |
| IBM | International Business Machine |
| ICT | information and communication technologies |
| IS | information system |
| ISO | International Organization for Standardization |
| ISP | internet service providers |
| ISs | information systems |
| IT | information technology |
| J&J | Johnson & Johnson |
| JIT | just in time |
| KMS | knowledge management system |
| KPI | key performance indicator |
| MAS | Malaysia Airlines |
| MIS | management information system |
| NHS | National Health Service |
| PC | personal computer |
| PDCA | Plan, Do, Check, and Act |
| PID | patient identification |
| PM | Panasonic Malaysia |
| POS | purpose of offer |
| PSC | Public Service Commission |
| RFID | radio frequency identification |
| SEM | structural equation modeling |

| | |
|---|---|
| SMEs | small and medium enterprises |
| SRM | student relationship management |
| TAFIS | Treasury Accounting and Financial Information Systems |
| TMC | Toyota Motor Corporation |
| TPS | transaction processing system |
| TQM | total quality management |
| TQMP | TQM program |
| USB | Universal Serial Bus |
| VOIP | voice over internet protocol |
| VPN | virtual private network |
| WLAN | wireless local area network |
| WWW | World Wide Web |

# PREFACE

Human resource (HR) management has broad roles and responsibilities in organizations, such as payroll, administration, and guiding, motivating, and mentoring employees. Manual HR systems may not be adequate in today's work environment. Systems such as management information systems (MIS) and human resource information systems (HRIS) provide support in an organization's daily activity.

This book explores human capital issues in human resource management (HRM), which is related to the redesigning of MIS. It highlights that when MIS is implemented or changed, the responsibilities of organizational members can be drastically altered. Employee resistance is one of the main challenges of this process. To solve this problem, several behavioral strategies are required, such as encouraging employees' involvement in adopting effective MIS to help them overcome resistance during organization changes.

The volume includes a literature review that focuses on this discussion. This book also highlights the changing skills requirements of the employees in the context of both MIS and HRM perspectives. It describes how the current trends have evolved in this fast-emerging market of competitive advantages and rapidly changing environments toward globalization. The literature review starts with an introduction of what HRM and MIS are all about and the linkage between them, followed by examples of some useful journals, articles, and websites to support the topic "Changing Skills Requirement: An MIS Perspective." Considering employees as assets of an organization who contribute effort, time, energy, and knowledge, the organization must accept the responsibility of sustaining efficiency through operation, execution, implementation, and the acceptance of changes, which involvesleaders and managers, as well as employees. Consequently, implementing and designing the MIS is paramount to bringing both benefits and challenges related to the financial costs, the quality, and the performance of the organization. Human resource information systems (HRIS) have aided many organizations by replacing traditional HR functions. Files, documents, and papers kept under HR are now replaced with electronic documents under a system that saves physical space and mitigates potential damages or losses to these files if the system is harmed. HRIS files are easily backed up and can be stored away from the original system. The challenges with HRISs are with the security of keeping the data and information within the system. Threats from hackers, information leakages, and natural incidents such as fire could disruptthe system and would jeopardize the data and information of the organization. There needs to be emphasis on security of HRISs to protect important data and information and prevent threats that could affect the system.

Finally, within the book, we have provided several case studies that have been used to strengthen our findings, and hence, we conclude that changing skills requirements in the MIS perspective are really important for improving the performance of employees in order tolead to greater efficiency and productivity of the organization.

# CHAPTER 1

# WORKFORCE PLANNING AND MANAGEMENT SYSTEMS: A HRIS APPROACH

## ABSTRACT

The advancement of technology has made a huge change in most of the professions and in all functions of management. Human Resource Management (HRM) is one of the areas impacted. It has been increasingly affected by the constantly developing new trends and has introduced e-HRM. The e-HRM is one of the information technology systems that includes the Web, and is designed for HR professionals and top management. It supports to manage the workforce, monitor changes, gather information needed in decision-making, and at the same time, the employees are able to contribute in the process and keep track of significant information. e-HRM enables easy interactions within employee and employers, where they contact these functions using the intranet or other web-technology networks. The implementation of e-HRM will consistently provide greater benefits toward the organization. The service and information quality impacts satisfaction toward the employees. e-HRM offers advantage of reducing the costs unlike the traditional HRM systems before. It can also lead to improved levels of satisfaction, motivations, and increased employees' performances. This eventually improves the ability of organization to compete more effectively and efficiently. Fully developed e-HRM can provide analysis capabilities and decision support systems for HR to pay, promote, assign, terminate, and reward the employees. It acts as a gathering data tool. It also can support the organization's goals.

## 1.1 INTRODUCTION

The advancement of technology has impacted many organizations nowadays. In a growing business organization that has expanded globally, its goals and objectives need to equally be achieved and communicated to subordinates. Decentralization process has been developed to assess this.

Human resources management (HRM) plays an important role in order to prepare the impacts of the changing technology as well as to sustain and achieve the organizational goals. They are required to make the employees well informed and make them understand the workflows of the organizations. Therefore, telecommuting capabilities have eased the organizations' flows. The implementation of management information systems (MIS) in human resources (HR) benefits the organizations as well as the employees.

### 1.1.1   RESEARCH AIMS

This chapter will be assessing and focusing on the use of MIS in HRM, specifically in decentralized work sites.

### 1.1.2   RESEARCH OBJECTIVES

There are three objectives of this chapter:

1.  To identify the importance of decentralization in the work sites and its impacts toward the employees and the organizations.
2.  To study the opportunities and challenges of decentralized work sites.
3.  To identify the types of MIS that were used in the decentralization organization and how it operates efficiently.

### 1.1.3   DECENTRALIZATION IN WORK SITES

Decentralization is the process by which the workforces in the organization are delegated toward the employees. The employees do not work together in the same office but collaborate in functional area.

#### 1.1.3.1   THE IMPORTANCE OF DECENTRALIZATION

Decentralization brings out work productivity of the employees as well as increases their quality of lives. Extension in delegations of works could make the employees authority and responsibility more on their assigning work scopes. This also gives lower level employees a chance for decision-making and individualization. Employees tend to be more motivated and creative.

The degree of delegations depends on the highest authority. It is also determined by the authority given. Meanwhile, for the organizations, divide the workloads according to the employees' work posts and skills.

### 1.1.3.2   COMPARISON BETWEEN BEFORE AND AFTER DECENTRALIZATION

Previously, decision-making was limited to a certain level only. It remained at the center. Transmitting orders and works were limited. This is crucial as the varieties of works are strictly limited and constrained.

As the world is rapidly changing with the advancement of technology, there are more opportunities and diversified workforces. Decentralizations make the workflows more diversified and qualitative as employees are not being pressured.

### 1.1.4   MANAGEMENT INFORMATION SYSTEM

The role of MIS is processing the data that are collected from different sources (input) and transmitting it to all the needy destinations as an output. The information collected is important for the operations throughout the organizations. It impacts the organization's performance and productivity. The good functions of MIS will support the management to be more efficient.

### 1.1.4.1   THE IMPORTANCE OF MIS

There are multiple functional systems running in the organizations. The system includes sales systems, call center systems, financial systems, inventory systems, and logistic systems. The purpose of MIS is combining all the information gained from multiple systems in which it contributes to better understanding of the HRM. Not only that, MIS helps HRM manager to take appropriate actions that can meet the needs of their customers. MIS also help in strategic planning, management control, operational control, and transaction processing. In addition, MIS plays an important role in information generation, communication, problem identifications, and decision-making processes.

## 1.1.4.2   THE FUNCTIONS OF MIS

The main function of MIS is to lead the manager to take right action and right decision-making. It can be types of questions and answers that are directly related to tactical or strategic goals of organizations. It is also used to review the past actions and make immediate changes if there is an increase in sales and meeting goals.

## 1.1.5   TELECOMMUTING

### 1.1.5.1   THE CAPABILITIES OF TELECOMMUTING

Telecommuting refers to the work styles in which the employees do their works outside the companies or work sites. They commonly work from home and communicate over the internet or phone call. It involves flexibility of work arrangement, for example, flexible schedule. Telecommuting has helped employees save time, fuel, and often increased their productivity.

For instance, a marketing researcher's job is to collaborate with his colleagues and clients. However, he can do his job by himself alone in order to give his best. This is to prevent him from being distracted. Therefore, telecommuting plays an important role as it helps him to do deep analysis and research, without disruptions from the office. Then he will be well prepared when present during work meeting.

### 1.1.5.2   THE IMPORTANCE OF TELECOMMUTING AND ITS OPPORTUNITIES IN BUSINESSES AND EMPLOYEES

With the rise of internet and affordable bandwidth, telecommuting benefits not only the employees as well businesses organizations.

First, telecommuting enables the employees to avoid long and stressful commutes. In other words, the telecommuters' working hours at home are shorter than working at their workplace. Based on the study on AT&T, there are 75% of telecommuters who commented that they felt greater satisfactions as they have more time to spend on their personal and family lives (Hamilton, 2017). In fact, employers find that this will help them to retain their valuable and highly talented employees.

Furthermore, telecommuting helps increase their employees' effectiveness. This is because the telecommuters feel lower distractions than working at hectic environment and they spend less time in meetings. Not only that, employees tend to think that their morale is becoming much more better because they tend to take less sick leaves and it lowers turnover rates of employers.

Moreover, the importance of telecommuting in the business organization is to help the firm reduce costs in facility cost or organization's overhead. This is based on the JALA International firm, whereby the company estimated that they have managed to save $11,000 per year for every employee who telecommutes 2 days a week (Hamilton, 2017). As the number of employees working at home or remote areas increased, there is a possibility for the organization to spend less investment and capital expenditure in office building, parking lots, and other physical capital. In addition, the evidence taken from anecdotes stated that the telecommuting allowed many roles to be done productively. This is reflected to the positions that have greater involvement in reading, writing, and other working activities that required high-focused concentration (Hamilton, 2017).

## 1.2   LITERATURE REVIEW

The continuous change in communication technology has changed the ways of people's works in all businesses functions including HRM. It is important for HRM to be more comprehensive, contribute better quality, and be fast and flexible in line with upcoming trends as it is one of the factors that can contribute to the success of businesses (Dorel and Bradic-Martinovic, n.d.). Effective HRM has allowed employees to provide effectively and productively in order to accomplish the overall organization direction and achieve the organization's goals and objectives (Manna Akter Lina, 2016). The greater challenge is to help the organization improve its effectiveness and efficiency in an ethical and socially responsible way. Thus, use of MIS has become essential for HRM. One of the current trends that HRMs are practicing has diversified their employees in the decentralized work sites (Aluvala, 2013).

Manna Akter Lina (2016) stated that the decentralized work sites are usually established in organizations and that knowing the telecommunication arrangement allowed them to find and having qualified employees without have to place business facilities. He has quoted (Khan and Taher, 2011) that the HRM decentralized work sites will require training not only for the employees but also for the managers in managing and controlling work, and establishing pay system that reflects their work arrangements.

Meanwhile, Dorel and Bradic-Martinovic (n.d.) stated that the systems that are commonly used in HRM to link between employers and employees are as follows: electronic human resource management (e-HRM), human resource information systems (HRIS), and human resource management systems (HRMS). Practically, it is importance to understand the differences between e-HRM and HRIS. The term of HRIS is to do with direct implementation in HR department and the users of this system are the employees within the department. HRIS has become one of the main modern tools in HR (Navytha, 2013). According to Hedrickson (2003), "Human Resources Information System (HRIS) can be briefly defined as integrated systems used to gather, store and analyze information regarding an organization's human resources."

Meanwhile, according to Tannenbaum (1990), "HRIS, one which is used to acquire, store, manipulate, analyze, retrieve and distribute information about an organization's human resources." On the other hand, e-HRM provides services which are available over internet or intranet and the users are not limited to HR department but also cover wider ranges of employees, potential employees, and management (Dorel and Bradic-Martinovic, n.d.). The output subsystem of the HRIS handles vital aspects of the HR management which includes workforce planning, recruiting, workforce management, compensating the employees, employee benefits, and preparing HR reports. This is how the output subsystems are determined.

Aluvala (2013) stated that the existence of telecommuting services has made many employees to be located outside of their workplace. He agreed that decentralized work sites provide better options for employees who are in need for diversified workforce, especially people with disabilities as these enabled the employees to do their work while they are at home. He added that there is a potential for employers not to rethink whether to locate their businesses close to their workforce. On the other hand, Haris (2013) stated that the diversified workforce would actually affect the balance between people's personal life and work. This is because employees tend to confuse between their work and life. There are possibilities for the organization to put employees working in longer hours.

However, Haris (2013) addressed on the importance of having knowledge workers, a group of people who are designated in job of acquisition and application of information. He also emphasizing the importance on technology as it supports organization in terms of productivity, creating and maintaining the competitive advantage, as well as it provides better and useful information. Apart from the technology's impact on decentralized work sites, it is the technology that has made the HRMs undergo staffing their employees, training and development, motivating knowledge workers, paying employees market value, and communication.

Meanwhile, Duggan (2017) mentioned about the employee development on decentralized system. He stated that implementing an HR system in decentralized system is more cost effective. This is because the decentralized system allowed HRM to conduct only the training courses rather than giving comprehensive course catalog or other employee development programs that it requires. He added that the main role of HR in the organizational structure is to satisfy the specific needs of people. By diversifying the employees' work sites including system, it will be easier and faster to check their employees' performance gaps when HRM is more familiar with works of its business functional unit. However, Duggan (2017) stated as decentralization is focused onto its specialization, it may lack attention on general professional skills such as presenting, speaking, and so forth.

Decentralized work sites also offer opportunities that meet the need of the diversified workforce. Diversity can be defined as acknowledging, understanding, accepting, and valuing differences among people with respect to age, class, race, ethnicity, gender, disabilities, etc. (Esty et al. 1995). Organizations need to take diversity positively and believe that it has a great potential to yield greater work productivity and competitive advantages (SHRM, 1995).

Apart from addressed benefits of decentralized work sites, there will be challenges that HRM team may face and one of the challenges is how to establish and maintain the quality of work life and on-time completion in diversified work site. Manna Akter Lina (2016) stated that the quality of work life actually reflects the overall organizational life of the HR. There are eight criteria provided by Richard E. Walton which are used in identifying the quality of work life in every organization. These include: adequate and fair compensation, safe and healthy working conditions, chances to use and develop human capacities, chances for continued growth and security, social integration in the work organization, constitutionalism, balanced role of work, and socially beneficial and responsible work. As decentralized work sites eliminate the traditional ways of work life, managers are required to find new ways to motivate their employees such as greater involvement of employees discretion to make decision that affect them (DeCenzo and Robbins, 2017). In addition, Aluvala (2013) stated that it is the responsibility of organization to ensure their full-time employees receive health and safety since they are working decentralized.

Safety is the major consideration for HRM. This is because it could affect the organization if the HRM failed to ensure that the workplace meets the requirement setting of federal and union standard. Worker protection issues include but not limit to protection of private employee information, use of "no fragrance" zone, and so forth.

Another challenge addressed is terms of compensation policy. It is difficult to search and retain the experienced employees. DeCenzo and Robbins (2017) stated that many organizations attempted to implement the extensive incentives and benefits which are rarely seen particularly by nonmanagerial employees. The lists of attractive incentives included signing bonuses, stock options, free health club membership, and subsidies for mobile phones. However, with the existing of technology, the compensations are now becoming more transparent. For instance, Glassdoor, which is an online source that helps the applicants compare salaries of employees in any organizations.

On the other hand, HRMs also have to consider the external factors that can affect the organization. In other words, it is the outside forces whereby the organization has no direct control over it. These external factors may positively or negatively impact the employee; these external factors include expectations of employee, changing demographics of the workforce, changes to the law of employment, greater highly educated workforce, and so forth.

## 1.3   RESULT AND DISCUSSION

From a business perspective, information system can be defined as an information technology oriented solution to organizational and management challenges (Melissa, 2010). The advancement systems can support the decision-making and implement the strategies.

In a growing number of people in the organizations, MIS system can eventually increase the effectiveness of managerial skills. The HRM manager can control the subordinates through telecommuting without any delays by tracking the employees' databases. In a situation of employees' absences at work, it can be checked whether the employees have logged into their computer or not. HRM can track it from the databases regarding these matters.

The implementation of MIS system can ease the operations in the organization. Moreover, when the organizations are situated in different geographical areas, the top managers will key in the data into the system. Then the system will collect, transmits, processes, and store the data. This information will be readily viewed by subordinates. This differs from the situation where if the organization is small, a memorandum will be used to pass down the news.

The vast amount of information stored through MIS system will ease the decentralizations without need to transfer the data by hands in addition to cost and time effectiveness without any hassles. However, the effectiveness of MIS system depends on the managerial skills. The managers need to know how to manage it and exploiting it according to the capabilities of the technology.

### 1.3.1 HUMAN RESOURCE INFORMATION SYSTEM

HRIS is a part of MIS that provides the information regarding the decentralized workforce in the organizations and helps HR in decision-making. HRIS is a system used to collect, record, and analyze data concerning an organization's HR. An effective HRIS is when it is able to provide detailed information on an organization's employees and is beneficial for helping HR in decision-making.

A well-developed HRIS helps the organization to store employees' information more accurately and securely. To be compared with in the past, organizations only recorded their data on papers and spreadsheets. However, with advancements in technology today, many organizations have realized the need to implement organized computerized systems such as HRIS. With the implementation of HRIS, companies are now able to update records regularly and increase its accuracy. This allows the organizations to project the growth of their companies as HRIS will also increase the efficiency of HR's decision-making in terms of its quality and increase the productivity of the employers and their employees.

#### 1.3.1.1 IMPORTANCE OF HRIS IN ORGANIZATIONS

The HRIS provides information to managers throughout the firm concerning the firm's HR. The input subsystems of the HRIS are the transaction processing system which provides input data, the HR research subsystem which conducts special studies, and HR intelligence subsystem which gathers environmental data that bear on HR issues. Meanwhile, the output subsystem of the HRIS handles vital aspects of the HR management which includes workforce planning, recruiting, workforce management, compensating the employees, employees' benefits, and preparing HR reports. This is how the output subsystems are determined.

### 1.3.1.2   FUNCTIONS OF HRIS IN HRM

An HRIS is effective when it is able to provide detailed data on an organization's employees such as the administration of all staff and rewards management. Figure 1.1 shows the model of a HRIS system. There are different types of HRIS output subsystems that are handling main aspects of the HR management which include:

#### *Workforce planning and management subsystem*

It is an information system that supports workforce planning through delegating the workforce accordingly such as into new office locations. The need of information on the quantity of the available workforce and their specialty is essential for the organization to plan their workforce.

#### *Recruiting subsystem*

The organization needs to design a proper recruiting plan in order to fill vacant positions at an organization and the requirement skills for interested applicants. Having a proper recruiting information system is essential to ease the recruitment processes for both the organizations and the employees.

#### *Compensation and employees benefits subsystem*

This system involves determining the employees' wages, salaries, and benefits which include the insurance benefits, retirement benefits, and medical payments. This information will be shared with the higher levels of management and it can be used to compare it with the budget plan of the organizations.

#### *Environmental reporting subsystem*

This system is used to prepare reports of the organization to be given to the government or labor unions regarding the employees.

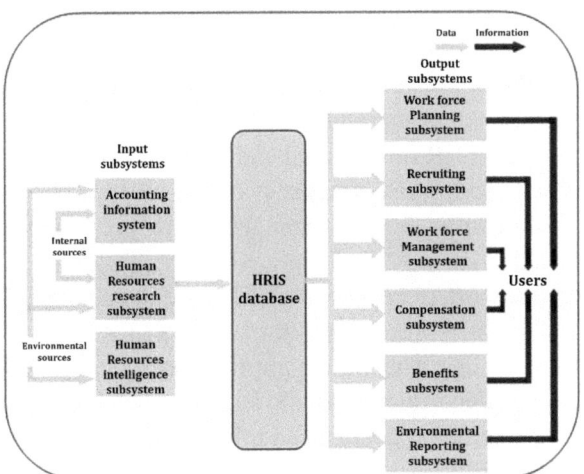

**FIGURE 1.1**    A model of a human resources information system (HRIS).

## 1.3.2    *ELECTRONIC HUMAN RESOURCES MANAGEMENT*

The advancement of technology has made a huge change in most of professions and in all functions of management. HR has been increasingly affected by the constantly developing new trends and has introduced e-HRM. The e-HRM is one of the information technology systems that includes the Web, and is designed for HR professionals and top management. It supports to manage the workforce, monitor changes, gather information needed in decision-making, and at the same time, the employees are able to contribute in the process and keep track of significant information. e-HRM enables easy interactions within employee and employers, where they contact these functions using the intranet or other web-technology networks.

The implementation of e-HRM will consistently provide greater benefits toward the organization. The service and information quality impacts satisfaction toward the employees.

e-HRM offers advantage of reducing the costs unlike the traditional HRM systems before. It can also lead to improved levels of satisfaction, motivations, and increased employees' performances. This eventually improves the ability of organization to compete more effectively and efficiently.

Fully developed e-HRM can provide analysis capabilities and decision support systems for HR to pay, promote, assign, terminate, and reward the employees. It acts as a gathering data tool. It also can support the organization's goals.

### 1.3.2.1   IMPORTANCE OF E-HRM IN DECENTRALIZED WORK SITE

e-HRM enables the organization's managers and employees to communicate on a geographical level to share information and create virtual teams. It also allows them to access HR information and increases connectivity of all parts of the department and organizations. It offers to improve services to HR activities for both employees and management by providing more transparency in the system.

e-HRM is an important support system in the management of HR and all other basic and support processes within the organization as well as in the decentralized work sites. The system improves efficiency and cost effectiveness within the HR department and allows HR to become a strategic partner in achieving organizational goals; in addition it creates more dynamic workflows in the decentralized work site process, productivity, and employee satisfactions. Furthermore, the system has standardized procedures that ensure the organization remains submissive with HR requirements, also ensuring more precise decision-making.

### 1.3.2.2   FUNCTIONS OF E-HRM IN DECENTRALIZED WORK SITES

With the achievement in implementation and outsourcing operational functions of e-HRM in an organization, the burden of the transactional HR activities has been released. The changes of HR services delivered by e-HRM can be illustrated through the functions presented such as electronic recruitment (e-recruitment), electronic training (e-training), electronic performance management (e-performance management), and electronic compensation (e-compensation).

### e-Recruitment

e-Recruitment is one of the most common applications of e-HRM, where in the traditional recruitment process it consists of several stages that are needed to be walked through by the applicants. The processes are such as identification of hiring needs, submission of job requisition and approval, job posting, submission of job applications, screening of resume, interviewing, pre-employment screening, and job offer and employment contract. This is a very time consuming process for the HR department.

Since the advancement of internet, e-HRM has taken the opportunity of having the idea of easily accessing it at anytime and anyplace. The benefits of e-recruitment are: it is time and cost saving, quick and flexible responses, has wide range of applicants, worldwide accessibility, efficiency, and is convenient for both recruiters and job seekers.

## e-Training

Another beneficial use of e-HRM system is training and development of the employees. e-training focuses on learning activities and supports through the information and communication technologies. It can be easily done and is accessible through the organization's intranet or web, aided by varieties of multimedia such as audio and video conferencing, and links to resources that offer more solutions of remote learning which relates to employees development.

There are several advantages of e-training as follows: firstly, the flexibility of how it allows the employees to learn through the systems at the chosen time and place of the employees themselves. Secondly, this change allows the organization to have cost reductions which includes the administrative costs, travel expenses, opportunity costs, and instructional cost. Lastly, this application can help understand and study the employees' needs in order to enhance their knowledge and improve their abilities by self-study to get promoted in their career level.

## e-Performance management

To track an employee's performance management may be a hassle and a difficult responsibility of the managers and top management of an organization. As performance management of an employee relates to the goal setting, performance planning, performance tracking, employee appraisal, evaluation, and feedback process, it sounds easier to imagine and talk about but causes difficulties to be implemented. With e-performance management application, the difficulty of the process can be minimized with the supporting tools available.

By having a web-based system, it has become beneficial for the HR professionals, line managers, and employees as the system increases the efficiency and consistency of the whole process. The supporting tools are firstly the specific tracking software that helps monitor and track an employee's job routines. Secondly, performance appraisal and feedback are made in computerized forms to save time and simplify the process.

Thirdly, the employees are able to track their own performance progress in the organization from a real-time performance data and receive fast and adequate evaluations from managers and top managements. Fourth, managers have opportunity to compose better quality appraisals as they need to frequently communicate with the employees regarding their performances, which helps in adjusting the employees' goals and performance to achieve organization's objectives. These activities are readily available in the internal network systems and are accessible to all employees, managers, and top management.

### e-Compensation

One of the strongest attractions and a way to retain competent employees in any organization is its compensation policy, especially in the fierce war for their talent. An excellent compensation system will very much help the organization in achieving its objectives by recruiting and possessing employees' loyalty, commitment, and performances. The compensation and benefits of employees can be sorted out and managed properly and with considerably less effort with the help of e-HRM system.

As it is a web-based tool, it somehow helps and enables the organization to list the rewarding tasks from gathering, storing, manipulating to analyzing, utilizing, and distributing compensation data in the organization. The benefits of e-compensations are as follows: firstly, the compensation information is available and accessible online despite the time and whereabouts of the employee as it can be accessed anytime and anywhere. Secondly, it helps tailor rewards and compensation as per individual employees' needs, as they can make their own choices and decisions regarding the choices of benefits or rewards to suit their compensation packages. This clearly shows that e-HRM enables employees to electronically select their preferred compensation plan, which hence reduced the HR's administrative workload.

There are various fields of e-HRM. These include e-training and development, e-performance management, and e-recruitment.

### e-Training and development

Nowadays, there are many companies which provide online training. Growing a business from different locations will let the employees learn new skills by staying at their work places. The participants will be able to complete the course whenever they have access to the computer and internet—via virtual classrooms and digital collaborations through e-learning. The employees will be equipped with knowledge of management strategies and other competencies. The training material will be available widely.

### 1.3.3   OTHER MIS SYSTEMS IN DECENTRALIZED WORK SITES

There are many examples of MIS system that are used by HR:

#### 1.  Enterprise resource planning

Enterprise resource planning (ERP) is a business process management system that allows an organization to use the system to manage the business and automate the back-office functions. HR stores employees' databases including their contact information, performance evaluation, and promotion of all employees. Most common subsystems under ERP are payroll system, time management, and self-development.

#### a.  HR payroll system

It is a software designed to organize attendance management for salary calculation. It keeps track of hours, delivering wages, checks, etc. The HR needs to key in employees wages information and hours.

#### b.  Time management

HR will keep track of how many hours that employee has worked, As well as the absence and leave management requested by the employees are stored into the system.

#### c.  Self-development

Training will be conducted online. This will reduce the time intensive in preparing the places and save every staff's time.

### 1.3.4   BENEFITS OF DECENTRALIZATION TO THE ORGANIZATION

#### 1.3.4.1   FLEXIBILITY

A growing business with expansions of operations can benefit from decentralization. When a new business opens up in different geographical area, it will be working independently. This means the workflow might be different from the main office as new rules will be managed by new manager or new employees. Flexibility takes place according to the specific needs of the areas.

It also benefits when in  case of emergency. A situation when the employees are unable to attend work, employees from other office can replace their work for awhile. The organization will not have a problem of nonsettled work or urgent documents.

## 1.3.4.2   DIVERSIFIED WORKFORCE

Diversified workforce is important for an organization's success as this workforce brings high values to the organization and they are able to share their ideas and creativity in finishing the tasks given to them. The differences will benefit the workplace by creating a competitive working environment and increasing the work productivity. Most working places, nowadays, are made up of diverse workforce; thus it is essential for the organizations to quickly learn and adapt to the differences in cultures.

It is important for the managers to acquire certain skills in order to create a successful working place with a diversified workforce. Firstly, managers need to understand discrimination and the consequences that it will bring to the organization and staff. Secondly, managers need to recognize the cultural biases and prejudices (Koonce, 2001). Diversity is not necessarily about differences among groups only, but rather about differences among the individuals. Each individual is unique and does not represent or speak for a particular group. Lastly, managers must be willing to change the organization if necessary (Koonce, 2001). Organizations need to learn how to manage diversity in the workplace in order to be successful in the future (Flagg, 2002). In a diverse workplace, employees are more likely to remain loyal to the organization when they feel respected and valued for their contributions at workplace.

## 1.3.4.3   EFFICIENT DECISION-MAKING

Some decision-making in a centralized and big organization can be time consuming, as the end decisions are needed to be approved or made by the most top management. This causes slow operations or more problems in dealing with customers or clients who expect answers or decisions made quickly. In a decentralized work site, the decisions made are more efficient as the manager can make decisions without having to wait for the top management command, as these fast actions can mean the difference between gaining or losing a customer or client. This clearly shows the engagement of making efficient decision will lead to better operations of the organization that will make a better turnover.

### 1.3.4.4   FEWER PEOPLE

By having fewer people in an organization it shows that an organization needs to encourage its employees to be more creative and motivate them to increase their productivity level as the number of employees is not as many as before. The use of HRIS helps the organization to save time, be more organized, well structured, helps in the immediate processing of the paperwork, and automate financial transactions. As the technology advances, tasks that were usually carried out by employees are now mostly being carried out by the computer systems. Thus, fewer employees are needed as compared with before the implementation of HRIS itself, which mainly needed written records.

### 1.3.4.5   COST EFFECTIVE

In addition, HRIS software may help HRM to orient, train, and develop employees in managing their career. Web-based or cloud-based service also enables the on-demand employees to have training and development. Online training and teleconferencing is also one of the ways that would save the HRM budget as it enables the HR to provide cost-effective training since it does not require large budget of HR department (DeCenzo and Robbins, 2017).

### 1.3.5   THE BENEFITS OF DECENTRALIZATION TO THE EMPLOYEES

### 1.3.5.1   FLEXIBILITY

Every employee wants work to be more flexible. Those people who have opportunities in working flexibility have an option in getting other matters done. Parents are able to continue their works while looking after their children. For those who have problems with transportations can be stress free from commuting. This allows them to spend more time with their families. This eventually increases up the work productivity as they are more happy.

Some people have personal diagnosis that working away from office enables them to give their best. They feel pressured when the managers were there. Some employees who have proven to be well self-managed can be entrusted on work flexibility.

## 1.3.5.2   DIVERSIFIED WORKFORCE

Through decentralization, the diversified workforce is able to work with a variety of backgrounds such as different in gender, race, age, culture, or religion. This diversity leads to the various contributions of workforce benefits to the employees which are as follows: firstly, by having different perspectives. By having employees from different backgrounds and cultures work together, they are able to share and show their different ideas and creativity. This allows them to brainstorm the possible solutions toward the issues or problems given to them. At this point, the employees have a better list of possible solutions and ideas. Secondly, the employees are able to increase their communication and language skills. Thirdly, diverse teams are able to perform better and increase their productivity. Fourth, there is greater opportunity for personal and professional growth. By having diverse colleagues, the employees are exposed to new skills and are able to develop an international network that can take their career to the next level.

## 1.3.5.3   EFFICIENT DECISION-MAKING

The efficient decision-making means that there are strong feelings of association among the employees. This also made them satisfied with their job, as their suggestions and recommendations are being implemented or put to practice. This also made the level of motivation among the employee increase which links to increase in productivity and improved quality.

## 1.3.5.4   FEWER PEOPLE

With decentralization, employees are able to improve their performance due to the equality in the workplace. This will encourage the workers from different backgrounds to feel confident in their ability and make them strive for the best. The chances of getting the best applicants is also high as the organization is accepting applicants of all backgrounds and this shows that they promote equality to their potential employees which is an appealing quality for the organization.

### 1.3.5.5   COST EFFECTIVE

Decentralized systems are more cost effective in providing training courses and education. However, the managers are still required to communicate and collaborate efficiently with their virtual employees whether they are full-time employees or contingent workers. These include understanding their soft skills but also choosing the right technologies to ease the desired interactions. Virtual employees are also required to have their own training. For instance, Dell provide their employees with training courses in a way of managing the career and progression through the company as a virtual employee.

## 1.3.6   THE CHALLENGES OF DECENTRALIZATION TO THE ORGANIZATION

### 1.3.6.1   WORK QUALITY

First, challenges of decentralization to the organization are to ensure and maintain the delivering of work quality and on-time completion. In other words, it is about how well the managers ensure that their employees are suitable and able to perform the particular job. For instance, Aetna assessed their employees based on three criteria: individual, job function, and home environment. Individual is ensuring the employees have the right capabilities and competencies, job function is ensuring if the job can be carried out at home, and home environment is ensuring the employee's home environment meets the standards of security. Meanwhile, Xerox company considered their employees based on their self-assessment (Morgan, 2015). This is because not all employees are able to be located away of their work facilities, for example, employees that worked in manufacturing facility.

### 1.3.6.2   ON-TIME COMPLETION

Supporting and maintaining different types of global workforce urged the employers to make changes in their workplace that enables employees to divide their time between career and family responsibilities. However, since they are outside working environment, there is an issue of ethical dilemma on how the organizations monitor their employees' behavior and ensures their productivity. Employers that are concerned about the employees' productivity observed their employees' computer use in order to prevent them from checking personal email and social media, as well as surfing the website while they are working.

### 1.3.6.3 COMPENSATION POLICY

One of the functions of HR is to look at the compensation and benefits for the employees. Compensation and benefits that are offered by an organization are seen as a key factor of having well-performed employees. A compensation system based on enterprise's success allows the employees to share the organizational success and see how their performance contributes as a whole (Fitzgerald, 1995).

Compensation policy was one of the issues that concerned the HR managers who have the difficulties in monitoring and explaining the compensation systems for those employees who are in decentralized work sites. Which may lead to time consuming, ineffective processes, and inflexibility. With the easily accessible e-HRM, and in order to survive in the decentralization world and with competitive market, the organizations need to enhance the designing and administering compensation programs in the e-compensation which helps achieve such goals with fewer efforts.

In the system itself, the HR managers and top management are able to review and monitor the employees in the decentralized work site's job routines and productivity. This helps them to decide whether the employees have fulfilled the required performance in order to entitle certain salary that is based upon their agreeable terms. The employees are also able to request through the system their desired day off from work, as well as to inform of any medical leaves. Through this, e-HRM systems are able to ease the HR activities in controlling the work routines of the employees in the decentralized work sites as this can minimize the problems that arise among the employees here.

### 1.3.6.4 HEALTH AND SAFETY

It is the organization's responsibility to ensure the health and safety of its decentralized workforce is being taken care of. Employers shall be responsible in providing a safe and healthy workplace to their employees and it is the duty of every employer to consult with their employees on health and safety matters. Furthermore, there are also other concerns in terms of security issues of the HRIS such as the privacy of employees' information. Irresponsible hackers can interfere into the HRIS, identify and obtain the confidential organizational data such as the data on the organizational performance and financial records. The hackers might misuse the data gained and even sell it to the possible competitors of the company itself. As for the employees' personal information, it is important to restrict indiscriminate access to the HRIS data on employees, as irresponsible ones might misuse the personal information of the employees for personal purposes. It is essential for the data to be kept confidential and to be accessed only when needed and only by trusted personnel.

An example of a company that experienced a breach in its security is Ameriprise Financial. In late 2005, a laptop that contained the clients' and the employees' personal information was stolen. Due to that, the company then determined that there was a need to increase the security inserted into the company's laptops as most of its employees bring it home with them after work. Ameriprise ensured all of its employees had the new security suite installed in order to securely keep all the information. This will help in building trust toward the company's workforce especially toward the HR employees who are handling the information.

### 1.3.7  CASE STUDY OF COMPANIES USING MIS SYSTEM

#### 1.3.7.1  LG ELECTRONICS

It is a growing business with more than 82,000 employees across 40 countries, where the HR faced challenges due to its giant scales of business. The challenges are: inefficient decision-making and high costs through manual processes. Therefore, LG implemented ERP system as a solution.

The ERP system included performance management system and staff portal benefits the organization through easy sharing of best practices across the different centers. Staff gains information through the portal and maximizes localized resources for employee learning. Online tutorials resulted in cost savings. This eventually increased employees' productivity.

#### 1.3.7.2  JOHNSON & JOHNSON

According to Johnson & Johnson's (J&J) CEO William Weldon, decentralizing management is a good way to trigger innovation and train corporate leaders. Currently, J&J has 200 companies that produce consumer goods such as shampoo for babies. Weldon said that the challenge of having over 200 operating companies is in need of 200 great leaders and to always develop great leaders who can run the business efficiently.

J&J has carried out a decentralized approach as they believe that decentralization helps in boosting the employees' mind to be more innovative and "it allows different people with different skills, different thoughts, to bring together different products and technologies to satisfy the unmet needs of patients or customers," he also says. J&J also uses "internal ventures" as a way to let employees come up with new ideas. Decentralization also helps with overseas operations such as Japan.

### 1.3.7.3   TOYOTA MOTOR CORPORATION

As for Toyota Motor Corporation (TMC), in order to improve productivity and for the development of its HR, it is essential for them to enhance technology usage for its business to be able to run smoothly. At first, the idea was to implement technology-enabled business process and it was then changed to improve the efficiency of the HR to meet the employees' needs. The estimated cost for the HR ERP system reached US$ 7 million. This includes the upgrade in its HR management system that would contribute to hiring, retention, and development of skilled and motivated employers and team members. This system contains the processes for recruiting, selection, and suitable placement of the employees. The system is very high in cost; however, it gives long-term benefits to the organization itself and also improves the structure of the TMC HR system to be more well structured and organized.

### 1.3.7.4   PANASONIC MALAYSIA SDN. BHD.

Panasonic Malaysia (PM) have signed a multimillion ringgit deal with Ramco Systems Sdn. Bhd. to develop their HRMS to a better quality with the use of digital system, namely, cloud-hosted Ramco human capital management (HCM) suite. "As our Malaysia operations expand, the complexity of managing HR and talent also grow multi-fold. We wanted to be a forerunner in Malaysia in leveraging on technology to build a centralized HR and payroll system," said the managing director of PM during the signing ceremony between both parties. Currently, PM has 21 HR and payroll systems across its 21 corporate entities in Malaysia for nearly 20,000 employees nationwide.

A cloud-hosted Ramco HCM suite will be incharge on the core HR, time and attendance, talent management, recruitment, planning, and payroll. The HR management system developed by India-headquartered Ramco Systems was commenced in April 2017 and the implementation will take 21 months to complete and its main intention is to provide ease toward the total management of HR. Previously, PM had different types of systems. Furthermore, Hiroyuki Imizu who was the deputy managing director of Panasonic Management believed that with the implementation of this cloud solution, Panasonic expects HR personnel's duties to shift from more administrative and bureaucratic tasks to doing more strategic work.

## 1.4  CONCLUSION

In conclusion, the implementation of MIS such as HRIS and e-HRM support system, specifically toward the HRM in decentralized work site, has successfully achieved its benefit operations for the employees in terms of the recruitment, training, performance management, and compensations. The great roles of both systems are as follows: firstly, to ensure that the employees and top management perform their tasks as required, which will be able to guarantee the quality and satisfaction of both. Secondly, the system gives positive effects on raising the organization efficiency and lays great importance toward HR.

Thirdly, the organization is able to achieve a competitive advantage through quickly adapting to the changes of working and processes guided with the advancement and trends of the information technology. Fourth, with this system it helps in managing the HRM reduces the cost, saves more time for the employee and managers, and it helps improve the procedures of processing of HR activity. Lastly, using this system the employees are able to manage private information, to upgrade their records when their positions change or gets promoted, and make various valuations on their hold, seeing HR experts if necessary.

MIS works as a main component of the organization and an excellent MIS will cater and provide important information on the desires and competencies of HR; this information will support the top management in forming the organizational missions and setting goals and objectives in motion. This shows clearly that MIS is not limited to the computer hardware and software applications that comprise the technical part of the system, it also includes the people, policies, procedures, and data required to manage the HR functions and other operations in organization.

## KEYWORDS

- **e-HRM**
- **human resource**
- **workforces**
- **strategic planning**
- **management control**
- **operational control**
- **transaction processing**

## REFERENCES

Aluvala, R. *Millennial Workforce: A Contemplation;* Zenon Academic: India, 2013.

Compare HRIS. Human Resource Information Systems. https://www.comparehris.com/human-resource-information-systems-/ (accessed Aug 23, 2017).

DeCenzo, D. A.; Robbins, S. P. *Fundamentals of Human Resource Management;* United States of America, 2017.

Deshwal, P. Role of E-HRM in Organizational Effectiveness and Sustainability. *Int. J. Appl. Res.* **2015**. http://www.allresearchjournal.com/archives/2015/vol1issue12/PartI/1-12-77.pdf (accessed Aug 25, 2017).

Dorel, D.; Bradic-Martinovic, A. *The Role of Information System in Human Resource Management;* Munich Personal RePEc Archive—MPRA: Munich, Germany, 2011

Duggan, T. *The Role of Human Resource Personnel in a Decentralized System;* 2017. http://yourbusiness.azcentral.com/role-human-resource-personnel-decentralized-system-25500.html (accessed Mar 2019).

Esty, K.; Griffin, R.; Schorr-Hirsh, M. *Workplace Diversity. A Manager's Guide to Solving Problems and Turning Diversity into a Competitive Advantage;* Adams Media Corporation: Avon, MA, 1995.

Fitzgerald, W. Forget the Form in Performance Appraisals. *HR Magazine,* 1995; Vol. 40 (12), pp 36–38.

Galuzka, P. *Johnson & Johnson CEO: Decentralization Works;* 2008. https://www.cbsnews.com/news/johnson-amp-johnson-ceo-decentralization-works/ (accessed Aug 23, 2017).

Green, K.; López, M.; Wysocki, A.; Kepner, K.; Farnsworth, D.; Clark, J. L. *Diversity in the Workplace: Benefits, Challenges, and the Required Managerial Tools;* University of Florida, 2015. https://edis.ifas.ufl.edu/pdffiles/HR/HR02200.pdf (accessed Mar 2019).

Ha, N. T. V. The Impact of E-HRM on the Roles and Competencies of HR. https://www.google.com.bn/url?sa=t&rct=j&q=&esrc=s&source=web&cd=6&cad=rja&uact=8&ved=0ahUKEwiJ_tHwufrVAhUFK48KHVGRDDIQFghOMAU&url=https%3A%2F%2Fwww.researchgate.net%2Ffile.PostFileLoader.html%3Fid%3D568662d35e9d976a468b45c5%26assetKey%3DAS%253A313047296479232%25401451647699542&usg=AFQjCNFXgcrbHPGEu2PkAD3o9wlx5DQAhw (accessed Aug 25, 2017).

Hamilton, E. *Bringing Work Home. Advantage and Challenges of Telecommuting*; 2017. https://www.bc.edu/content/dam/files/centers/cwf/research/publications/researchreports/ Bringing%20Work%20Home_Telecommuting (accessed July 2019).

Haris, M. Strategic Implications of a Dynamic HRM Environment (Chapter 1). *Human Resource Management;* Nov 5, 2013. https://www.slideshare.net/tiens4pk/hrm-chp-1 (accessed Mar 2019).

Joseph, C. The Advantages of a Decentralized Organizational Structure. http://smallbusiness. chron.com/advantages-decentralized-organizational-structure-603.html (accessed Aug 24, 2017).

Khan, T. I.; Jam, F. A.; Akbar, A.; Khan, M. B.; Hijazi, S. T. Job Involvement as Predictor of Employee Commitment: Evidence from Pakistan. *Int. J. Bus. Manag.* **2011,** *6* (4), 252.

Koonce, R. *Redefining Diversity: It's not Just the Right Thing to do; it also Makes Good Business Sense. Train. Dev.* **2001,** *55,* 22–33.

Laxmikant, S. *E-HRM;* Apr 7, 2015. https://www.slideshare.net/laxmikantsoni92/e-hrm-slideshare-ppt (accessed Mar 2019).

McLeod, R. Jr.; DeSanctis, G. A Resource-flow Model of the Human Resource Information System. *J. Inf. Technol. Manag.* **1995,** *6* (3). jitm.ubalt.edu/VI-3/article1.pdf (accessed Feb 2019).

Melissa, B. *Centralized Versus Decentralized Information Systems in Organizations*; 2010. https://melissakay83.files.wordpress.com/2010/11/centralized-versus-decentralized-information-systems-in-organizations.pdf (accessed Mar 2019).

Morgan, J. *Five Things You Need to Know About Telecommuting*; May 4, 2015. https://www.forbes.com/sites/jacobmorgan/2015/05/04/5-things-you-need-to-know-about-telecommuting/#44c93653418f (accessed Apr 2019).

Navytha, K. A Study on Emerging HRIS Trends in Coorporateorganisations. In *Human Resource Management Horizons;* Aluvala, R., Ed.; Zenon Academic Publishing: India, 2013.

Principle of Management. *Design a High-performance Work System.* https://open.lib.umn. edu/principlesmanagement/chapter/16-7-designing-a-high-performance-work-system/ (accessed Aug 24, 2017).

Salaman, G.; Storey, J.; Billsberry, J.; Eds.). *Strategic Human Resource Management: Theory and Practice;* Sage, 2005.

Sakthivel, R. S. *Role Impact and Importance of MIS*; Sept 1, 2014. https://www.linkedin. com/pulse/20140901121616-270946654-role-impact-and-importance-of-mis (accessed Aug 23, 2017).

Shobhit, S. *LG as a Case Study of a Successful Enterprise Resource Planning System*; n.d. http://www.investopedia.com/articles/investing/111214/lg-case-study-successful-enterprise-resource-planning-system.asp (accessed Aug 24, 2017).

Toyota Motor Manufacturing North America, Inc. Drives Growth and Targets US$ 7 Million Annual Payback by Investing in Its People. https://www.cisco.com/c/dam/en_us/about/ ac79/docs/success/Toyota_Final.pdf (accessed Aug 23, 2017).

Wallace, G. S. *Panasonic Malaysia Moves to Cloud with HR Management System*; 2017. https://www.digitalnewsasia.com/business/panasonic-malaysia-digitalise-unify-hr-management-system (accessed Aug 23, 2017).

# CHAPTER 2

# STRATEGIC HUMAN RESOURCE MANAGEMENT: AN ICT POSSIBILITY SHIFTING PARADIGM OF HUMAN RESOURCE MANAGEMENT

## ABSTRACT

Technology innovation would enable human resource to put more focus on the value added in achieving technology's full potential and business strategy. It can be seen that software and interface are used by employees to perform their jobs. Here, IT can potentially bring such benefits such as an increase in productivity, minimization of response times, and lower administrative costs. In addition, HRIS is capable to create an IT-based organization. To implement HRIS in an organization, there are variety of factors to take into consideration from internal and external factors. Implementation of HRIS will require an organization to inform stakeholders and to provide them evidence on the benefits of HRIS. The presence of technology and systems such as HRIS and MIS has allowed them to work systematically, structurally, and coherently. For instance, the HR department can keep track records of employees' details such as name, contact number, home address, training records, and achievement records. From these records, HR department will enable to review what sort of training can be given if necessary. In addition, with technology, HR can reduce carbon copies. Communication between HR department and employees can be conducted as well. Telecom, visual conferences, and e-mail enable both HR department and employees to communicate wirelessly.

## 2.1 INTRODUCTION

Due to an inadequate involvement of employees, the productivity of an organization will not increase, in other words, the overall performance of the organization will be affected, and this may result in not being able to meet expected goals of an organization or collective. Thus, understanding the impact of management information systems (MIS) on the performance of employees is very crucial for all types of organizations, as it potentially can improve the performance of the organization and individuals. However, the value of understanding the impact of information systems (IS) on user performance has not received sufficient attention.

In this 21st century, there are various competitors with mass business strategies that are entering the market. Therefore, to sustain and to achieve competitive advantage among the rivalries, an organization must be alert of their own competitive advantages. One of the valuable resources is the people in the organization. Without doubt, to understand employees, the human resource management (HRM) plays a significant role to manage their employee's needs and problems. For instance, in relation to the revolution of technology, the HRM has to ensure that their employees have no difficulties with the new changes specifically in using MIS and skills required.

### 2.1.1   STATEMENT OF PROBLEM

Considering the competitive pressures in the business environment, the top management of the organization is urged to achieve the maximum productivity of the people, processes, and IS. Questions such as "how can the top management configure which type of IS will lead to the highest performance of its employees" need to be addressed as most organizations would invest the significant resources in the adoption of IS. Thus, this chapter intends to assess the impacts of IS on the performance perceived by the employees. In conducting this literature research, the following questions will need to be answered:

1) How would the involvement of employees' participation and voice contribute to effective MIS in an organization?
2) How will IS influence the performance of the employee in an organization?

### 2.2   LITERATURE REVIEW

### 2.2.1   EMPLOYEE INVOLVEMENT

Bredin and Soderlund (2001) describe employee involvement as the degree of employee influence in the decision-making processes as well as their individual influence and working condition. Moreover, they believe that giving motivation to employees and encouraging them to have an opinion in the decision-making process is beneficial for both the employees and the organization. Chatleska and Sofijanova (2013) indicate that presently, there are rapid changes in the business environment, and one of the possible ways in sustaining the business is through employee involvement in both decision-making and problem solving as it is essential and gives impacts toward an organizational innovation and effectiveness. Additionally, employee involvement can potentially result in job satisfaction and improvement in financial performance, as employees feel empowered to perform their job not only to achieve their self-goals but to also achieve the organizational goals. Prida and Grijalvo (2008) emphasize that *"the involvement and participation of the employees not only improves their quality of life, but has also proved to be very useful when it comes to facilitating organizational flexibility and improving productivity and product quality"* (p. 347). To elaborate in brief, employee involvement or participation and the right delegation are critically important. To some extent, this allows employees to participate in the organization's decision-making and it also gives independence to employees to make their own decision-making. This way, it permits employees to become more efficient as they understand and acknowledge their own responsibilities with minimal interference and less control by managerial staff for directions. Glew et al. (1995) assert that involvement of employees can increase the input of employees' input into decisions.

This also has been supported by Elias and Hassan (2007) research. They highlighted that to strive effectively, successfully, and for businesses to grow, the organization should allow employees to be involved in an organization as well as in managing and sharing of knowledge to other employees. Eventually, it will develop a mind of expert and will assist in the sustainability of an organization as employee involvement required the knowledge from the organization and the employee itself to make the right decisions and for sustainability.

On the other hand, there may be limitation on employees' application in technology. Weiss (2012) observes that Generation X have a difficult time in using the technology or software provided in the organization, due to different perceptions. In addition, Weiss (2012) pinpoints that Generation X argued that what matters is the job to be done quickly rather than to make the task more effective and efficient; however, from the top management's view, the availability of technology is to make employees to be more effective and efficient. At the organizational level, there are several barriers in implementing new technologies such as lack of capital, lack of skills, setup cost, maintenance cost, lack of support and commitment from top management, lack of human knowledge, and lack of application for users.

## 2.2.2  MANAGEMENT INFORMATION SYSTEM

Considering the wide use of IS in organizations nowadays, IS are very much needed to get employees stay involved to achieve the organizational goals or objectives, through communication, delegation, leadership, and teamwork. MIS is a structured system meant for managing data and to make use of the data of ERP system for the need of decision-making and analysis (Michálek, 2013). Looking into the components of MIS which are consisting of data, hardware, software, telecommunications, people, and procedures (Oz, 2009), this indicates that IS still requires people to function it. ERP is known as enterprise resources planning or "operation system." It is refers to the systems used for planning and to organize the enterprise resources such as accounting, human resources, stock detail, and others. Therefore, the collaboration between MIS and ERP will enable to produce better planning and analysis. However, the MIS which links to ERP needs to be properly designated in correspondence to the situation of the context. Without MIS, it is difficult especially for managers to make right decisions as there are lacking supporting data.

In the Sixth Edition Book of Management Information System written by Oz (2008), there are types of IS that can be beneficial to organizations such as transaction processing systems, supply chain management systems, customer relationship management, business intelligence systems, decision support systems, expert systems, and geographic information systems. Additionally, IS revolve around business functions in accounting, finance, marketing, and human resources.

Many researches have been conducted on the approaches such as, techniques and technologies for the design and development of MIS. However, there are few articles that covered the impact of MIS on the employee involvement and decision-making. From the literature above, MIS enables the exchange of experiences and information transfer to any of the hierarchy levels to sustain competitive advantage. Therefore, this has been supported by Barachini et al. (2009) that it is imperative for the organizations to continuously motivate their employees to share valuable information for their intellectual capital can be leveraged.

Hence, with the aid of MIS, employee can be efficient in organizing the resources of an organization, yet there are also some disadvantages of MIS driven from people itself. MIS could deteriorate the organization's budget if the employees do not practice MIS. Aside from that, there is a need in providing maintenance for continuous improvement as MIS acts as a provision of the raw data input (Markgraf, 2017).

### 2.2.2.1   THE ROLE, IMPORTANCE, AND BENEFITS OF MIS

In the study of Pilarczyk (2016), he stated that the role of MIS is to make an organization to achieve its objectives and performance as well. Pilarczyk (2016) also addresses the importance of MIS as a life saver for the organizations to have excellent performance on employees.

The roles of MIS are as follows:

1)   Easy accessibility to data: Manager and employee will be able to access to data globally or locally.
2)   Time availability: Efficient and effective.
3)   MIS provides solutions without both delayed the risk of mistakes arises.
4)   MIS provide benefits in planning, reporting, and analyzing.

Argris (1991) describes MIS as a system that is using formalized procedures to administer all levels in the management with relevant information from internal and external sources which may assist them in effective decisions for planning, directing, and control. Thus, in his statement Argris (1991) also points out how MIS can be beneficial to the organizations.

There are different results shown by top management and human resources regarding employee involvement. First, to identify the main causes of employee performance that has interconnected with their involvement in the organization. This has been studied by different researchers and some researchers believe that employee involvement revolves around human resources practices, leadership, and delegation.

### 2.2.2.2   IMPACT OF MIS: HRMS VIEW

Within the technology advancement, the usability of manual HR systems may not be adequate in today's trends, when the IT Emerging Technology become factors of enabler and driver of business processes reengineering in managing human resource (Beckers and Bsat, 2002). Broderick and Boudreau describe HRIS as a combination of computer applications, hardware, software and databases to collect, record, manage, store, deliver; and manipulate data for human resources. Shrivastava et al. (2003) stated that technological innovation could be supported by HRM and would enable HR to focus on value addition in achieving technology's full potential and business strategy.

Broderick and Boudreau (1992) added that activities such as payroll, administration, and transaction can be automated by information technology (IT) in which HR has the opportunity to focus more on the organizational strategy. In relation to this, the invention of HRIS facilitates professionals to accomplish better performance and also allows professionals to participate in consultancy in an organization (Bussler and Davis, 2001). Moreover, according to Othman and Teh (2003) HRIS is capable to create an IT-based organization. To some extent, IT can potentially bring such benefits such as an increase in productivity, minimization of response times, and lower administrative costs (Snell et al., 2002). Therefore, the implementation of HRIS will require an organization to inform stakeholders and to provide them evidence on the benefits of HRIS (Lengnick-Hall et al., 2003).

## 2.2.2.3   IMPLICATIONS OF TECHNOLOGY IN THE ORGANIZATIONS

It is undeniable that the implementation of technology in an organization also brings implications, particularly for employee and financial. Downsizing and cutting staff have also been the implications. Some factors have been identified by Dickson (1970) in which there are implications in implementing MIS in an organization:

1) The implementation of MIS can lead to the changes in the organizational units
2) Employees who are the users of MIS should be included as there will be modification on job design and processes
3) MIS team and users will need to communicate as users will need to be aware of the objectives of the implementation
4) Job performance management will also be changed to a new measurement or may not be applicable.

Taking into a case study in Nigeria by Kenneth and Laudon (2001), they stated that the implementation of IS in the country is expensive and is difficult to develop, not all tasks can be solved by IS, expectations of the manager on IS can be unrealistic and computers can be attacked by viruses and get sabotaged.

### 2.2.3   HUMAN RESOURCE MANAGEMENT

According to Hellriegel et al. (2009), HRM is the process of managing human resource needs of an organization in ensuring satisfaction of its strategic objectives. To perform MIS, it requires people. HR management provides many purposes such as selection, analysis, pension plans, payrolls, and information of employees (Oz, 2008). Therefore, it can be seen that people who are referring to employees are the drivers and users of the system. In reality, people may have difficulty in familiarizing with the new job designs. For instance, previously, employees would just save their files on Microsoft Excel without sharing it to others. In contrast to today, employees would need to save the Microsoft Excel files and save them into a database. To do so, it would require employees to go on respective software. Another example, previously people communicated through written notes and letters. Today, the presence of email and instant message has made communication easier and quick without having employees face-to-face to interact. Thus, HR in an organization has a role from guiding, mentoring, motivating, providing training, and addressing issues of the employees.

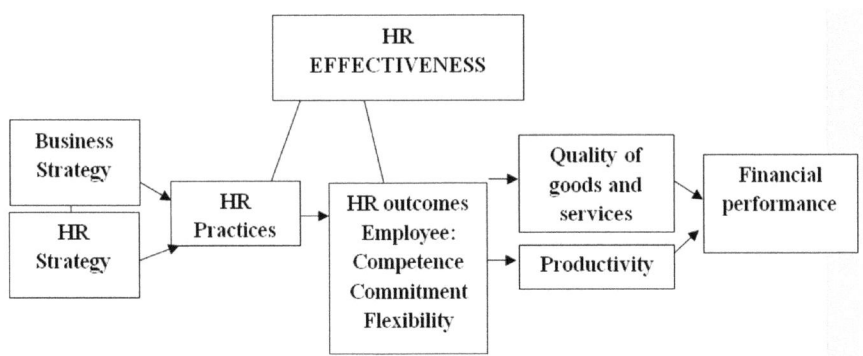

**FIGURE 2.1**    The relationship between strategic HRM and performance (Armstrong, 2006, p. 75).

In Figure 2.1 above, it can be seen that there are connection between HR and performance. Therefore, it can be said that HRM can influence employees' behavior and their job performance. Additionally, when HRM detects that there is a need for employees to undergo specific training on IT, HRM should consider why training is necessary, to what extent the training will be effective to broaden the skills of employees, and also how much it will cost to the organizations. To some extent, Kumar (2009) clarifies that employees are keen to do assigned jobs willingly in a supportive environment rather than in a controlled environment. Furthermore, employees would also demand for better pay and working conditions (Hendry et al, 1990). Thus, the role of a manager has now shifted to not just directing employees and expecting results, but also as enabler and driver for successful employees by increasing their productivity within organization

## 2.2.3.1   LEADERSHIP

Yukl (2006) defines leadership as a process of influencing others to comprehend and concur on what should be done, how to do it and it is also a process of facilitating people in fulfilling shared goals. With the changes in business patterns, the occurrence of technology in the organizations, a leader plays a crucial role in guiding and making employee feel secured. With this, based on the literature review above, discussion for leadership will cover the types of leaders along with the flow of time and benefits of having great leadership correlated with the assistance from MIS.

There are new characteristics possessed by leadership accordingly to the business trends such as a performance role as a coach instead of giving order (Johnston and Marshal, 2016). They stated that to have an effective employee it all starts from the leadership in an organization. In comparison to traditional leadership, Keulder (1997) argues that traditional leader is more into the old system with one voice and orders are given. To some extent, traditional leadership is based on power gained and held by leader.

Dubrin (2007) describes leadership as the ability to support and inspire confidence of people who are in need to achieve the organizational goals. This indicates that leaders can help to elevate employees in an organization to perform the IS and understand changes in their job designs and processes. As leadership relates to achieving goals, this demonstrates that the role of a leader is also to direct people toward the accomplishment of tasks. Thus, Kotter (1998) highlighted that both strong management and leadership are important in an organization. However, Rowe (2011) states that there are contrasts between leaders and managers as *"Managers are people who do things right and leaders are people who do the right thing"* (Bennis, and Nanus, 1985, p. 221). Nonetheless, in this report, we will discover how leadership brings advantages to employees. Interpersonal skills, analytical skills, experiences, and knowledge possessed by a leader can be embedded in ensuring the people to be capable to perform jobs better.

## 2.2.3.2   COMMUNICATION

Regardless of any nature of businesses organizations, communication remains essential as it allows people in an organization to interact, socialize, exchange information, share knowledge, request information transfer, planning, give directions, and receive information. Organizational communication is a process whereby employees capture information on what is revolving around the organization including changes, as defined by Kreps (1990). Relevant communication facilitates employees at all level to achieve common objectives (Barrett, 2002). Through communication, there can be a decision-making process. Thus, Duran and Corral (2016) examine that employee involvement is concerned with the capacity of employees to influence decisions as individuals rather than through representatives. Additionally, through communication, it gives employees the interpretation and understanding on why there are changes and why should they enhance skills as employees may have lack of understanding on the benefits that IS can bring to their daily routines. Hence, communication as a tool can close the gaps. Therefore, effective communication assists in informing employees that certain changes especially in the application of IS. For instance, a telecommunication organization is initiating to expand its operating in other regions of a country. Looking into the size and nature of the business, it would be inefficient for the telecommunication organization to do offline. Before deciding whether or not to decide to expand its operation, first it has to consider various aspects. Top management will need to communicate with specific hierarchy level especially the finance department as the finance department is responsible for calculating the related costs. Therefore, in planning and decision-making, it requires interaction and communication. Smidts et al. (2001) stated that effective communication fortifies workers' relationship with the organization, which adds value to organization's financial performance and success. Therefore, communication will allow employees to acknowledge what MIS can bring to their job. Thus, Abraham et al. (1999) suggest that communication has to be well managed to avoid confusion during any change in process and it can be done through honest and clear messages aided by media with high coverage.

## 2.2.3.3   TEAMWORK

Teamwork is essential in an organization especially in the world of technology and the presence of competitors in today's business environment. Cohen and Bailey (1997) stated that promoting trust and care by team leaders with high level of teamwork can affect an organization's competitiveness. Moreover, Belbin (1981) asserts that effective teams consist of eight team roles such as shaper, chairman, monitor, evaluator, investigator, finisher, and team worker. In addition, the output of team in meeting the quantity, timeliness, behavior, and attitude define the effectiveness of a team (Hackman, 1990).

## 2.2.3.4   MOTIVATION

Organization can be more efficient when they have motivated employees for both short-term and long-term success. Shanks (2003) asserts that personal and organizational goals are the two factors that link to why employees should be motivated. With today's context and the existence of technology, employees are advised to use new software and technology and organizations are to figure out ways to increase their employee's confidence in the application (Kian and Fauziah, 2012). However, Ramlall (2004) argues that employees also want their work that they are doing is meaningful. Employee involvement in any activities or changes in an organization can support in a quality decision-making. It is through effective management system that can motivate employees to think in creative ways for them to be involved in the suggested process (Spangler et al., 2003). Thus, this will allow employees to accept changes.

   In accordance to our research, we believe that Theory X and Theory Y also play major role in improving employees' low performance regardless of older or younger workers. Using this motivation, it will directly and indirectly increase their employee involvement in an organization and productivity. Therefore, with theories X and Y, it relates with a complete interaction and their own work, respectively. This model is created by Douglas McGregor which has been developed and contributed to effective outcomes (Moona and Mohamad, 2013). In general, there are two theories, X and Y yet each of the theory brings its own advantages and disadvantages. In the study of Tim (2008), he states that Theory X is stressed by top management and HRM is to see the changes on their employee's productivity and demand for their output. As for this theory, employees tend to be monitored and followed up.

On the other hand, Theory Y is far more relaxing. Carson (2005) indicated low-performing employees are given self-direction and self-control to be improvised from time to time.

### 2.2.3.5   EMPLOYEE TRAINING AND DEVELOPMENT

Swart et al. (2005) assert that employees are assets of an organization who provide various knowledge, skills, and talents that need to be retained by providing them motivation and training for the enhancement of their skills and the organizational success. Employees may face obstacles while facing changes in their jobs. For instance, employees are expected to perform their work with IT. Considering there are senior staffs who have been working with an organization for quite a long period of time, they may not be familiar with new processes and workflow. Therefore, appropriate training can be provided (Sanchez, 2003). However, in Germany, Zwick (2011) highlights older workers may be less interested in training due to low-financial incentives in comparison to the young workers and less social pressure. On the other hand, training motivation is declines with age (Warr and Birdi, 1998). Nonetheless, IBM (2011) states the value of training perceived is having a skilled workforce that brings benefits to the organization from sales, marketing to customer services. In addition, Zahra et al. (2014) recommend training to be planned thoroughly by aligning the training's objectives with organizational objectives.

## 2.3   METHODOLOGY

In answering a given statement, questions were established. In answering the questions, information was extracted from books and journal articles. Information was also gathered from online database and quality of the data was checked.

## 2.4   DISCUSSION

From the literature review above, it was stated that employees are the assets of an organization. Therefore, employees are the users and people who will use the IS. Thus, in discussion, it will revolve around how the involvement of employees can contribute to effective MIS in an organization. Thus, HRM has the role in ensuring employees to be familiar with technologies, guiding employees, and motivating employees to apply MIS.

It is undeniable that technology is a tool that makes people's life easy and more efficient. In other ways, it actually automatically changes the pattern of the employees' productivity and job design. Employees would first wonder as "How technological change could influence their current job design?" For instance, technology can increase the productivity of low-skilled manual labor with the use of better tools and machinery. Eventually, technologies begin to hit high on skilled workers because manual jobs are now computerized. Thus, complex and time-consuming work are being replaced with new IS. Therefore, it demonstrates that technology acts as complement in making tasks easier. Overall, technologies influence job design from time to time.

### 2.4.1  THE ROLES OF HUMAN RESOURCE MANAGEMENT AND EMPLOYEE INVOLVEMENT

According to Hellriegel et al. (2009), HRM is the process of managing HR needs of an organization in ensuring satisfaction of its strategic objectives. To perform MIS, it requires people. Therefore, it can be seen that people who are referring to employees are the drivers and users of the system.

To further elaborate on the role of HRM, looking at the relationship between strategic HRM and performance by Armstrong (2006, p. 75), it will allow readers to further understand why employee involvement can have impacts on the implementation of MIS in an organization as now there has been an extension from HRM to Strategic Human Resource Management.

An example will guide readers to further understand. It is shown in Figure 2.1 above that the effectiveness of HR will result in outcomes such as increase in employees' competency, commitment, and flexibility. Part of these outcomes may also revolve from employee involvement as HR may practice employee involvement in every aspect of decisions. This is because, for HRM, employees are the assets and knowledge; therefore, to achieve both qualities of goods and services as well as productivity, HRM will need to consider the people in the organization first. Indirectly, the success of HR practices will result in financial performance. This explains why HR in an organization has a role from guiding, mentoring, motivating, providing training, and addressing issues of the employees. In addition, both internal and external environment influences revolve around HRM's role. Internal environment encompasses of employees, products, services, values, culture, leadership style, and management; while external environment includes factors such as socioeconomic, political, technological, legal, and competitions. Thusly, both internal factors and external environment factors should be taken into consideration in implementing MIS.

Bredin and Soderlund (2001) describe employee involvement as the degree of employee influence in the decision-making processes as well as their individual influence and working conditions. If employee involvement is provided in implementing effective MIS and new job design, employees will get to understand better. For instance, employees may encounter numerous customers on customer service. Feedback and complaints from customers are instantly received by employees first than the manager. Therefore, employees may come up with solutions on how to minimize the problems faced by customers. For example, customers are not satisfied with their purchase as wrong amount in invoice was sent to them. From this scenario, employees would want to suggest the organization to provide a specific file in a database for the records of feedback and complaints from customers. This way, it would be easier for employees, manager, and top management to review. Employees would want their suggestions to be heard. Therefore, Bredin and Soderlund (2001) believe that giving motivation to employees and encouraging them to have an opinion in the decision-making process is beneficial to both employees and the organization. Consequently, employee involvement can potentially result in job satisfaction and employees feel empowered to perform their job. Furthermore, Prida and Grijalvo (2008) emphasize that *"the involvement and participation of the employees not only improves their quality of life, but has also proved to be very useful when it comes to facilitating organizational flexibility and improving productivity and product quality"* (p. 347), it has been proven that HRM and employee involvement play a role in MIS.

### 2.4.2  MANAGEMENT INFORMATION SYSTEM

As described by Michálek (2013) MIS is a structured system meant for managing data and making use of the data or ERP system for the need of decision-making and analysis. This proves that it assists both employees and the management to perform tasks. Thus, here again is proven that HRM has the role to achieve the organizational goals or objectives through people in which communication, delegation, leadership, and teamwork can accelerate the process. Gradually, it is imperative for the organizations to continuously motivate their employees to share valuable information so that their intellectual capital can be leveraged (Barachini et al., 2009). For instance, employees who have been working in an organization for a long period of time may have less knowledge on how to operate MIS. However, looking into the importance of MIS as a life saver for the organizations to have excellent performance on employees (Pilarczyk, 2016), demonstrations to employees on how to operate MIS may be beneficial. For instance, in supply chain, employees would not need to go to the physical location to retrieve information of suppliers, information on products, and contact details of suppliers as technology such as e-mails, clouds, and databases will allow communication and transfer of information between employees and suppliers. Hence, if employees are consistently informed about the advantages, they would perceive changes and improvement positively.

| Humans | Computers |
|---|---|
| Think | Calculate and perform programmed logical operations faster |
| Have common sense | Store and retrieve data and information faster |
| Able to make decisions | Able to perform complex, and arithmetical functions accurately |
| Able to learn new methods | Able to perform routine tasks inexpensively |
| Able to accumulate expertise | Able to be programmed and reprogrammed |

**FIGURE 2.2**    How human and computers contribute to synergy (Oz, 2008).

From Figure 2.2 above, it can be seen that both human and computers can contribute to synergy. MIS provide opportunities to both organization and employees to apply technology. For instance, features on technology gives convenience to users to perform job better. For instance, Microsoft Excel files which contain a variety of valuable information can be stored in databases. Files, needed by employees or users will be able to retrieve data at a faster rate.

Thomas J. Watson, the founder of International Business Machines (IBM) stresses that to build a business it needs people to work on it rather than mechanize as machines cannot make decision and are programmed rather than human. This view has been supported by Chatleska and Sofijanova (2013). Even now, technology has taken over in giving support in the tertiary industry from the existence of IS and management information yet the system still require employees to make the final output. Let us have a look on the achievement by one of the organization that is still successful, Nike. Previous and latest inventions of new shoes are succeeding due to technology and people. In the past, Nike faced a technical failure due to lack of research yet the team developer managed to settle the problem. At that time, Nike faced criticism on the price launch of Nike 'Back to the Future' shoe as they costed a thousand dollar each pair. This showed that Nike had a weakness on data management with prices of shoes. However, from the event, employees learned from their mistakes and considered voice from consumers for the next new products. This also shows that employees have the ability to make a right decision based on experiences and knowledge with the assistance from MIS.

To some extent, it is found that employee involvement increases the confidence level of employees and productivity as the output produced by employees is the result from the top management and human resources. This is represented with Richason's research. Richason argued employees' productivity either positive or negative are determined through the behavior and action from top management. If top management did not trust and give a job's full responsibility toward their employees, employees will exhibit negative and low performance. Thus, top management action with minimal supervision and less in restriction increased their capacity. Additionally, Richason's studies gave a clear idea to be learnt by all of us and organizations that in order, for top management and human resources, to have committed and hardworking employees in organizing their tasks to be done efficiently and effectively, individuals should not be exploited, isolated, and treated badly. It is one of the reasons that management should allow employee to do their work in their own ways. Consequently, it can result in high job satisfaction and motivation. Therefore, Richason's research has revealed that employees' productivity is highly depended on relationships between employees and management as it reflects on employees' output.

## 2.4.2.1   HUMAN RESOURCE INFORMATION SYSTEM

As discussed before, the impacts of IT Emerging Technology as a fundamental rethinking and reengineering of HR business processes may lead to shows that manual HR system may not be adequate for further usage of managing human resource (Beckers and Bsat, 2002), as a consequence, combination of computer applications, hardware, software and databases to collect, records, manage, store, deliver and the manipulation of data for human resources is extremely needed Boudreau (1992). This shows that technological innovation could be supported by HRM technology and technology innovation would enable HR to put more focus on value-added in achieving technology's full potential and business strategy. It can be seen that, software and interface are used by employees to perform their jobs. For instance, when an organization is to contribute payroll to employees, HR department can liaise with other department such as finance department and IT department. For example, there may be an increase in the payroll of particular employees as they have work overtime. Therefore, HR department would want to liaise with IT/finance department for amendment of information. Thus, it demonstrates that activities such as payroll, administration and transactions can be automated by IT in which HR has the opportunity to focus more on the organizational strategy which has been added by Broderik and Boudreau (1992).

Moreover, Bussler and Davis (2001) highlight that the invention of HRIS facilitates professionals to accomplish better performance and Human Resource Information System (HRIS) also allow professionals to participate in consultancy in the organization. HR department has a broad roles and responsibilities from payroll, administration, motivating, guiding, and mentoring employees. Thus, Snell et al. (2002) state that IT can potential bring such benefits such as an increase in productivity, minimization of response times and lower administrative costs. In addition, Othman and Teh (2003) mention that HRIS is capable to create an IT-based organization.

To implement HRIS in an organization, there are variety of factors to take into consideration from internal and external factors. Lengnick-Hall et al. (2003) highlighted that the implementation of HRIS will require an organization to inform stakeholders and to provide them evidence on the benefits of HRIS.

To further understand why there is a need to consider various factors, a scenario will be included. Large organizations, such as, telecommunication, hospital, and university have variety of stakeholders. To some extent, the number of employees can be extensively high. Therefore, if an organization was to do all work manually, it would be a burden. However, the presence of technology and systems such as HRIS and MIS has allowed them to work systematically, structurally, and coherently. For instance, the HR department can keep track records of employees' details such as name, contact number, home address, training records, and achievement records. From these records, HR department will enable to review what sort of training can be given if necessary. In addition, with technology, HR can reduce carbon copies. Communication between HR department and employees can be conducted as well. Telecom, visual conferences, and email enable both HR department and employees to communicate wirelessly. This has not only given convenience to HR but employees can also be motivated in many factors. Employees would not need to be face-to-face with HR when they wish to ask how many leaves they have. Aside from that, with a system, it enables employees to take leaves online.

## 2.4.2.2   *IMPLICATIONS OF TECHNOLOGY IN ORGANIZATIONS*

Most literature reviews cover the implication on technology that revolve around the implementation of technology. Dickson (1970) identified several implications in implementing MIS in an organization. From the study of Dickson (1970) communication, users and job performance management will affect within the duration. This indicates organizations need to consider employees or the users of IS as these are the people who will be using the IT. For financial implications, a study of Nigeria by Kenneth and Jane Laudon (2001), the implementation of IS are expensive. For instance, large organizations may require comprehensive IS, suitable computers, software, training, professionals, and consultants. All these involve costs.

## *2.4.3   LEADERSHIP*

Employees can be motivated by many factors and to be guided and mentored by leaders is one of them. With the changing of business landscape, there is an increase in number of innovative competitors; with the help of new technologies, an organization must retain their competitive advantages. Therefore, one of the ways is to have a great leadership to guide employee and to give support when there are changes in the organization. Related back to LR Yukl (2006) and Dubrin (2007), they highlighted that having a great leadership motivated employee to perform outstandingly as the role of a leader is to inspire others to be strong and follow up with trends in businesses. As there are changes in the technology in organizations, employee involvement in organization must be balanced while using technology to make tasks efficient and effective in which it begins with the leader. With the characteristics of leadership mentioned by Johnston and Marshal (2016), they can run the organizational objectives smoothly and can encourage their employee to use MIS.

The emergence of new types of leadership consider their subordinates' voice, equal delegation of the task, keen in creating learning experiences, and focused to the personal needs of their employees. As been described by Johnston and Marshal (2016), the intellectual knowledge of the leader indirectly motivates employee involvement in an organization to accept challenges and changes in meeting the objective of the organization. To be an innovative leader, leader must learn and enhance their interpersonal skills to create an organizational climate to support their employees and to develop new products and services. It is about growing a culture of innovation and to support each other with less control. The studies from Keulder (1997) has not been used and practiced along with the changes in business patterns, as it has weakness and is perceived as an old method with high chance of failure. In opposition to Keulder (1997), Kotter (1998) believes that leadership is the key for employee involvement to become productive. Unfortunately, based on our findings, in reality leadership is not one of main factor alone that drives them to be productive in the organization, as it can also be from peers.

Regardless of the types of leadership, it still plays a role in shaping the way how the organization will run and how to ensure the organizational objectives are to be reached through people. Moreover, leadership is responsible to transform potential solutions into reality. To some extent, leaders act as guidance. Therefore, HRM also needs to support leaders to go for training to enhance their interpersonal skills. Apart from that, HRM has to encourage and give support to leaders and its importance cannot be isolated. Without leader, it would be difficult for organization to survive. Thus, leaders are to influence and inspire employees to perform task effectively, efficiently, and competently.

### 2.4.4 COMMUNICATION

Along with the advancement in technology, IS, communication between employees and employers also changes. Communication enables people in an organization to interact, socialize, and exchange and share information. For Koontz (1994), communication is essential as communication allows in generating social bonding and connecting employer with employees. Today, communication between people in the organization can be performed through software and technologies such as social network, email, visual conferences, and electronic devices. With them, any changes in an organization, communication is part of the solution implemented by HRM as a tool to close the gaps of problem faced by employees. Technically, there are two ways of communication between top management and employees. First, centralized communication where authority makes decision and planning solely without employee involvement. Second, decentralized communication is where employees from different levels are given chances in making decisions and to involve in meetings for any ideas and suggestions. By comparing both types of communication, employee involvement would be in the decentralized communication.

Decentralized communication will also give employees the opportunity to make decisions in their own tasks. For instance, when employees encounter a minor problem with customers, employees do not necessary have to refer to their superior. However, communication in the organization will also depends on the size of the organization. Duran and Corral (2016) highlighted that decentralized communication is one of the backbones which can increase employee productivity. Furthermore, it gives employees the privilege to speak up and to give ideas. To conclude, communication connects people regardless of any occurrence of problems in an organization. Therefore, this relates to the study of Smidths et al. (2001) that an effective communication makes the relationship between employees and organization stronger in today's rapid business environment.

### 2.4.5   TEAMWORK

Based on our findings, we find out the importance of teamwork in the organizations. To have a great team, a leader has to have a start or the initiative to support and guide from the start for an effective management of team. Cohen and Bailey (1997) state that promoting trust and care by team leaders with high level of teamwork can affect an organization's competitiveness. The importance of trust placed on team members from a leader of the team can give the members the empowerment to complete their task and increase their involvement in the organization. A good leader of the team would also give an opportunity to his members to perform at its optimum level and to demonstrate their potential. By doing so, it indirectly shows that the organization will have an effective leadership and can prove that the organization has developed and recruited right candidates. Thus, a great teamwork starts from the leader itself and the effective management is based on right selection of a leader and a successful organization is correlated with the setting goals and objectives of the organization in creating their succession plan.

In relation to the study of Hackman (1990), to generate a great performance with outstanding results in performing tasks, it is important for a team leader to delegate, assign a role and responsibility. This will allow team members to know the objectives and goals of assigned tasks. This shows that if there is an absence of teamwork, members of the team would encounter difficulty in completing tasks assigned. For instance, a marketing department is to provide a report on their monthly sales. The team is consisting five members. Each member has a specific role:

Team member 1: To collect all contact and sales information

Team member 2: To input all contact and sales information into a system

Team member 3: To check the accuracy of information

Team member 4: To ensure that team is completing the task in a specific amount of time

Team member 5: To prepare a report from resources provided by the other members.

From the scenario above, it can be seen that each of the member has a different role. The reason for this is to ensure the quality of the report that they will have to submit to their department leader. Team member 5 is responsible for preparing a report. Therefore, without collective efforts from all team members, it would be impossible for team member 5 to do so. Thus, every team member has a function and a goal. Consequently, this demonstrates how important teamwork is especially in marketing department as the department is not only dealing with customers but also sales predictions, abundance customer information, and huge amount of money.

This shows that if there is an absence of teamwork, members of the team would encounter difficulty in completing assigned tasks. Thus, the allocation of job roles is essential for each member to recognize their own responsibility. As previously mentioned by Belbin (1981), each member has specific responsibility to make tasks to be done efficiently and effectively. If there are tasks assigned and designated for each member, they would become clueless. Eventually, the will feel discouraged.

Allocation of roles can result in productivity and performance. In relation to today, there are numerous competitors and potential problems faced by the organizations. Thus, through teamwork, it can lessen the effects from problems in achieving competitive advantages. Consequently, it will allow team members as well as employees to generate more idea, increase learning experience with each other, enhance communication skills developed, and share knowledge. Teamwork is applicable for any type of organization. Teamwork also exists in the organizations that involve in constructions. Eventually, different organizations have different styles in managing their teamwork. In general, each assigned task is connected to other employees' responsibility, for instance, as mentioned in the example above. To some extent, to retrieve information, the team would also require to approach the finance department as there would be information that they do not have access to. Thus, the output of team in meeting the quantity, timeliness, behavior, and attitude define the effectiveness of a team as asserted by Hackman (1990).

Often, tasks given to employees have a duration of time to be completed. Also, often times, tasks assigned are to be worked in a team. Thus, it requires every member to contribute ideas, thoughts, input, time, and effort in accomplishing the tasks. This has been demonstrated in scenario above. However, here, it is highly emphasized that teamwork is part of motivation. For instance, employees also are motivated by the support given by their peers. Another example, there may be new employees in a team of five. Thus, team work setting would be a great setting for him to discover various information, skills, and learning. New employees may also be involved in a group of experienced employees who have been working for so long in an organization. The valuable knowledge gained from the experienced employees can be fruitful to the enhancement and development of job performance.

However, teamwork also has its disadvantages. As the world is revolving around technology, employees tend to work alone along with the presence of MIS in the organization. Employees would state that MIS is outstanding as it reduces their workload. Unfortunately, even with the existence of technology, it still requires team members to come up with solutions and input data collected into the system. Apart from that, teamwork also can bring negative effects to slow employee involvement in an organization due to crisis led by unprofessional employees as they do not favor the assigned role. Miscommunication may also arise frequently, since big number of team that may interpret wrongly, specifically in e-mail as there is an absence of verbal communication to ensure what the points and purpose weal with Employees would rather communicate online than face-to-face. To conclude, to have an outstanding teamwork with the assistance of technology, it will still depend on how team leader and members are performing. Apart from providing empowerment to members, it is essential for team leader to get to know their employees and ask feedback.

### 2.4.6 MOTIVATION

Employees tend to receive motivation in various aspects. There are several motivations that can be implemented in correlation with how employees can be motivated to use MIS. From a research done by Kian and Fauziah (2012) based on Malaysia perspective, they argue that motivation provided by an organization must be accepted by employees as it will help to produce the organizations to be innovative. Failures in meeting the requirement from these aspects of motivation may lead to low performance which may result in the decline of HRM effectiveness toward motivating employees. On the other hand, unfriendly working environment will lead employees to having difficult times in accepting technology. In relation to this report, both intrinsic and extrinsic motivation, Douglas McGregor's XY Theory will be included. For older employees, they may encounter difficulty in operating IT and involvement in the organization as older employees prefer intrinsic rewards which can result in high level of motivation in comparison to Generation Y (Mohani et al., 2010). Based on a study conducted in Japan in an electrical and electronic manufacturing company, their study has gained attention and was supported from a previous research four years ago namely, Ringer and Garma (2006). On the other hand, Zwick (2011) found that older workers are keen and are attracted to extrinsic motivation and it is also applied to Generation Y. Under a research of Jang (2008), he pinpointed that Generation Y tend to move other organizations if they are offered better extrinsic motivation.

Thus, it can be seen that employees can be motivated by many factors not just intrinsically but also extrinsically. To find out how to motivate employees, an organization must first understand the behavior of the personnel and create a culture to drive employees' creativity in the organization. To obtain and utilize employees' creative ideas, they implement effective management systems. With the rapid changes in the business environment, it is highly depended on HRM and organizations to encourage and motivate their employee to use MIS in their job routine. Doing so, effective management system will allow motivated employees to generate creative ways in finishing their task and will tend to be involved any activity held by organization (James et al., 2003). Therefore, HRM must use their interpersonal skill in motivating their employee to use MIS in their task. Additionally, giving emotional support to employees to reach organizational objectives will also require actions from HRM. This is supported by Jamaes et al. (2003) that according to them, HRM plays a role in motivating their employees to engage with the changes to meet their organizational goal.

To increase the productivity of the workforce, based on our research we found out that motivation by McGregor's XY Theory Management Style can increase employees' confidence level and productivity. By implementing reward systems, it will automatically boost employees' creativity and input in an organization. Thus, it is important to be aware that creativity is always the main actors in competitive markets. Ever since before 2000s, IT and MIS have already grown drastically to become crucial parts of our life.

How does motivation encourages employees to utilize MIS? It can be done through intrinsic and extrinsic motivation. Both are involved with employees' comfortable and happiness. On the other hand, intrinsic motivation involves with employees satisfaction which is related to the outcome from their responsibility and work. Table 2.1 states the differences of rewards.

**TABLE 2.1** Intrinsic and Extrinsic Motivation.

| Intrinsic | Extrinsic |
|---|---|
| **Acceptance:** Employees feel belong with the environment and are accepted by other workers | Awards |
| | Benefit packages |
| **Honor:** To be respected | Bonuses |
| **Independence:** Feel to be unique and capable | |
| **Power:** Need power that can influence in making decision | |
| **Socials status:** Feeling important | |

Looking into Table 2.1 above, a scenario is given for further understanding. For instance, an electrical engineer who has been working for 10–20 years in an organization, awards may not be favorable as for him he wishes to be honored for the tasks he has completed. Another example, employees who only have been working for 2–3 years, they would want awards for their achievements. This way, they will receive recognition and get motivated.

According to Moona and Mohamad (2013), in the context of Malaysian perspective, it is argued by using Theory X and Y, it allows people to grow, develop, and work hard on their efforts. Therefore, an organization with the help from HRM has to designate which theory is suited to be used in the situation of employees. According to literature review above on Theory X and Y, in reality, Theory X is not favorable for employees as they are being controlled by top management (Tim, 2008). From our research, it is agreed that Theory X can slow down organizational performance which can lead to turnover of employee toward the organization. In addition, Theory Y will be the most suitable approach to be used for employees that have low performance and to guide older employees according to the changing patterns of business. Consequently, employee becomes more productive when more trust is placed and tasks are delegated to them instead of being control and monitored.

## 2.4.7 *EMPLOYEE TRAINING AND DEVELOPMENT*

As employees are part of the core assets in an organization, it is vital for Human Resources Management to ensure employee to perform better as highlighted by Hellrieg et al. (2009). Additionally, as HRM has roles to manage the needs of an organization and employees, it needs to detect their employees' deficient skills. For instance, if employees have less knowledge and have insufficient skills in IT, trainings on such Microsoft Office and software can be given. Weiss (2012) mentioned that, Generation X has knowledge in the usage of technology. Consequently, they encounter frustration and low job performance. However, Sanchez (2003) argues that the solution lies in training provided by HRM and organizations. In addition, other researchers such as Zahra et al. (2014) also address and support Sanchez (2003) that to overcome their low performance, the researchers recommend implementing an effective training according to the context of the situation. Therefore, the training is to be planned thoroughly by aligning the training objectives with organizational objectives. As for Swart et al. (2005), they highlighted that employees are part of useful asset regardless of their age thus the organizations need to retain them due to their variety of skill, experience, knowledge, and talents. As they are still valuable to a company it is essential for HRM to keep them motivated by sending them for further trainings.

Considering different gaps in employees' age, HRM has to consider the types of training and development for them. In the research of Bova and Kroth (1999), training for employees has to be suited along with their ages. According to their study, older employees did not prefer face-to-face trainers and they preferred to spend time with their family as they value time more than learning. They would prefer short training and visual stimulation. Unfortunately, the older people still need to be familiar and practice the IT. Consequently, this would lead them to be demotivated to attend training. Furthermore, rewards play a role for older employees to participate in training. This can be shown in the case study of Zwick (2001) that in Germany, he mentioned that older workers are less participated in training due to low reward offered by the organization. In general, not all older workers are less engaged with training as it depends on the culture of an organization or approach established by HRM. Importantly, interaction with older employees may differ from interaction with younger employees. Therefore, reward may not necessary be the solution.

On the other hand, age is the factor that affected the focused of Generation X to be involved in training provided. In the study of Warr and Birdi (1998), in correlation with the age of people, HRM does not have any power in controlling the productivity of an older people's mind and perception. Again, it is related to Weiss (2012) perspective. Moreover, it is not only the organization's support or types of training yet the trainer itself needs to be engaged with the training and be enthusiast in teaching older employees. This has been supported by Alleencomm's research; he stated that training is essential in an organization and to enhance and improve their skills. Unfortunately, in reality Aleencomm's study, he did not take effective trainers into consideration. Also, Aleencomm's study was not aware that the performance of the employee is correlated with their trainers as they play a major role in creating more skilful and knowledgeable workers. Thus, training will be ineffective if the trainers do not full do their part. This can potentially lead into waste of an organization's budget. Thereupon, training brings benefits and value if it is aligned with the training's objectives and organizational objectives. In addition, IBM (2011) states that training would cost a lot yet in the long run, training brings benefits and would be perceived as a tool in enhancing and boosting their skills.

## 2.4.8   KNOWLEDGE OF EMPLOYEES ALIGNED WITH THE CHANGING IN BUSINESS TRENDS

As time goes and organizations facing various changes, knowledge, and skills have to be followed up. There are two types of knowledge, *Tacit and Explicit*. Tacit knowledge consists of employees' experiences, skills, and interactions. While explicit knowledge consists of tangible sources such as documents, books, and written records. All these knowledge are known as intellectual resources of the organization as they belong to organization and employees. For an organization to sustain and remain competitive, it can be done through investing and retaining employees. To conclude, knowledge has to be shared as different people have different skills. Our findings found that Elias and Hassan (2007) research is useful as they believed that for an organization to expand and meet objectives, HRM should manage and organize knowledge for employees to be expert. Few of the approaches to sustain employees are through sharing section, job rotation, to encourage employee to store any information in a document, etc. Apart from that, information and knowledge are stored in a database. This is to ensure that information is documented as employees may retire or leave an organization and they may leave their skills and experience with them. For instance, an IT expert may leave an organization after working for a year. The IT expert may have a variety of information on technical solutions. To utilize his skills and knowledge, he can pass the knowledge to other employees or document the solutions in the form of a document.

## 2.5   CONCLUSION

Considering business challenges, changing business practices and technological changes, employees still remain one of the important topics in the management area as they are part of the assets of organizations and users of MIS. In relation to implementing MIS in an organization, employees also will need to be taken in consideration. Employees would not be familiar with the new job designs and processes. However, HRM has a role in guiding, motivating, coaching, mentoring, and providing training to ensure that employees are well equipped, computer literate, and prepared to accept changes. Changes that revolve around today may not be avoided as opportunities can still be tapped into. For instance, previously, an organization may have to do all tasks manually. Presently, the presence of computer as well as internet, a wide range of information is accessible. In addition, problem solving and decision-making are made easier. For instance, it is not necessary for a leader of an organization to have a face-to-face meeting with another leader of an organization in another country as there are facilities such as teleconference, conference calls, etc. Another example, is point of sales for retail selling, this allows restaurants and retail stores to review their sales and inventory. HRM practices such as leadership, teamwork, motivation, and communication may assist in the successful implementation of MIS and the operation of MIS. Support from leaders will enable employees to perform work better with the assistance of technology. Aside from that, teamwork allows employees to work toward common goals. For motivation, employees can be motivated by various factors from giving them intrinsic and extrinsic values. Employees may not necessarily motivated by money but rather delegation and guidance given to them. Communication remains one of the powerful tool as to ensure every person in the organization to work toward the organizational goals.

## KEYWORDS

- ERP system
- IT-based organization
- decision-making
- low-performing employees
- HRM

## REFERENCES

Abraham, M.; Crawford, J.; Fisher, T. Key Factors Predicting Effectiveness of Cultural Change and Improved Productivity in Implementing Total Quality Management. *Int. J. Qual. Reliab. Manag.* **1999,** *16* (2), 112–132.

Argyris, C. Management Information Systems: The Challenge to Rationality and Emotionality. **1971,** *6,* B-275.

Armstrong, M. *A Handbook of Human Resource Management Practice,* 10th ed.; Cambridge University Press: London, 2006.

Barachini, F, Cultural and Social Issues for Knowledge Sharing. *J. Knowl. Manag.* **2009,** *13* (1), 98–110.

Barrett, D. J. Change Communication: Using Strategic Employee Communication To Facilitate Major Change. *Corp. Commun.: Int. J.* **2002,** *7* (4), 219–231.

Beckers, A. M.; Bsat, M. Z. (). A DSS Classification Model for Research in Human Resource Information Systems. *Inf. Syst. Manag.* **2002,** *19* (3), 41–50.

Bennis, W. G.; Nanus, B. *Leaders: The Strategies for Taking Charge*; Harper & Row: New York; 1985.

Bredin, K.; Soderlund, J. (2011). Human Resource Management in a Project-Based Organization. Palgrave MacMillan.

Broderick, R.; Boudreau, J. W. HRM, IT and the Competitive Edge. *Acad. Manag. Execut.* **1992,** *6* (2), 7–17.

Bova, B.; Kroth, M. Workplace and Generation X. *J. Workplace Learn.* **2001,** *13* (2), 57–65.

Burke, R. J.; Cooper, C. L. The New World of Work and Organizations: Implications Human Resources Managememt. *Hum. Res. Manag. Rev.* **2006,** *16* (2), 83–85.

Buchbinder, S, B.; Shanks, N. H. *Management and Motivation.* Jones; Barlett, Eds.;Chapter 2, 2007; pp 23–35.

Bussler, L.; Davis, E. Information Systems: The Quiet Revolution in Human Resource Management. *J. Comput. Inf. Sys.* **2001,** *42* (2), 17–20.

Carson, C. M. A Historical View of Douglas McGregor's Theory Y. *Manag. Decis.* **2005,** *43* (3), 450–460.Emerald Group Publishing Limited. Exploring Different Management Styles. ManagerialSkills.org.

Sofijanova, E.; Zabijakin-Chatleska, V. Employee Involvement and Organizational Performance Evidence from the Manufacturing Sector in Republic of Macedonia. **2013,** 23–35.

Dickson, G.; Simmons, J. The Behavioral Side of MISL Five Factors Relating to Resistance. *Bus. Horizons* **1970,** *13* (4), 59–71.

Dubrin, A. *Leadership: Research Findings, Practice, and Skills*; Houghton Mifflin: New York; 2007.

Durán, J.; Corral, A. *Employee Involvement and Participation at Work: Recent Research and Policy Developments Revisited.* Retrieved from European Observatory of Working Life: https://www.eurofound.europa.eu/observatories/eurwork/articles/working-conditions-industrial-relations/employee-involvement-and-participation-at-work-recent-research-and-policy-developments-revisited (accessed, October 25, 2016)

Elias; Hassan. *Knowledge Management*; Dorling Kindersley: India; 2007.

Grates, G. F. Are Your Employees Working With the Volume Off? *Edelman Memo to Management;* 2006; Vol. *1* (4). www.edelman.com/expertise/practices/employee_change/documents/Edelman_MemotoManagement_vol1issue4.pdf

Glew, D.; O'Leary-Kelly, A.; Griffin, R.; Van Fleet, D. Participation in Organizations: A Preview of the Issues and Proposed Framework for Future Analysis. *J. Manag.* **1995,** *21* (3), 395–421

Hellriegel, D.; Jackson,S.; Slocum, J.; Staude, G. Managing Human Resources. *Management*, 3rd ed.; Oxford University Press, 2009, [ISBN 9780195982169].

Hendry, C.; Pettigrew, A. Human Resource Management: An Agenda for the 1990s. *Int. J. Hum. Res. Manag.* **1990,** *1*, 17–43.

Hindle, T. *Guide to Management Ideas and Gurus;* John Wiley & Sons, 2008, Vol. 42.

IBM. (2011). The Value of Training. Retrieved from https://www-03.ibm.com/services/learning/pdfs/IBMTraining-TheValueofTraining.pdf

Jichul Jang. The Impact of Career Motivation and Polychronicity on Job Satisfaction and Turnover Intention Among Hotel Industry Employees. Unpublished Master's Thesis, University of North Texas, 2008.

Johnston, M. W.; Marshal, G. W. *Sales Force Management: Leadership, Innovation, Technology*; Routledge: New York; 2016.

Keulder, C. *Traditional Authorities and Regional Councils in Southern Namibia*; Friedrich-Ebert-Stiftung: Windhoek; 1997.

Kian, T. S.; Yusoff, W. F. W. Generation X and Y and Their Work Motivation. In *Proceedings International Conference of Technology Management, Business and Entrepreneurship,* 2012, December, pp 396–408.

Kotter, J. P. What Leaders Really Do. In *Harvard Business Review on Leadership.* Harvard Business School Press: Boston; 1998, pp 37–60.

Kovach, K. A.; Cathcart, C. E. Human Resource Information Systems (HRIS): Providing Business with Rapid Data Access, Information Exchange and Strategic Advantage. *Public Pers. Manag.* **2002,** *28* (2), 275–281.

Kreps, G. L. *Organizational Communication,* 2nd ed.; Longman: United Kingdom; 1990.

Lengnick-Hall, M. L.; Moritz, S. The Impact of e-HR on the Human Resource Management Function. *J. Labor Res.* **2003,** *24* (3), 365–379 (accessed Dec 2018).

Markgraf, B. (2017). *Common Problems in Management Information System.* Retrieved from http://smallbusiness.chron.com/common-problems-management-information-systems-63376.html

Marchington, M.; Wilkinson, A. *Human Resource Management at Work: People Management and Development*, 3rd ed.; CIPD: London; 2005, p 176.

Mohani Abdul, Hashanah Ismail, Noor Ismail Hj. Jaafar. Job Satisfaction Among Executives: Case of Japanese Electrical and Electronic Manufacturing Companies. Malaysia. *J. Global Bus. Manag.* **2010,** *6* (2), 165–173.

Mohamed, R. K. M. H.; Nor, C. S. M. The Relationship Between McGregor's XY Theory Management Style and Fulfillment of Psychological Contract: A Literature Review. *Int. J. Acad. Res. Bus. Soc. Sci.* **2013,** *3* (5), 715.

Sanchez, N.; Arago, A.; Arago, B. I.; Valle, S. Effects of Training on Business Results. *Int. J. Hum. Res. Manag.* **2003,** *4,* 105–122.

Othman, R.; Teh, C. On Developing the Informated Work Place: HRM Issues in Malaysia. *Hum. Res. Manag. Rev.* **2003,** *13* (3), 393–406

Oz, Effy. (2008). Management Information System. 6th Edition.

Pilarczyk, K. Importance of Management Information System in Banking Sector. Annales Universitatis Mariae Curie-Skłodowska. Sectio H. Oeconomia 2016, 2, 69-80

Ramlall, Sunil. *A Review of Employee Motivation Theories and Their Implications for Employee Retention Within Organizations. J. Am. Acad. Bus.* **2004,** 52–63. Print.

Richason, O. (n.d.). *What Are the Benefits of Practicing Employee Involvement & Empowerment?* Retrieved from Chron: http://smallbusiness.chron.com/benefits-practicing-employee-involvement-empowerment-1842.html

Rowe, W. G. Creating Wealth in Organizations: The Role of Strategic Leadership. *Acad. Manag. Execut.* **2001,** *15* (1), 81–94.

Ringer, A.; Garma, R. Does the Motivation to Help Differ Between Generation X and Y? In *Australian & New Zealand Marketing Academy.* Conference (2007: University of Otago), University of Otago, School of Business, Dept. of Marketing, Dunedin, N. Z., pp 1067–1073, 2006.

Shrivatsava, S.; Shaw, J. B. Liberating HR Through Technology. *Hum. Res. Manag.* **2003,** *42* (3), 201–222.

Smidts, A.; Pruyn, A. T. H.; van Riel, C. B. M. The Impact of Employee Communication and Perceived External Prestige on Organizational Identification. *Acad. Manag. J.* **2001,** *5,* 1051–62

Snell, S.; Stueber, D.; Lepak, D. Virtual HR Departments: Getting Out of the Middle. In *Human Resource Management in Virtual Organizations,* Heneman, R. L., Greenberger, D. B., Eds.; Information Age Publishing: Greenwich, CT; 2002, pp 81–101.

Swart, J.; Mann, C.; Brown, S.; Price, A. Human Resource Development: Strategy and Tactics Elsevier Butterworth-Heinemann Publications: Oxford, 2005.

Warr, P.; Birdi, K. Employee Age and Voluntary Development Activity. *Int. J. Train. Develop.* **1998,** *2,* 190–204.

Weihrick, H.; Koontz, H. (1994). Menedžment, Mate, ISBN 953-6070-08-1, Zagreb.

Zahra, S.; Iram, A.; Naeem, H. Employee Training and Its Effect on Employees' Job Motivation and Commitment: Developing and Proposing a Conceptual Model. *IOSR J. Bus. Manag.* **2014,** *16,* 60–68. Retrieved from http://www.iosrjournals.org/iosr-jbm/papers/Vol16-issue9/Version-1/I016916068.pdf

Zwick, T. (2011) Why Training Older Employee is Less Effective? Discussion Paper No. 11-046. Retrieved from https://ub-madoc.bib.uni-mannheim.de/3200/1/dp11046.pdf

# CHAPTER 3

# MANAGING INFORMATION SYSTEMS FOR HUMAN RESOURCE MANAGEMENT: A TECHNOLOGICAL EVOLUTION

## ABSTRACT

As a business grows, there will be an increase of workloads and processes, which may impact to become complex and require more resources and this could create some employment opportunities to the market. In other hand, recruitment is the "most critical human resource function for organizational success and survival" and it is the "foundation of the organizational performance." Successful recruitment practice may result in strategic competitiveness of the organization. By applying a MIS to HRM practice, it may have solved the problems on the growing numbers of employees in an organization. This study found that MIS is essential in HRM, the size of the firm indicates the choices of the firm to decide whether to implement the MIS or just use the internal recruitment and selection sources. In reality, a relatively small firm would commonly prefer the simple approach, whereas big multilevel organization would require a reliable MIS for their HRM, since by having an IS it would require a credible HR practitioner and teams with IT tools and skills to run the system.

## 3.1 INTRODUCTION

Human resource simply defines the human workforce. There are several definitions of human resource management (HRM) defined by various scholars and HRM practitioners. In general, HRM is a planned, integrated, and comprehensible approach to the employment, development, and well-being of the working in organizations. According to Beer et al. (1984), "human resource management involves all management decisions and action that affect the nature of the relationship between the organization and its employees—its human resources." Guest (1987) on the other hand

believed that HRM is a set of guidelines designed to maximize organizational incorporation, employee obligation, elasticity, and quality of work.
Added more, HRM involves the physical, psychological, and social aspects
of an employee and the organization (Chand, 2014). Each and every definition agreed that HRM is involved in the process of hiring, job analysis, and
recruiting either newly appointed employee or existing employees.

## 3.2    MANAGEMENT INFORMATION SYSTEM

The technological evolution keeps on growing, from manual papers and
filing system the world has move on to the computerized management information system (MIS). In general, MIS is the study of people, technology, and
organizations. Nowadays, each and every part of life essentials or routine
involves the usage of technology. For example, bill payment is now well
known using either via applications or on the job systems, these systems
will record on time transactions. MIS is also used as a tool to improve the
performance of the organization and to make an effective decision where it
is necessary.

Several scholars or writers gave varieties of definitions according to
their understanding of what management information is all about. MIS is
the department-governing hardware and software system used for business-
related critical decision-making within an organization (Rouse, n.d). In addition, Beal (n.d) believes that MIS broadly refers to a computer-based system
that provides managers with the tools to organize, evaluate, and efficiently
manage departments within an organization. MISs professionals help firms
realize maximum benefit from investment in personnel, equipment, and
business processes.

## 3.3    CURRENT TREND OF HRM

Working experience is vigorously changing. In general, HRM is part of
organization's operational business function. However, HRM falls under the
administration department or governance. It is crucial for HRM to understand the needs of the employees to accommodate the rapid change of the
world technological aspect. Business or organization today is widespread,
ranging from various sectors and levels. This results in globalization and

workforce diversity where HRM needs to ensure the appropriate mix of the employees in terms of their knowledge, skills, etc. (Andrew, 2016).

## 3.4   OBJECTIVES OF HRM

Each and every departmental organization has its own objectives in line with its organizational mutual mission, vision, and objectives. Primarily, HRM is to ensure the effectiveness of having the right person for the right job with the suitable qualification in order for them to achieve their main goals efficiently. In other words, their main objective is to accomplish the goal effectively and efficiently by providing skilful and diligent employees'. From the main objective we can break it down into several subobjectives. It is to ensure the employees' job satisfaction and employer's expected positive performances (Patel, 2016). In addition, it is to maintain the quality of work of the employees along with their ethical behavior within the organization.

Werther and Davis (2016) classified the objectives of HRM into four different categories, namely, social, organizational, functional, and personal objectives. Each objective serves its functional purpose. For example, societal objectives serve the union management relations where employees are allowed to voice their opinion through their labor union group.

## 3.5   HRM IN MIS PERSPECTIVE

Human resource information system is one of the global technical approach practices. With the aid of information system, HRM can be analyzed effectively when it comes into deducing the possibilities of employees with lack or better skills, their knowledge capability, and work efficiencies.

## 3.6   LITERATURE REVIEW

### 3.6.1   IMPORTANCE OF HRM

For a few decades, scholars and human resource practitioners started to acknowledge the importance of HRM for overall organizational performance appraisal (Kehoe and Wright, 2013). Below are several common HRM importance in general practices:

### 3.6.1.1    FURTHER INCREASE THE ORGANIZATIONAL DEVELOPMENT EFFECTIVENESS

As previously mentioned above, HRM is a tool to increase the development of organizational progress. By having HRM, an organization has the ability to control its employees to act according to their ethical code of conduct and achieve the mutual goals of the organization itself. HRM is responsible for the recruitment of the right person for the right job, training the existing employee to accommodate the needs of the current world trend in work-related field, and more to ensure the smooth progress and positively increase the organizational development process.

### 3.6.1.2    MAINTAIN DIVERSITY'S RELATIONSHIP

A study has been conducted in Oman by Moideenkutty et al. in 2011 for the relevance of HRM practice to the organizational performance. The result is positively encouraging and accepted in the context of Islamic world. Oman is one of the Islamic countries that has a lot of expatriate and locals. Businesses are widely available and these require a high involvement of HRM to manage all the employees from various nations. This is to ensure they are able to synchronize the standard performance measurement among employees despite the fact of their nationality, religion, and races.

### 3.6.1.3    EMPLOYEE'S ATTITUDES AND BEHAVIOR

Employee's personal behavior and attitude are differed from one person to another. Psychologically each human being is prone to different things. One person might be motivated in a positive way, the other might be the other way around. HRM is responsible to monitor their employee's progress and stability.

### 3.6.1.4    JOB SATISFACTION

Job satisfaction is one of the main contributors to the improvement of employee's performance and also the end result of a good motivation at work. HRM needs to understand the needs of the employees and know how to differentiate the various behavioral acts of each employee. They are responsible to keep the employee motivated by managing their well-being, aware of what they are lacking and what skills need to be improved, etc. Positive vibe shown by HRM team may lead to incredibly well job satisfaction by the employee as they are highly appreciated and welcomed to the organization's community.

### 3.6.1.5  JOB COMMITMENT

Commitment in general is the willingness of the employees to something they have been appointed to or something they believe in (Cambridge dictionary, n.d). High job commitment toward the work is what the employees are expected to give in return of their paid salary. High job commitment may lead to a positive change in life and ability to adapt any changes easily to their working circumstances. HRM team has a role in this particular case as an agent to drive the employees for a better work commitment. They are to ensure the well-being of the employees as what have been mentioned on the previous point. Both are related to one another; job satisfaction may lead to better commitment.

### 3.6.1.6  MAINTAINING WORK ATMOSPHERE

An individual performance of the employee is largely influenced by the positive work atmosphere or work culture (Administrator, 2017). A good working condition is one of the vital benefits for the employees to be more productive and innovative. They can be expected to be more actively engaged and help their team members. Health safety executive (HSE), a safe, clean, and healthy working environment may lead to healthy minds which hugely affect their working habits and outcome. HRM is responsible for the pleasant working environment to enhance the intellectual capability and physical skills of the employees' in their everyday operations.

### 3.6.1.7  MANAGING DISPUTE AND DEVELOPING PUBLIC RELATIONS

It is normal for an organization to experience a rough path among the employees and employer when they are not on the same wavelength. It often occurs when one party is not satisfied with the other. As previously mentioned, it is the responsibility of human resource department to manage the employees' satisfaction to their job and environment to minimize the possibility of crisis in the workplace. As a human resource officers, they are the very first to hear the complaints and grievances. They are expected to at least come up with solutions though it is only for temporary matter (Administrator, 2017).

To some extent, HRM is also responsible for the company image and relations with all stakeholders including the customers. Human resource department plays some important roles in organizing the business meetings, agendas, covering issues, and clearing the misunderstanding occurring to the organization (Administrator, 2017).

### 3.6.2   HRM TRENDS IN MISs

Over the years, the rapid development of technologies has forced an organization to constantly evolve simultaneously with the current trends of HRM. From a traditional approach to the most advance information systems (ISs) using the latest IT tools, it has solved most problems that encountered using the HRM traditional ways.

### 3.6.2.1   RECRUITMENT

As a business grows, there will be an increase of workloads and processes. It will become more complex and require more resources and this could create some employment opportunities to the market. According to Taylor and Collins (Phillips and Gully, 2015), recruitment is the "most critical human resource function for organizational success and survival" and it is the "foundation of the organizational performance." Therefore, successful recruitment practice may result in strategic competitiveness of the organization.

By applying a MIS to HRM practice; it may have solved the problems on the growing numbers of employees in an organization. According to Russell and Chamberlain (2016), "Sourcing is a skill that is to some extent emulated by expert finding recommender systems, where machine learning is used to select the best suited individual to perform a particular task." Their study found out that 27% of HRM specialists' time is actively consumed on searching the right applicants. Therefore, it may save the energy and time for HR department to get the qualified candidate by having the right IT skills. The system would automatically extract the shortlisted candidates that meet the job description.

Another article that has the similar opinion on using computer-based HRM is by Tripathi (2011) who stated that by implementing MIS would help top management to control and make the right Human Resource decision-making. Furthermore, he also mentioned that MIS could provide accurate, timely, and reliable information before making a crucial decision which could directly affect the performance of the organization. By having MIS in HRM, it becomes easy for HR department to filter the applicants and match the right qualification in the shortest possible time.

However, Barber et al. (cited in Phillips and Gully, 2015) found that though MIS is essential in HRM, the size of the firm indicates the choices of the firm to decide whether to implement the MIS or just use the internal recruitment and selection sources. In reality, a relatively small firm would commonly prefer the simple approach, whereas big multilevel organization would require a reliable MIS for their HRM. This is because by having an IS it would require a credible HR practitioner and teams with IT tools and skills to run the system.

As a whole, MIS does play an important role to help organization to remain competitive, productive and effective, in managing a diverse workforce.

## 3.6.2.2   TRAINING

Developing employees' skills is also important for an organization in order to maintain the productivity and quality of its workforce. To remain competitive, an organization has to update the systems regularly and this will require HRM to conduct training programs to support the systems.

According to Karikari et al. (2015), the use of human resource information system in organization encourages the necessities in identifying the insufficient of employee's skill gap that needed to be fulfilled, it helps to enhancing their capability in providing suitable for the job with proper training for the effectiveness of training programs to meet the standard requirements. With the integrated system being implemented in human resource, it is easier for the management in planning on the long-term purposes and forecasting the opportunity of organization development by gathering, storing, processing of information on employees' qualifications, work performances, recruitments for training program needs.

Human resource information system is the effective method for training workforces in computerized system which is much more relevant in acknowledging the availability of positions for the right people to occupy in an organization. They claimed that the new advancement and development in the era of 21st century have affected the work environment in creating innovative way of obtaining knowledge, skills, and requirement of employees (Lee et al., 2009). They used one example in relevance to their research paper on "Training Older Workers for Technology-based Employment," it has stated that the advancement and reliability of technology nowadays had change the structure of employment's processes, task, and skills requirement. The stronger dependence on computer-based technology nowadays indicates increasing number of internet and electronic application users at work that will encourage higher demand for skilful and educated employees in the future. Hence, the participation of older people who do not have know-how of using technology and have to adapt the new changes of work environment are in needs of continuous support in training program in order to be able competitive in the organization.

Maurer et al. (2003, cited by Lee et al., 2009), claimed that the involvement in training and engaging with activities came from the inner self of the employees which is influenced by their personal reasons that they needed to make improvements in themselves and seek the favors of the programs. Throughout the research they have conducted they found that participants who responded their questionnaire have shown that they were lacking in technology skills. As a result, for an effective training program, top level management needs training to understand the significance of senior workers by giving them the opportunity in the workplace. Furthermore, senior people are advised to have their technology training programs with the appropriate standard requirement. Regardless of the age differences, the involvement of workforce is a key to MIS in the organization on the productivity; their knowledge can be transferred to younger generation and vice versa with the availability of information technology as guidance.

### 3.6.2.3  COMPENSATION

Within every organization, the human resources department is always considered as the most crucial department for the whole organization. Its various functions serve as the backbone for the company serving entirely from recruitment and selections, training, evaluation, compensation and benefits management, promotion and employee advancement opportunities, and more. Since human resource is the highest expense for majority of the organizations, human resources department assists the companies in acquiring the greatest value of this important asset.

Compensation and benefit management are simply monetary values given to the employees in return of their services. In the book of HRM, Gary Dessler defines compensation in these words "Employee compensation refers to all forms of pay going to employees and arising from their employment" (Feb 12, 2010).

It is an important component in HRM as it helps to motivate the employees and reward them with benefits based on their performance at the workplace. The greater the compensation and benefits granted by the employer, the higher the level of loyalty and motivation to work and perform better. However, companies that give lower salaries can see higher rate of staff turnover and less productivity from the employees. All these elements make compensation and benefits as critical factors of managing human resource.

Therefore, Human resources department must have the right tools and software in order to operate more efficiently. According to Dave (2017), a human resources information system (HRIS) is a type of software program that can be utilized within the department to help human resources employees and managers improve their productivity and the results of their efforts.

By implementing Human Resource Payroll Systems, it supports the human resource identification, evaluation, and selection of HRIS software that will best suit to the organizational needs. The system includes personnel tracking, employee self-service, applicant tracking, time and attendance, performance reviews, payroll systems, and more.

Investment in the information system helps to achieve the organizations' operational excellence objectives which include the productivity and efficiency, innovative new product and services, improving the customer relationship with better decision-making, achieving the organization competitive advantage and ensuring survival.

As stated by Papia et al. (2015), middle management uses human resources systems to monitor and analyze the recruitment, allocation, and compensation of employees. Operational management uses HR systems to track the recruitment and placement of the employees. HRIS can also support various HR practices such as workforce planning, staffing, compensation programs, salary forecasts, pay budgets, and labor/employee relations.

Teo et al. (Papia et al., 2015) said in article "Adoption and Impact of Human Resource Information Systems (HRIS)" that HRIS was used mainly for administrative purposes like payroll and employee record keeping, rather than strategic applications like succession planning. He found a tremendous amount of unrealized HRIS potential as very few respondents are using the HRIS strategically to directly improve their competitiveness; HRIS was simply viewed as the use of computer hardware and software applications to perform HRM activities.

This is also supported by Farhat Ali Syed (Papia et al., 2015) who stated in a study of HRIS in Indian banking scenario (issues and challenges) that as Indian banking is going through structural changes, this concept is proving a boon for its overall development. The manual system adopted by the banks was the cause of procedural delays, outdated as well as inaccurate information, higher expense of storing files and their safety. He added that banks are able to integrate different HR functions by using third generation of feature-rich, broad-based, and self-contained HRIS.

## 3.6.2.4  HEALTH AND SAFETY

Another trend of MIS for HRM is on the health and safety (H&S) of its employees. H&S is one of the HRM function. Managing H&S is very important and it is one of the major concerns in every organization. "Every 15 s, a worker dies from a work-related accident or disease. Every 15 s, 153 workers have a work-related accident" (International Labour Organization, 2017). This statement was followed with a result of reported number of deaths from 215 countries from 1989 to 2013 on fatal occupational accident or work-related of 6300 people daily. This equals to more than 319 million deaths per year.

The main cause of these deaths was inadequate and weak occupational H&S practices at work. The consequences of accidents at work may not just cause significant loss of life but will cost the organization to pay the compensation, medication, legal action, and even the cost of finding new employee for replacement. This will include the cost of recruitment and selection and training the new employee for replacement. Therefore, it is the responsibility of the organization to prevent hazards and protect their employees and stimulate with a safe working environment at the very best way possible.

According to the Ministry of Labor Department in Alberta, Canada (2017), "health and safety management system is a process put in place by an employer to minimize the risk of injury and illness." This can be done by identifying, assessing, and controlling risks to all employees in any functional area. They also stated that one of the factors that could make H&S

management system to be effective was management system administration. Therefore, it is important to have professional H&S personnel to fully perform the safety working environment duty according to the type of workplace and operations carried out.

The role of MIS in relation to H&S will comprise the availability of recorded data on reported injury or incident at work. With MIS, the organization can monitor the total number of injuries that had happened and can trigger to the organization if there is a gradual increase in the number of accidents. MIS would also help analyze when, where, and frequency of the incident. By knowing these statistics, HRM can take preventive action or initiate policies and practices.

Overall, it is clear that MIS plays a major role for all HRM processes such as in recruiting, training, compensation, and H&S, in order to acquire an effective human capital and develop the skill of current employees in an organization.

### 3.6.3   CHANGE MANAGEMENT IN WORKPLACE

Organization may come across obstacle to accept changes in their culture and work-related matter. This is especially according to their employees' point of view. Therefore, CEOs have to take the initiative to change their employee's mind setting to accept the changes occur in order for them not to held back behind the fast development of technology in everyday life (Price, 2003).

#### 3.6.3.1   PURPOSE TO BELIEVE

When the organizations take a step into changing their organizational norm to fit in the current trend of the world, they have to believe on what they are doing. Believe by doing so it will open another opportunity for the organization. Changing an organization is not easy yet doable in certain circumstances. When human brain believes on something, automatically their brain does accept the change sooner or later. What they are waiting for is the switch to trigger their thinking that change will bring positivity.

#### 3.6.3.2   REINFORCEMENT OF SYSTEMS

Positive reinforcement of acceptable system leads to a positive acceptance of change. It does motivate the people to go through the transition from one level to another. Organizational designer, design the structure, management and operational process of the organization. It is their initiative to reinforce an easy-going system. However, the acceptance might not be stay still when maintenance is not done accordingly. People always anticipate a positive change yet they forget to keep on conveying the fun side of the change.

### 3.6.3.3    THE SKILLS REQUIRED FOR CHANGE

Change required the right skills for the right program of change. When people are transiting from one phase to another they have to either born with the skills or developed the skills through training, etc. This is in order to prepare them for the new exposure, so they will not feel alienated when doing the work. When an organization conducts clusters for employees' merger from different operational departments that require different background of expertise to complete a project within the timeline. Employees would have the tendency and willingness to share their knowledge as part of members' exposure from communication and teamwork. Sometimes employees tend to hinder themselves from confronting as they might be lacking in certain areas which not only affected the employees' performance but also the productivity. The knowledge transfer often reflects to how the application of the information can be extracted from and can be used. However, to restructure the retain skills to line up with the recent advancement approach, employees need to have a training and workshop that ensure with the organization's requirement for certain jobs and positions. Employees are ensuring the need to fulfill their skill discrepancies through getting the right training in specific field. A person's eagerness to learn new skills is depending on the situation through self-readiness to improve their expertise and when dealing with a situation that encourages a person to adapt and be able to absorb the skills in any circumstances.

### 3.6.3.4    CONSISTENT ROLE MODELS

This is a very crucial aspect of change. Leadership leads their subordinates to either a positive or negative change or simply neglecting change at all cost. Leaders are meant to be followed, and they are the main reference for the subordinate to see the future of new changes in their workplace. Positive aura will attract people to accept changes.

## 3.7   RESULT AND DISCUSSION

### 3.7.1   IMPACT ON THE ABSENTEEISM OF HRM

There are several logical impacts on the absenteeism of HRM in an organization.

#### 3.7.1.1   CHAOTIC OPERATION

As known by all, HRM is very important in shaping the organizational structure, designing the recruitment, and selecting suitable applicants, etc., how can an organization be in a good operational strategy when they have no one to monitor and control the increasing number of employee within an organization. Their daily productivity depends on the efficiency of the HR department. For example, in work allocation, are they actually placing the right person for the right job? Or in terms of job hour, are they actually allowing the employee to work less or more than the standard working hour? These should not be overseen by the CEO. As per the study conducted in Oman, despite the fact their diversity it proves that HRM is very important for the organization in all aspects. HRM helps to unite the employees into one tree and work together to achieve the mutual goals of the organization.

#### 3.7.1.2   EMPLOYEE'S ERROR

In the absence of human resource department, there is no one to manage the employee recruitment, training, and selection. The organization may pick random person without the proper experience, knowledge, and qualification. This may lead to decreasing of job performance of the employees as they are not handling something they are familiar with.

#### 3.7.1.3   UNABLE TO ACHIEVE THE ORGANIZATIONAL GOALS

Organizational goals must be in pars with departmental objectives. When HR is not in function to organize the objectives together with the employee's outcome, it is impossible to work together to achieve the goals. For example, the organizational goals are to be the very best beverage provider globally, yet their employees only targeted local market without taking into consideration the other opportunities. They are not on the same wavelength.

### 3.7.1.4   BAD PERFORMANCE OF BOTH EMPLOYEE AND ORGANIZATION

Performance plays some important roles in attracting more investors and shareholders. When the operations are bad they cause a withdrawal effect to the investors. HR should start from the inside as a happy employee may lead to a very good outcome and further performance of the organization.

### 3.7.2   E-RECRUITMENT

Many organizations have successfully implemented MIS into their HRM processes. One of the online HR systems that currently used by the Brunei government is the Public Service Commission (PSC) Recruitment (Fig. 3.1). This system "was created to replace the existing system which is Public Service Information System (SIMPA, System Informasi Pelayanan Awam) used by the PSC since 1999. SIMPA is no longer able to meet the current needs of the growing and changing demands and technology according to time. Online Recruitment System has more functions than SIMPA, which uses a web-based platform and enables access at any time anywhere online" (PSC Recruitment, 2015). It was implemented under the Public Service Commission Department commencing February 2015.

**FIGURE 3.1**   An image of PSC recruitment system website.
*Source:* https://www.recruitment.gov.bn/en-us/Pages/welcome.aspx.

The advantage of this system is it allows PSC Department to advertise the job vacancy online. This will reduce the processing time and reduce the paper cost for advertising the post. Secondly, it allows candidates to apply for job vacancy online by searching the available vacancy in government sector via the portal and entering their personal information details directly into the system. By doing this, the system will have the personal record of candidates and provide report on the number of applicants and listed down applicants that fulfilled the job descriptions. Thirdly, this will also resolve the problems on limited space for storage in PSC as previously every job application form must attach with applicant's copy of relevant documents.

Another e-recruitment system found in Brunei Darussalam is the Jobcentre-Brunei or Agensi Pekerjaan Tempatan Dan Pembangunan Tenaga Kerja (APTK) as shown in Figure 3.2. This system was conducted and monitored by the Energy and Industry Department, Prime Minister's Office, Brunei Darussalam. This system is similar to PSC Recruitment but it only focuses on candidates who are interested in applying job vacancy under private sectors in Brunei. Using this system, it will instantly display the numbers of jobseeker, numbers of job vacancy available, and numbers of companies that offering the job. It also stated the expected age and salary offered by the companies. Therefore, it is easier for the system to filter the candidates according to the requirement.

**FIGURE 3.2**    An image of Jobcentre-Brunei website.
*Source:* http://www.jobcentrebrunei.gov.bn/.

The main limitation for these systems is the internet connection in Brunei Darussalam, as it requires candidates to have an internet connection and a computer or smartphone in order to access the systems. Another limitation is how fast or frequent the system will be updated if the vacancy has been filled.

The third example of MIS for HRM process in Brunei Darussalam is the I-Ready. It is a 3-year apprenticeship program and acts as a platform to expose graduates to various industries in both public sector and private sector and gain real working experiences (Energy and Industry Department, Prime Minister's Office, 2017). This program was first introduced in April 2017 as part of the government efforts to reduce unemployment rate in the country. As of March 4, 2017, the number of unemployed local were 12,966 people and 2328 people out of this were graduates who have degree and above qualification.

Majdi (2017) reported with I-Ready, a monthly allowance of $800.00 per month will be funded by the government to the selected candidates to allow them gain the reskilling opportunities up to a maximum period of 3 years; however, the companies have the option to permanently employ the apprentice at any time. The process of selecting candidates is similar to recruitment process in HRM. Therefore, with MIS, apprentice can easily access the I-Ready system online, and the respective host organization with JBC representative is responsible in screening, short listing, and interviewing the selected candidates. Successful candidates will be offered an apprenticeship contract and the host organization is responsible in coaching and mentoring them by giving continues support.

In another example, MIS has been used globally over the years. The current trends used by the world are the Applicant Tracking System. This online system consists of Candidate Resource Management which allows candidates to update their details, qualifications, and experiences instantly into the system. For example, Zoho Recruit. It is an automated online recruiting software for staffing agencies and human resource department which can efficiently track resumes and talent and streamline the hiring process. There are two packages in order to subscribe the software. The first package called Enterprise costs $50.00 per month, whereas the standard package costs $25.00 per month. Economically, the organization will get a low cost online software and will not give much effect on the financial implication. In general, MIS will not only smoothen the hiring process but also hire the best skilled labor and reduce the risks of skill deficiencies issues.

### 3.7.3 TRAINING

Civil Service Institute (CSI) under Brunei's Prime Minister Office is an institute that delivers consultation, providing courses and training programs

required for public servants. The objective of CSI is to provide continuous learning and development in human resource aligned with the aims
of national development. As human resource become part of the important function management in the organization, CSI is responsible to ensure
the objectives are achievable by enhancing the potential, capabilities, and
competence of civil workers with the emerging contemporary learning
associated effectively. Thus, it provides a framework that links the training
activities in consists of research, consultation, and development programs.
One of the initiatives, for example, to increase human resource capabilities
within the government agencies is conducting the training, workshop, and
programs provided by Public Service Department and Ministry of Education. Current approaches being conducted are competency-based training in
utilizing public servants' skills, such as in information tracking and ICT,
through Leadership Development Program Framework, which compiles
some of the development programs for advance executive, senior executive,
and administrative officer. Few of the difficulties they have to confront are the
long-term training and development based on organizational strategic plan,
and the result of training assessment in human resource were not feasible
effectively. To adapt the situations, selected strategies application can be
implemented through enhancing the use of technology by public servants
such as Government Employee Management System and encourage Human
Resource experts being part of the team in strategic planning. Being able to
step ahead with full dependency on current MIS in comprising the combina-
tions of management in system, hardware, people, and networking nowadays is valuable to the human resource in providing strategies to fulfill the
workforce skill gap. The major impact of changing skills in public servants
has dominated the needs of information system in human resource to be able
to be a competence workforce as one of the competitive strategies.

As a training approach for learning, it has become a trend of gathering
information system especially beneficial for the human resource in organization. The support of information system in functional departments in organization will help the management run effectively in gathering the information
credibility. For example, Brunei E-Government National Centre (EGNC),
Prime Minister Office is offering a training course "Digital Literacy Certification Programs for civil servants." It provides online training courses
in digital web application using iPad, mobile phones, and other electronic
devices.

The course duration held for 8 days with given three stages for applicants
to complete once registered. For all applicants, they were given easy access

to additional resources with Microsoft e-learning courses for free, depending on applicants' preferable time. The courses were divided into three level of modules that will help the users to understand the fundamentals of using computers, Microsoft office e-learning courses application commonly used everywhere such as Microsoft Word, Excel, and PowerPoint, beneficial for the learners' competency in communication, networking within the application of internet and web. The objective of the training is to provide relevant knowledge and skills on ICT skills assessment usage to synchronize their potential equivalent with the current information in preparing themselves working with government (e.g., Google Training).

### 3.7.4  MIS IN BALANCE SCORECARD

Balanced scorecard framework (BSC) is widely used by all the organizations to measure strategic planning and for evaluation of performance and communication. It is an effective practice in identifying the key performance indicators and other factors that needed initiatives for improvements. It identifies four major aspects of finance, customers, internal processes, and learning that provide clear direction to evaluate the performance of the organization. According to Kettunen and Kantola (2005), Kettunen has stated that the application of BSC as a strategic planning might be sufficient for organization but it does not give much attention to development of the strategy into the next level. Based on their research, they found that balanced scorecard was lacking and inapplicable way to link the operational function with organizational strategic plan and ensure the measures of success strategy as a result of occurrence of overlap information and inaccuracy calculations. They also justified that with MIS IT implementation can help to redesign the management effectively to meet the requirement.

With MIS involvement in the organization, it is easier to check which area of weaknesses to be compared when using BSC alone. Record tracking system, efficiency in checking human errors, reducing workload among members in different level of management are some of the potentials of how efficiently it can provide with the application of IT as a tool. It is useful for operational departments in which it helps to reduce complications of manual workload and improve the performance and productivity. An example for human resource, attendance can be effectively registered by using fingerprint scanner which is hassle free from attendance recorded manually as it directly goes into the system. It makes the organization paperless society in cutting cost of using paperwork where the system can store the information and helps reuse it for future forecasting and strategic planning. With the new system being implemented, people can always track their attendance and productivity on how efficient they can be in producing better work within the timeframe in the organization. All the information is more precise and accurate for key performance index of employees, whether they are in a state of doing great performance and measures for bonuses and promotion applicable in fastest way. Feedback and complaints through system are quite reliable to get aside from difficulties in finding where and to whom to go. Employees counseling can be easily handled with the system by identifying the concerned problems encountered by employees, this application will help to analyze and minimize the difficulties by providing the right training for instance ethic issues.

## 3.7.5 COMPENSATION

Information technology in an organization particularly in human resource assists to manage compensation such as salary, bonuses, commission, daily or hourly pay more efficient and effectively. Implemented system program simply calculate hourly, weekly, monthly pay based on the organization's annual salaries and automatically generate payslip or direct deposits, which are known as "electronic fund transfer" (EFT) from the company's bank account to the employee's. This software can be accessed through intranet or internet.

For an example, HSBC Bank in Brunei implemented a self-service system for employees to access their account and monthly salary information. The system allows them with an option either to print or view only for the monthly statement. This can be accessed 24 h using the internet. The system has cut down the workload of the HR staff and reduced the bank's overhead costs through eliminating printing employee's monthly payslip and statements.

One of the initiatives of His Majesty's Government to implement the e-government strategy, the Treasury Accounting and Financial Information Systems (TAFIS), was introduced in March 2002. According to Azhar Ahmad (n.d), TAFIS project is one of the initiatives in line with the e-government strategy of the Brunei Darussalam National Information Technology Council. It is aligned to His Majesty's directive to move the nation toward a knowledge-based economy (p 3). The TAFIS is implemented for the Brunei Darussalam civil service involving all the ministries, departments, and government agencies.

As stated by Mohiddin (2008), the objectives of the implementation of TAFIS are to improve efficiency of government financial transaction processes. The transaction processes to include daily rated payroll, gratuity and pension, and allowances. Based on the survey showed in his slides, few improvements had been achieved. By transforming from manual process, shorter time is taken for payment approval and invoice payment. Less number of days to pay leave allowance and education allowance and a higher number of payments to employees and vendors automated via EFT.

Other objectives are to improve monitoring and management of government funds and thereby optimize utilization, to transfer of technology-based knowledge, skills and abilities, to move toward a "paperless" government by eliminating printing process besides reduced paper and printing cost, and finally, is to set up TAFIS information technology networking centre. Therefore, it is convenience for the vendors and civil servant to deal or make any inquiry with regards of any payment issues.

With the integration of TAFIS system, few advantages and disadvantages have been learned and indirectly give an impact to the people, working processes, and technology infrastructure in Brunei Darussalam.

First is the impact on people, the online approval, and gaining of new skills, this had developed the workforce to achieve higher productivity and reduce duplication of work. Second is the impact on the workflow and working processes of PC-based system integrated into TAFIS system. Elimination of majority manual processes such as manual spreadsheet for cash control file, Journal statement and balance sheet, manual reconciliation using cash book, ledger control system and budget control system. Currently, the manual processes are replaced by computerized reconciled data onto reporting views payment is automated via EFT. It is paperless driven by workflow, data entry, and retrieval is done online, automatic reconciliation with single point of data entry in order to avoid duplication of data entry.

Finally, the impacts on implementation of TAFIS system on technology infrastructure, it improved and developed network connectivity and establishing further interfacing with other application systems and highly focusing on network and information security.

### 3.7.6   E-HEALTH AND SAFETY

An organization will be beneficial if safety data can be transformed into actionable insights. With MIS, it could help attain the goals to H&S compliant and efficient organizational operation by automated H&S processes. One of the example using MIS in H&S is available online using a software called Environmental, Health and Safety (EHS) MIS. EHS is a software system that provides solution and constant guidance of best safety practices to any business to improve its H&S environment. EHS MIS ensured that their program is "A great safety program helps companies prosper, and makes sure compliance is achieved and workplace and environmental risks are reduced" (EHS Insight, 2016). All H&S personnel could understand, measure, analyze, and report data using this program. Furthermore, based on historical data points, EHS MIS may assist in proposing solutions and forecast the potential performance of the organization.

There are a few advantages of this EHS MIS which could help to enhance the business safety culture and direct to improve the employee morale. Secondly, HRM can focus more on employees' health and business operations because redundancy can be automated and the paper-based tasks help to conserve time. And thirdly, by preventing and eliminating the hazardous incidents, it could help the organization to save cost.

### 3.7.7 CHALLENGES OF MIS

#### 3.7.7.1 FINANCIAL

The development of information technology may lead to organization's success; however, developing MIS should be based on the cost–benefit analysis. The value of making a decision after developing the MIS should be more than the cost of developing it. New system should be accompanied with the system administrator in order to continue maintain and run the new system. Therefore, large organization may need large financial resources to build a more complex system. That is why some firms are reluctant to change to the new technology due to financial constraints.

#### 3.7.7.2 AVAILABILITY

Not all ISs are widely available and user friendly. For example, in Brunei we might be using what the other country is currently using but not every other country will follow. The availability of each and every information system depends on its producer, copyright, and product distribution.

#### 3.7.7.3 SECURITY

In terms of security, every production of information system is produced with good security measure. Toward what extent does an information system could protect the information stored in the system. Take example for employee information system, employee might feel violated when their personal information is leaked or breach by hackers. In another case, when financial information is accidentally stolen by competitors. Therefore, a great measure of security needs to be taken into consideration to ensure the end user to feel secure and confident in using the system.

### 3.7.8   LIMITATIONS OF MIS IN HRM

#### 3.7.8.1   REDUCE FACE-TO-FACE COMMUNICATION

The increasing number of technology used in the workplace cause the less verbal communication among the workers. Though they are communicating via online ISs that they have shared within the organization, it is still lack of emotions and sentimental value of the conversation. Other than words and way of speaking, emotions may help in deducing the conditions of the employee. Let us say they are having trouble in their work and asking for a guidance from their superior via text messages, the superior might take it the wrong way that the employee is notifying them of something instead of asking for help because he or she could not hear the voice notation or their facial expressions.

#### 3.7.8.2   MISUSE OF THE IMPLEMENTED SYSTEM

Based on personal experienced, these systems do have some drawbacks as for example in the APTK system not all employers are actively engaged with the jobseekers. Some are just for the sake of posting an advertisement to fulfill the requirement of registered employers. On another occasion, their reasons for rejecting the jobseekers are not acceptable as they said the position already filled in by the walk-in jobseekers before they even have a look on the online applicants. They are only taking advantage of the readily available platform without actually using it accordingly.

#### 3.7.8.3   LACK OF SUPPORT FROM TOP MANAGEMENT LEVEL

This is the main limitation of implementing a system. When the higher up is not keen to accept the change how can the subordinates be willing to change or excited to keep on using the system. When the higher up is already neglecting the change due to their comfortableness in the older system, it does not help their employees to be more productive and up to date with the current trend system used.

### 3.7.8.4 *RESISTANCE TO CHANGE ESPECIALLY FOR THE SENIOR EMPLOYEE*

Common problems occur to the majority of senior employees. They are very hesitant to change their way of work. Adapting technology in their old fashion work style caused them the need to learn many new things and understand many new jargons to improve their work performance. They do believe in benefits of changing their work style from manual HR process into using information system, but the process of learning the new technique that stop them from moving forward. Being in their comfortable zone makes them feel secure so that they could minimize the failure or error in using the new system (Reynold, n.d). They have also set their mind that the new upcoming generation is sufficient enough to cover up their lack of effectiveness in the workplace.

In addition, a greater reliance on the implementation of MIS system allows workforce to be within their comfort zone. The maintenance support of human activities in handling specific task will encourage themselves to resist in changing own status quo, such tendency of laziness may occur and person's skill development might be deteriorating in work place throughout the time if there are no measures to be taken for such situation. Therefore, a proper planning and a suitable technique would be required in making sure that employees are equipped and fully aware and know how to operate effectively if there is a system failure. Contingency plan is needed to be prepared in case of emergency and to avoid unforeseen circumstances by providing training procedures.

### 3.7.8.5 *IMPROPER PLANNING COORDINATION FOR SUITABLE MIS IN THE ORGANIZATION*

The initial stage of planning for selecting MIS to an organization is very crucial. The management is not only focusing on choosing the right tools for the organization in order to increase the performance and productivity, the organization is taking the risk of investing in the suitable software that matches with the organizational needs. The cost of implementing a software is usually very high. Mismatch software implemented can be a waste for the company.

## 3.7.8.6   IMPROPER SELECTION OF MIS

Without well strategic planning on application use in management may create challenges in communication among workers especially misuse of technology information, users of WhatsApp and telegram in sending direct information became prominently popular in the workplace. Employees can easily have lied about tracking attendance with MIS through WhatsApp text when the user is not currently present at the office. If there is no suitable selection, MIS implemented appropriately with standard requirement in every level of operation management. The information credibility will be taken for granted as the information does not apply to the right users. For instance, in finance department provide own accounting system format while the applicants were exposed in training for different accounting systems which creates confusion of skills requirement equivalent with the job description.

## 3.7.8.7   LACK OF SKILLS AND KNOWLEDGE IN USING MIS

Implementation of new software in the organization may require the user to be fully equipped with skill and knowledge in using the MIS. When user faced difficulty with the new function, employees will be not productive as they are not familiar with the system. This may lead to other problem such as demotivation.

## 3.8   CONCLUSION

MIS plays an important role in human resources management process in both public and private sector. The process includes recruitment and selection, training, compensation, and H&S. Implementation of a perfect and good utilization of information system can lead to a smooth and effective workflow to an organization in the long run in order to achieve organizational goals. But this can only be achieved with proper planning coordination and careful consideration before engaging with HRIS venture. Research are limited to the availability of secondary resources via online research and primary data base of past experiences of the researchers.

In conclusion, the systems mentioned throughout this chapter that have been implemented by Brunei Government are positively effective. Changes can be seen in the current way of handling the job recruitment processes and more straightforward. Applicants can simply click the job they wish to apply for and get notification within the good period of time either to be accepted or not. It is proven that MIS acts as indicator that driven the human resource department into an efficient and productive department.

**KEYWORDS**

- multilevel organization
- job satisfaction
- job commitment
- work atmosphere
- human resources information system

**REFERENCES**

Andrew, R. A. 14 Current Trends in Human Resource Management. 2016. https://www.linkedin.com/pulse/ 15-current-trends-human-resource-management-rutaihwa-aristides-andrew (accessed Aug 23, 2017).

Armstrong, M. *Armstrong's Handbook of Human Resource Management Practice,* 11th ed.; Kogan Page Limited: London, 2009.

Beal, V. MIS—Management Information System. n.d. http://www.webopedia.com/TERM/M/MIS.html (accessed Aug 21, 2017).

Chand, S. Human Resource Management: Meaning, Objectives, Scope and Functions. 2014. http://www.yourarticlelibrary.com/hrm/human-resource-management-meaning-objectives-scope-and-functions/35229/ (accessed Aug 20, 2017).

Commitment Meaning in the Cambridge English Dictionary. n.d. http://dictionary.cambridge.org/dictionary/english/commitment (accessed Aug 26, 2017).

Craig, R. The Importance of Applicant Tracking Systems: An Interview with Talent Tech Lab. *Forbes* 2017. https://www.forbes.com/sites/ryancraig/2017/04/28/the-importance-of-applicant-tracking-systems-an-interview-with-talent-tech-labs/#36ab94383a81 (accessed Aug 22, 2017).

Dave, R. HR Payroll System. What are the Benefits of HRIS. 2017. https://www.hrpayroll-systems.net/hris-benefits/ (accessed Aug 24, 2017).

E-Government National Centre. Digital Literacy Certification Programme for Civil Servants. 2017. http://www.egnc.gov.bn/Shared%20Documents/Digital%20Literacy%20Certification%20Programme/EGNC%20BOOKLET%206-4-13(LQ2).pdf (accessed August 25, 2017).

EHS Insight. *EHS Management Information System.* 2016. https://www.ehsinsight.com/ehs-management-information-system (accessed Aug 27, 2017).

Eller. What is MIS? n.d. https://mis.eller.arizona.edu/what-is-mis (accessed Aug 28, 2017).

Ghazzawi, K.; Accoumeh, A. Critical Success Factors of E-Recruitment System. *J. Hum. Resour. Manag. Labor Stud.* **2014,** 2 (2), 159–170. http://jhrmls.com/journals/jhrmls/Vol_2_No_2_June_2014/10.pdf (accessed Aug 26, 2017).

Guest, D. E. Human Resource Management and Industrial Relations. *J. Manag. Stud.* **1987,** *24* (5), 503–521. DOI: 10.1111/j.1467-6486 .1987.tb00460.x.

Human Resource Excellence. Importance of Human Resource Management. 2017. http://www.humanresourceexcellence.com/importance-of-human-resource-management/ (accessed Aug 26, 2017).

International Labour Organization Safety and Health at Work. 2017. http://www.ilo.org/global/topics /safety-and-health-at-work/lang--en/index.htm (accessed Aug 27, 2017).

Jobcentre Brunei. Manpower Policy and Planning Unit, Energy and Industry Department, 2017. http://www.jobcentrebrunei.gov.bn/ (accessed Aug 25, 2017).

Karikari, A. F.; Boateng, P. A.; Ocansey, E. O. N. D. The Role of Human Resource in the Process of Manpower Activities. *Am. J. Ind. Bus. Manag.* **2015,** *5,* 424–431. https://file.scirp.org/pdf/AJIBM_2015062913581793.pdf (accessed Aug 25, 2017).

Kehoe, R. R.; Wright, P. M. The Impact of High-performance Human Resource Practices on Employees' Attitudes and Behaviors. *J. Manag.* **2010,** *39* (2), 366–391. DOI: 10.1177/0149206310365901.

Keramati, A.; Salehi, M. Investigating Website Success in the Context of e-recruitment: An Analytic Network Process (anp) Approach. *Appl. Soft Comput.,* **2012,** *13* (1), 173–180. http://www.sciencedirect.com/science/article/pii/S1568494612003687?via%3Dihub  or  http://www.irantahgig.ir/wp-content/uploads/40065.pdf (accessed Aug 20, 2017).

Kettunen, J.; Kantola, I. Management Information System Based on the Balanced Scorecard. *Emerald Insight J. Campus-Wide Inf. Syst.* **2005,** *22,* 263–274. http://www.emeraldinsight.com.ezproxy.ubd.edu.bn/doi/pdfplus/10.1108/10650740510632181 (accessed Aug 28, 2017).

Lee, C. C.; Czaia, S. J.; Sharit, J. Training Older Workers for Technology-based Employment. *PMC US National Library of Medicine National Institute of Health* **2009,** *35* (1), 15–31. DOI: 10.1080/03601270802300091. https://ntl.bts.gov/lib/30000/30100/30165/810915.pdf (accessed Aug 25, 2017).

Majdi, D. Get Ready to Join I-RDY. *Borneo Bulletin* 2017. http://borneobulletin.com.bn/get-ready-join-rdy/ (accessed Aug 28, 2017).

Manpower Policy and Planning Unit, Energy and Industry Department, PMO (2017). *i-Ready (i-RDY) Apprenticeship Programme.* http://www.energy.gov.bn/Shared%20Documents/i-Ready%20Company %20Briefing%2010032017%20943am.pdf (accessed Aug, 28, 2017).

Marouf, L.; Rehman, S. U. Organizational and Human Resource Aspects of IT Management. *Electronic Libr.,* **2005,** *23* (4), 383–397. DOI: 10.1108/02640470510611454.

Ministry of Finance , Brunei Darussalam (n.d). Treasury Payroll Unit. http://www.mof.gov.bn/index.php/payment-section (27, 2017).

Ministry of Labour, Alberta, Canada (2017). Health and Safety Management Systems. https://work.alberta.ca/occupational-health-safety/health-and-safety-management-systems.html (accessed Aug 27, 2017) (accessed Nov 2018).

Mohiddin, M. *Brunei Darussalam: Towards Better Public Service Delivery Through E-Government.* 2008 [PowerPoint Slides]. http://unpan1.un.org/intradoc/groups/public/documents/eropa/unpan033 714.pdf (accessed Aug 20, 2017).

Moideenkutty , U.; Al-Lamki, A.; Murthy, Y. S. L. HRM Practices and Organizational Performance in Oman. *Personnel Review* **2011,** *40* (2), 239–251. https://doi.org/10.1108/0048348111110610 (accessed Nov 2018).

Papia, S, N.; Naidu, J. G. HRIS Efficiency and Its Impact on Organizations, 2015. https://www.academia.edu/19666974/HRIS_efficiency_and_its_impact_on_Organization (accessed Aug 26, 2017).

Parry, E.; Stavrou, E.; Lazarova, M. *Global Trends in Human Resource Management.* 2013. https://link.springer.com/content/pdf/10.1057%2F9781137304438.pdf (accessed Aug 24, 2017).

Price, E. L. The Psychology of Change Management. 2003. http://www.mckinsey.com/business-functions/organization/          our-insights/the-psychology-of-change-management (accessed Aug 24, 2017).

Promerantz, S. How Many of the Top 10 Most Common Organizational Challenges Plague Your Company. *Forbes.* 2017. https://www.forbes.com/sites/forbescoaches-council/2017/02/24          /how-many-of-the-top-10-most-common-organizational-challeng-es-plague-your-company/#1ae5da691e79 (accessed Aug 21, 2017).

PSC Recruitment. *About Us.* 2015. https://www.recruitment.gov.bn/en-us/Pages/AboutUs.aspx (accessed Aug 25, 2017).

Reynolds, M. What Are the Disadvantages of a Human Resource Management System? n.d.          http://yourbusiness.azcentral.com/disadvantages-human-          resource-management-system-16368.html (accessed Aug 21, 2017).

Rodger, J. A.; Pendharkar, P. C.; Paper, D. J.; Molnar, P. Reengineering the Human Resource Information System at *Gamma Facilities* **1998,** *16* (12/13), 361–365. DOI: 10.1108/02632779810235681.

Rouse, M. What is MIS (Management Information Systems)?—Definition from WhatIs.com. n.d.          http://searchitoperations.techtarget.com/definition/MIS-management-information-systems (accessed Aug 20, 2017).

Russell, T. R.; Chamberlain, J. Searching for Talent: The Information Retrieval Challenges of Recruitment Professionals. *Bus. Inf. Rev.* **2016,** *33* (1), 40–48. DOI: 10.1177/0266382116631849.          http://journals.sagepub.com.ezproxy.ubd.edu.bn/doi/pdf/10.1177/0266382116631849 (accessed Aug 25, 2017).

Tripathi, K. P. Role of Management Information System (MIS) in Human Resource. *Int. J. Comput. Sci. Technol.* **2011,** *2* (1). http://www.ijcst.com/vol21/tripathi.pdf (accessed Aug 19, 2017).

Vui-Yee, K. The Impact of Strategic Human Resource Management on Employee Outcomes in Private and Public Limited Companies in Malaysia. *J. Hum. Values* **2015,** *21* (2), 75–86. DOI: 10.1177/0971685815594257.

Weidong, Z.; Haitao, L.; Weihui, D.; Jian, M. An Entropy-based Clustering Ensemble Method to Support Resource Allocation in Business Process Management. *Int. J. Know. Inf. Syst.* 48(2), 305–330. https://link.springer.com/article/ 10.1007/s10115-015-0879-7 (accessed Aug 24, 2017).

What is MIS? Management Information Systems. (n.d.). http://mays.tamu.edu/department-of-information-and-operations- management/management-information-systems/ (accessed Aug 20, 2017).

# CHAPTER 4

# CHANGING SKILL REQUIREMENT: IMPROVING MARKET EXPECTATION OF HUMAN RESOURCE TROUGH MANAGEMENT INFORMATION SYSTEM

## ABSTRACT

Human resource (HR) is deliberated as one of the important assets of business organizations nowadays. The business-dealing process tier of management information in HR capacity is related with daily activities such as records of attendance, participation of employees, and expense estimate. The functional position activities also involve keeping up of records of employees, which is used as a basic foundation for critical layers. With the increasing significance of HRM and growing capacity of the organizations, perpetuation of the employee connected data and creating suitable reports are the important prospects of any company and organization. This study is to feature the execution of IS in HRM based on the employee's skills, through knowledge identification, composition, modernization, procreation, and growth of the abilities are some of the management practices as fundamentals of economic event have fluctuated crucial methods in past years due to the power of globalization, affluence of information technology (IT), the opportunity of information, and the growing environments of organizational pattern. There are a numbers of practical researches and studies, which aim only on definite trait of knowledge management but not on the entire management system.

## 4.1 INTRODUCTION

This study emphasized two aspects of emerging information technology.
The first aspect is the management information system (MIS) perspectives within organization point of view. The second aspect is market expectation of future information systems (IS) professional as well as which knowledge/

skills are predominant players in this competitive advantage industry. The other section explains to what extend human resource management could help by changing and improving skills requirement for employers in an organization. Both MIS and human resource management (HRM) play major role in the management, administration, and operations of an organization. This will be further explained in the literature review and discussed under the result section. In addition to that, this study is further supported with several references from academic articles, journal, and case studies from trusted resources.

MIS is an IS that merges data from every section of an organization's work and gives proper administration and business concern with the information they need. MIS is grown to boost system of business for organizations' managers, giving useful business information which contains economic, accounting, and business work. Besides that, MIS also changes out-of-date, nonunified reporting and data systems available in the organization. MIS is a versatile regimen connecting modern technologies, employees, actions, and organizational system. Without any doubt, the achievement of MIS is determined by its distinctive and organizational accomplishments. Nevertheless, expanding new information technologies in an organizational framework needs a wide area of capability. The MIS requires to be fully experienced at making business connections, achieving the crucial technology, and adapting the organization for alteration (Jamwal and Singh, 2011).

One of the very crucial guidance in the growth of MIS ability is MIS human capital. There are two main important signs that we focused on: skills and precision that they have. Skills applies to the range which the employees have the precondition mechanics and business skills, whereas precision applies to the range to which MIS employees have organization-specific expertise such as a belief of the values of the people and habitual activity of an organization. The activities of MIS are mainly required in knowledge-exhausted areas and need particular industrial skills. In addition, applicable business and social skills are required to efficiently provide services of MIS to the ultimate users (Teo and King, 1997).

Human resource (HR) is deliberated as one of the important assets of business organizations nowadays. The business-dealing process tier of management information in HR capacity is related with daily activities such as records of attendance, participation of employees, and expense estimate. The functional position activities also involve keeping up of records of employees which is used as a basic foundation for critical layers (Tripathi, 2011).

With the increasing significance of HRM and growing capacity of the organizations, perpetuation of the employee connected data and creating suitable reports are the important prospects of any company and organization. As a result, computer-based HRM systems are practiced by many organizations. This study is to feature the execution of IS in HRM based on the employee's skills (Tripathi, 2011)

The knowledge identification, composition, modernization, procreation, and growth of the abilities are some of the management practices followed worldwide nowadays. The fundamentals of economic event have fluctuated crucial methods in past years due to the power of globalization, affluence of information technology (IT), the opportunity of information, and the growing environments of organizational pattern. There are a numbers of practical researches and studies which aim only on definite trait of knowledge management but not on the entire management system (Syed and Xiaoyan, 2013). Knowledge management undertaken is directly connected to organizational accomplishment which is also directly connected to monetary and financial performance (McKeen et al., 2006). Knowledge management system (KMS) can boost up performance of a company by significantly reducing administrative costs and can also boost up the output. (Feng et al., 2004).

As the primary resources of MIS are composed of people, technology, and information, it is stated by Kumar (2012) that MIS has an essential part and role in the HRM. Known as a computer-based system with the capability of providing the HRM with tools for the tasks which involves a process of organizing, evaluating, and ensuring the efficient run of the department and large organization, MIS could help the department of HRM to figure out the past, present, and future through providing not only with the software for decision-making and data resources such as database, but also with the hardware resources of the system, decision support systems, people management, project management application, and any other high-tech processes that could ease the HRM to be more efficient in completing the tasks. Through this linkage, a system named as human resource information system (HRIS) is established where it refers to the intersection of the systems and processes between IT HRM and MIS (IS and IT) to help HR managers perform HR functions in a more effective and systematic way using technology.

The systems and subsystems involved in the HRIS are: first, staffing which includes the subsystem of personal record keeping, employee skill inventory, and forecasting requirement; second, training and development that involves the performance appraisal planning and succession planning;

third, the compensation which contains the salary forecasting and incentives planning; and lastly, the government report. The data resources of HRIS are composed of the personal application and appraisal form, appointment letter, attendance and leave record, wage and salary agreement, record of sources of requirement, industry data on manpower skill performance, biodata, and production data. As a result, it gives report output such as report on forecasting manpower requirements, report submitted to the government agencies, performance appraisal, training and development program that has been conducted which reveals about the success and failure (Kumar, 2012).

## 4.2  LITERATURE REVIEW

### 4.2.1  REASONS THAT DRIVE ORGANIZATION TO EVOLVE

According to Martin and Healy (2008), there are four major sets of reasons that drive organization to evolve: the flexibility demand, the autonomy demand, the consumer focus demand, and lastly the knowledge demand.

a) The *flexibility* demand is determined toward the progression of consumer markets and expansion in economic competition as well as globalization where the associations can never again use the usual production method for producing similar types of goods and services. The organization must now have the capacity to quickly adjust to this evolving demands and erratic competition by quickly modifying their products or services to meet those expectations. Thusly, the organization has to come up with new alternatives. In the adaptable working environment, the organization and its people should have the capability to move from one point to another as the necessity emerges. Furthermore, the work should be redesigned with the goal that individuals work in flexible team, completing the required employment, and generally sort out the task efficiently among the team members.

b) The *autonomy* or self-determination demand: the company sees that they should rebuild their organizational structure to be able to stand out among its competitors and improve its productivity by providing new strategies for ensuring that its people do what are they required to do. The organization wants its people to be more self-ruling, using their own perception in making the judgment precisely on what could be done next, and have a capability to concentrate on accomplishing an ultimate result that is ideal for the company.

c) The *'consumer focus'* demand: the expanding importance of consumer relations is another measurement to the developing aspect of why organization should evolve. The companies are focusing to satisfy their customer by providing the best customer services they could deliver as they are one of the key players to help the company achieve their initial goals.

d) The knowledge demand is another key factor that drives the company to evolve because knowledge and information are seen as the new important elements in the economic advancement which correspond to the effect of the latest information and advances in communication. Works should be composed in a proper manner so that knowledge can effectively flow and can be delivered as well as shared inside the workplace where the employees can learn from each other.

### *4.2.1.1   MIS PERSPECTIVES*

### *4.2.1.1.1   IT Career Prospect and Trends*

It is significantly important for the future employees, especially the undergraduates, to understand the current job trends chiefly, the job prospects for IT or IS industry. This is to help the prevailing arrangement of MIS, for instance, to tackle the problems that arise such as the absence of clarity about IT positions and diversified IT or IS positions, running from the specialized to business introduction. The future employees have to understand what does each specific IT or IS worker do and what are its requirement or qualification, with the title of the job given; some may know and some may not know what are the key qualification that are matching with the job position, or some may mislead the title of the roles with other roles.

For instance, one example that is stated in the Yew's journal is regarding the requirement of working under the computer and mathematical occupations. The future employees might think that one should have a strong mathematics or computer science education skills to apply for the positions, however, indeed quite a number of the IT posts does not require much advanced mathematics education as long as one had the right skills; this is further explained in the next sections. Thus, several publications that give essential description of career information about the trends in job market were pretty much needed. Hence, one of it was published in 2008 by the US Department of Labor, Bureau of Labor Statistic (BLS) stating relevant career instructions about the job advertisement around the US regions (Yew, 2008).

Moreover, researching for the latest evolution of the job market helps the future employees to view on the fastest growing job opportunities. An example to that was an universal classification system, named the US Standard Occupational Classification (SOC). It was used to take into account correlation of occupational statistical information and to monitor the advancement of occupations from year to year (Yew, 2008). This aided the US future employees understand the current trends of job prospect and prepare themselves before getting into the related workforce and be as part of a team.

In addition to that, a more extensive review at the general patterns is necessary for molding the IS educational programs in assessing the demand of critical abilities in the IT industry. The expanding utilization of IT tasks is an issue for the undergraduates; hence, the program review processes are much in need to prepare the graduates for the job market. As a result, MIS program review with four-phase flowchart was introduced (Yew, 2008) as follows:

- Phase I: Distinguish significant concerns, assemble work data for IT occupations, and table of skill weight in MIS programs.
- Phase II: The literature review of the program surveys; trends and knowledge or abilities for IS or IT workforce, discourses on the work duties suitable for understudies.
- Phase III: Tactics for enhancing work attractiveness and survey.
- Phase IV: Latest programs in supply chain management, enterprise resource planning, and internship; advancement of internship and center courses to meet workforce desires; listing out the programming instruments need for educating; and lastly, to investigate IT certification program.

### 4.2.1.1.2   Guide to Fulfill Industry Expectations

There will always be a conflict of meeting the requirements of industry expectation especially in this sudden change of IS due to education program specifically may not constantly keep up with the changes, hence, an educational development is displayed to precisely measure the demands of the industry . The curriculum development process guided future IS professionals to prepare themselves and meet this requirement by helping them identify the shift in MIS workplace and types of skills and knowledge demanded. One example of curriculum development process can be seen in Figure 4.1 (Noll and Wilkins, 2002).

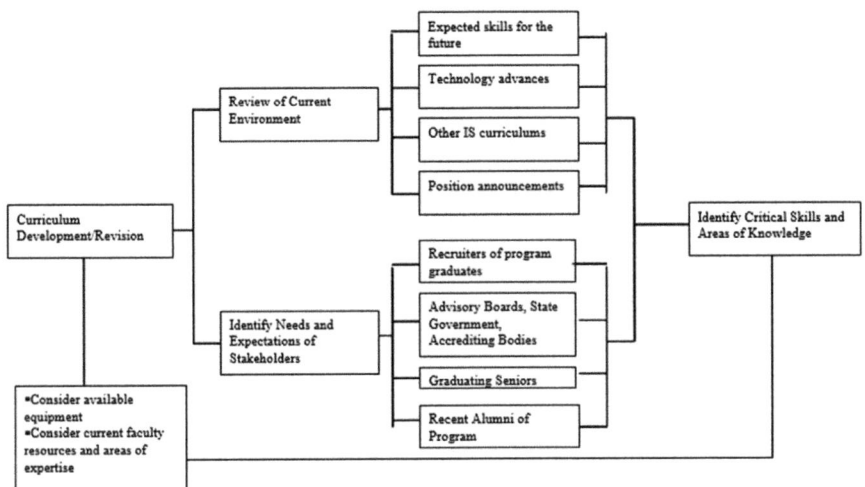

**FIGURE 4.1**   Curriculum development model.
*Source*: Noll and Wilkin (2002).

Lee et al. (1995) indicate that the industry want those IS experts who can manage organizational IS exercises not only just by the technological abilities but also as a whole package of business operations, administration, and interpersonal skills (Noll and Wilkins, 2002). Besides that, Ehie (2002) revealed meaningful information about two industry expectation toward the future workforce: (1) the industries are searching for people with a solid introduction and have a good comprehension of an integrative business value chain and (2) people who have general knowledge of programming and have sensible capability in at least one programming language (Yew, 2008).

According to Todd et al. (1995), job advertisements provide insight on valuable information of technical and business knowledge/skills needed for specific organization and the result varies depending on the requirements of a particular industry; some require more technical skills and the others may request for more business skills. For example, in services industry, they request for individuals who are specialized in the area of being a programmer and analyst; hence, those who have the qualification of having technical skills get higher chances of being recruited. Todd et al. conclude that technical skills, information of the business, critical thinking abilities, and individuals who have broad administration abilities are important knowledge/skills in a particular marketplace (Yew, 2008).

In addition to that, study that led by Noll and Wilkins in 2002 also successfully underlined the significance of IS skills that IS professions based on three staffing groups of programmers, analyst, and the user support should have, by summarizing, the core knowledge or skill such as:

- Knowledge of business practical operation and knowledge of particular industry and its condition.
- Capability to solve business issues and establish proper solution.
- Capability to work as a team in collaborative working situation, and to design, organize, and lead the task.
- Capacity to create and deliver effective, informative, and powerful oral introductions.
- Capability to design, construct, and compose technical manuals, transcription, and records.

### 4.2.1.1.3  Knowledge/Skills Requirements in a Rapidly Changing Industry

In response to the worries of industry administrators in regard to the changes in the knowledge or skills prerequisites of IS/IT profession and necessities to the academic preparation for future IS/IT laborers, numerous studies were conducted by academic researchers to investigate and to help determine the job performance of future graduates workers. Each study presented different cases of industry assumption of knowledge and skill one should have.

Investigations expressed that the fast changes in knowledge base and work requirements are not driven by the technological developments but rather additionally because of the change in business and social needs, that is, the management and organizational skills. Below were three categories of basic skills that the businesses are looking for in each individual: personal attributes and attitudes, personal skills, and technical skills (Dench, 1997).

- **Personal attributes**

Personal attributes are generally inborn; they are identified as a man's quality or characteristic, states of mind, and perspective of the world, for example, trustworthiness, respectability, determination, and have a good sense of teamwork and fit nicely in the work given. These sorts of attributes have for quite some time been significant to employers; even nowadays, employers are still looking for the people who have strong personal attributes because they want to hire people who will fit in their current workforce.

- **Personal skills**

Personal skills and personal attributes may sound the same but it proposes more than the presence of specific qualities in an individual, for example, make sound decisions, capacity to interact and works with others, and self-confidence. These personal skills can be created in different areas and individuals will accomplish different levels of skills. Due to the changing organizational structures and services and competitive pressures, employers are searching for individuals who are able to work effectively and efficiently.

- **Technical skills**

Employers want individual who can work confidently using the technology and who are fast learner, able to adjust, and keep up with new or updated ideas, and also those people who have much greater understanding of the "know-how" to use the technological systems especially related to software or programming. These technical skills can be learned through educational training or from previous experiences.

In addition to that, below are among important knowledge/skills needed in current industry of rapidly changing workforce because to be able to stay ahead of the competitors, organizations should always look at the quality which could lead the organization further along the changes of emerging globalization. In order to easily understand what types of qualities or abilities are needed, the knowledge or skill are grouped into following four classes (Yen, Lee and Koh, 2001):

1. **IS technology knowledge/skills:** IS technology (hardware-related and networking software), IS management (IT strategy and other technological trends), and IS development methodology (works as analysis, design, and performing new techniques or procedures)
2. **Organizational and societal knowledge/skills:** related with particular areas, ventures, and general working condition of the company.
3. **Interpersonal knowledge/skills:** related to interpersonal conduct skills, such as emotional intelligence, communication capability, and interaction with other people.
4. **Personal traits and/or knowledge/skills:** means personal encouragement and creative, productive, and critical thinking skills.

## 4.2.1.1.4   *From Technical to Nontechnical Skills Among Information System Suppliers*

The IT and IS management literature implies that there is transformation from technical to nontechnical skills (Aasheim et al., 2012; Gallagher et al., 2010). In addition, IS education and training curricula are progressively accepting courses in nontechnical skills, with specific focus on management (European e-Competence Framework, 2014; Ahmad et al., 2011; Davis and Woodward, 2006).

A research was conducted by Benedicte Branchet and Pierre-Yves Sanseau based on investigating the reason of the transformation of skills in the IS suppliers. The objective of this research is to answer the main question: what are the rising skills in the IS service suppliers sector? The aim of the research is to look into the changing technical to nontechnical skills and if this change shows the actual situation in companies, and if the adjustment in IS/IT curricula fit the growing market demand. For both skills and curricula, this study implements IS and IT reciprocally, because the two fields overlap each other and both of them basically need the same basic skills. Moreover, it practices the term critical skills to describe the skills people have that are precariously crucial for their companies (Branchet and Sanseau, 2017).

This study investigates the three important challenges from the viewpoint of IS suppliers :

1. IS services are immensely technical, therefore, technical skills are more reasonable to stay essential in this field, taking into account that the development in the field of technologies are very fast and boundless (Chia-An and Shih, 2005; Lang and Urbancic, 2009; Scott, 2007; Alsudairi and Dwivedi, 2010).

2. IS skills are vital for value creation in IS suppliers and also their business, however, it is less important for companies in other sectors that regulate their own IS. The shift in the demand for IS service skills is typically based on the skills available. But there is lack of skills in this field, so they are in urgent need. One of the achievable solution is outsourcing and offshoring (Lai et al., 2006; Gulla and Gupta, 2012; Rosebush et al., 2012).

3. There is never ending progress for the global IS services market in the face of recent technologies, international competition, and evolving markets, as a proof of large new developments (Fichman et al., 2011; Mamaghani, 2006).

IS supply business must constantly implement both their organization and their strategies, and such revolution are most probably involving in their structures, processes, internal practices, resources, technologies, and cultures (European e-Competence Framework, 2014).

Penrose (1959) and Andrews (1971) used the term skills to incorporate the concepts of resources and capabilities. They hypothesize that all firms are depending on resources, but will incorporate them in various ways, resulting in their building unique expertise (Andrews, 1971; Espino-Rodriguez and Padron-Robaina, 2006; Ethiraj et al., 2005; Mata et al., 1995). They pointed out that a firm's unique expertise are not just technological, but based on three skills:

1. Technological skills: these are the base of the company's technological development and computerization (Branchet and Sanseau, 2017).
2. Process skills: to facilitate firms on developing "sequences of operations" in technological or administrative work progress that may cut across organizational operations and framework (Branchet and Sanseau, 2017).
3. Business skills: to facilitate firms on doing specific business functions, and can partake in both individual and group competency (Branchet and Sanseau, 2017).

### 4.2.1.1.5 Mobile Customer Relationship Management as a Tool That Could Contribute to Personal Performance and Changing Skills

Customer relationship management (CRM) has been described as one of the most assuring technological advancement in the business realm for having the enormous possibility to build up employee's work efficiency and promote interaction with customers (Li and Mao, 2012). In the era of mobile communications, modernization and change have been seen to take the central role in business success, and in this situation, the mobile innovation brought by the smartphones serves as the latest archetype of enterprise management (Mirusmonov et al., 2015). Because an underlying aspect of mobile devices is the vigorous movement, there is a rising demand for mobile customer relationship management (m-CRM) to incorporate online and offline CRM because of its aspects such as mobility and pervasiveness ( Mirbagheri and Hejazinia,2010). However, IS achievement model may become as the guideline of organization infrastructure for showing elements that may influencing m-CRM personal performance (Delone and Mclean's, 2003) (Fig. 4.2).

The study aims to resolve the link between m-CRM and personal performance in an organizational perspectives referring to the framework of the model of Delone and Mclean's IS achievements, which has shown the link between IS attribution and net benefits (Garrido-Moreno and Padilla-Melendez, 2011). According to the literature review, this study form variables for information quality, system quality, and service quality, which are the main aspects of m-CRM. Moreover, the study gives an experimental research on the outcome of information quality, system quality, and service quality on personal performance based on m-CRM (Mirusmonov et al., 2015).

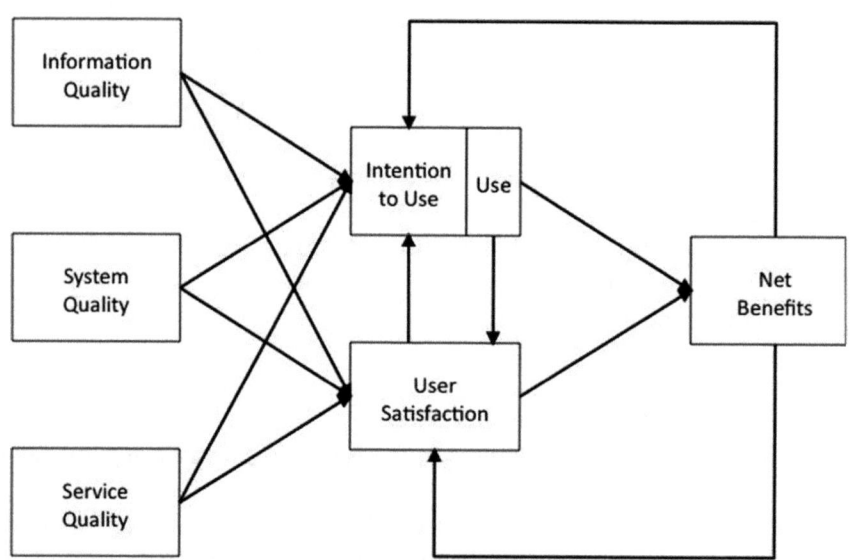

**FIGURE 4.2**    Delone and McLean IS success model.

The attributes of information quality depends on the database and value analyses for customer segmentation and the combination of customer data and information. Meanwhile, for system quality, it depends on the system extensibility where it is the extent to which system modules, databases, and networks can be extended. Nonetheless, system flexibility is also one of the attributes that are under the system quality. System flexibility measures the level of convenience to access the information of customers and employees. For service quality, it depends on the immediacy, personalization, user satisfaction, system use, and personal performance (Mirusmonov et al., 2015).

This study forms different hypotheses about the link between the major variables proposed in previous research. Previous research proposed the need to design a customer database for firms based the competency of m-CRM in gaining customer information (Sinisalo et al, 2007). Moreover, customer segmentation supports the personalization of products and services by the confirmation of resource allotment (Aeron et al., 2012). It was proven from the previous studies that customer segmentation can considerably increase staff's satisfaction.

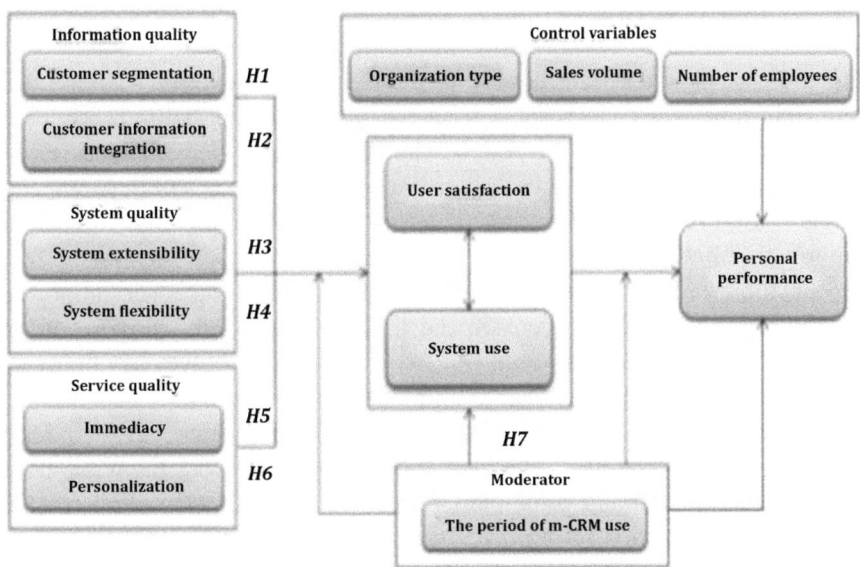

**FIGURE 4.3**   Research model.

There are several hypotheses in this study that the researchers believe that will lead to employees' personal performance. Hypothesis 1: The outcome of customer segmentation on personal performance is caused by system use and user satisfaction. Hypothesis 2: The outcome of customer information integration on personal performance is caused by system use and user satisfaction. These two hypotheses emphasize on the information quality of m-CRM (Fig. 4.3).

Hypothesis 3: The outcome of system extensibility on personal perfor-mance is caused by system use and user satisfaction. Hypothesis 4: The outcome of system flexibility on personal performance is caused by system use and user satisfaction. The third and fourth hypotheses focus on the system quality of m-CRM. Hypothesis 5: The outcome of immediacy on personal performance is caused by system use and user satisfaction. Hypothesis 6: The outcome of personalization on personal performance is caused by system use and user satisfaction. The last two hypotheses highlighted the attributes on service quality (Mirusmonov et al., 2015).

## 4.2.1.2   HRM PERSPECTIVES

### 4.2.1.2.1   *Strategic Workforce Planning of HR in an organization*

Nowadays in the HR community, the commitment for company and organizations to have the right people, in the right place, with the right skills at the right time is very important. In every business, preparation it needed to make sure that you have the people who have the knowledge and resources in the organization to bring both short- and long-term goals of the company/organization is quietly disputing. The term "workforce planning" differs of approval (*Workforce planning Right people, right time, right skills*, 2010).

On the list of things to do, strategic workforce planning is very important in the organization, however. On the list of things to do, strategic workforce planning is very important in the organization; however, one of the ability trainer/operator of tenable high accomplishment is the capability of the company/organization to make procedure that equity both their long- and short-term preference. The next generation of HR work likewise shows knowledge of planning as a vital procedure that assists to improve the company/organization (*Workforce planning Right people, right time, right skills*, 2010).

This ongoing research established that companies/organizations were committed in workforce planning not only for the right size for the ongoing drop but also to have the ability to perform in place to do well in the future and to ensure they persist to draw attention and make ability to accord with a dimension of future sequence of events (*Workforce planning Right people, right time, right skills*, 2010).

What is the definition of workforce planning? There are actually many different definitions from literature, research, journals, and interviews because the definition itself is likely to take into account a number of activities. Indeed, the study found out that many companies and organizations are performing more in terms of workforce planning than seem at first sight, but it is basically an accumulation of practices such as progression training rather than an orderly work to check and make the whole labor force (*Workforce planning Right people, right time, right skills*, 2010).

Other than that, workforce planning is actually not being used regularly. The public sector, however, will discuss about the "workforce planning," whereas in the private sector, it is being used correspondingly with other terms such as resource planning (*Workforce planning Right people, right time, right skills*, 2010).

In Figure 4.4, there will be a recommended method bringing together some of the essential feature of workforce planning based on interviews and researches.

# Process model for business planning (CIPD)

| Business strategy | | |
|---|---|---|
| Organisational strategy | People strategy | Operations plan |

| Analyse and discuss relevant data | |
|---|---|
| Input information from data collection exercise | Input resourcing information from HR business partners and business managers |

| Agree objectives of the plan | |
|---|---|
| Review labour supply data both internal and external | Review workforce capability to deliver the plan |

| Agree actions and implement plan | |
|---|---|
| Agree assessment and evaluation criteria | Regularly review outcomes |

**FIGURE 4.4**  Recommended process model for business model.

There are four stages in total:

Stage 1. Business Strategy: A very good planning must come with a good business strategy. The first stage of workforce planning which is the "business strategy" will be affected by three essential features such as organizational strategy, people strategy, and the operations plan that is shown in Figure 4.4. In many companies and organization they interviewed so far, workforce planning begins with strategic preferences from which the society associations are derived. Workforce information provides an overview of awareness of ongoing case and what needs to be adjusted to achieve these strategic necessities in time to come. Information such as numbers, location, and skills requirements are included in the case (*Workforce planning Right people, right time, right skills*, 2010).

Stage 2. Analyze and Discuss Relevant Data: The second stage of the planning process includes analysis and conference around the appropriate and accessible data. At this second stage, both HR and managers may give in their inspection and forecast around resourcing necessity which will then be deliberate alongside the information explained above (*Workforce planning Right people, right time, right skills*, 2010).

Stage 3. Agree the Objectives of the Plan: These conference should result in concurrence about what the scheme is trying to accomplish, which will need to be inspected opposite to available assets so administrator needs to analyze: (1) inventory of worker both external and internal, (2) the possible ability to perform of the workforce to make new skills, boost of efficiency and effectiveness, and accept the act necessary to initiate and accommodate to change, (3) recognize and advice enrollment and activity of development that may be needed in the future (*Workforce planning Right people, right time, right skills*, 2010).

So basically, any skills difference or inadequacy in the training plan will be needed to be considered. Many of the expert explained that at this third stage, they acknowledge the point of view of the managers on the number of people they accept they need to continue functional activity (*Workforce planning Right people, right time, right skills*, 2010).

This third stage may also be essential to comprehend between the conditions of the activity giving an account of numbers of people, jobs, and skills, and that giving an account of those arrangement of workers and skills to ensure that they are fully useful for the company and organization (*Workforce planning Right people, right time, right skills*, 2010).

The final stage of the process (agree actions and implement plan) includes in cultivating, enrollment of workers on the numbers and levels, coaching and preparation—goal, knowledge, and growth—realizing promise, organization arrangement, and design-directing work groups and act and arrangement—advising management methods to use the availability of HR (*Workforce planning Right people, right time, right skills*, 2010).

This will then be broadcasted to organizational managers to allow them to recognize the conduct they need to take and to form their behavior, for example, in determining how they occupy individual position so that it is suitable for them. This will then be broadcasted to organizational managers to allow them to recognize the conduct they need to take and to form their behavior, for example, in determining how they occupy individual position so that it is suitable for them. It is also needed to take into account on how other HR procedures such as cultivating workforce competence and skills advancement to achieve organization goals as final stages (*Workforce planning Right people, right time, right skills*, 2010).

In workforce planning, there are number of different methods that are somehow interrelated to other HR activities. The practices were necessary in some cases and it was more about administering the outflow of knowledge from one place to another. Whatever the connection looks like, the important point is that there is an adjustment between different kinds of activities and that they are backing up each other. This will need skills of connection and collaboration with colleagues in a larger organization (*Workforce planning Right people, right time, right skills*, 2010).

There is no confirmation from latest literature that workforce planning is one of the important parts of HRM. It also contributes proof that importance of workforce planning is boosting and that it can also help cultivate business alertness and strategic intuitiveness that distinct positive management of people. In some literature, workforce planning is explained as a current active action developing good understanding to help decision-making. It also supports irresistible controversy in assistance of workforce planning as a means to recognize, understand, and manage with future requirements (*Workforce planning Right people, right time, right skills*, 2010).

### 4.2.1.2.2  *Training*

The problems posed by the fast pace of alterations in the world of information must be investigated in order to answer some of the fundamental questions posed to librarians and other information workers by this revolution as they strive to come up with answers to the new centuries of information vitality. Librarians and information workers in general are more likely to speed into the era of automation, and it is becoming more important to look into their training needs and development of information skills (Odini, 1999).

It is significant for those who partake in the training of librarians and information workers to dedicate substantially more time and energy not only to the procedure of change but also to authorize it and gain acceptance of it. Enabling change means to supervise and construct the direction, then modify in some way the outcome of change (Odini, 1999). According to Aina (1993), the percentage at which computer technology and communications technology are being brought into Africa is increasing, and the correlated use of these technologies with information is becoming more solid.

The evolving information environment needs a workforce which accustoms with the rise of information and communication technologies (ICT) which will be needed in the new world of information. Librarians must hook themselves with the required skills in the use of different software packages, the internet, CD-ROMS, and printed sources. The workforce should have the nature and ability to incorporate traditional and ICT services to convey the library service of the new era (Odini, 1999).

The best learning and training and the most adequate management structures have to guide them. In favor of satisfying the information needs of present-day users, the library staffs should be introduced with the newest innovation in the field of information service and information science. Hence, it is important for training institutions to form their curriculum to the essential education and training at the relevant level (Odini, 1999).

The drawbacks of establishing information science curricula in African countries are the distressing impact of the current status of IS and the shortage of a distinctly defined national information. But the main drawbacks of teaching the information sciences are feasibly copied from the recentness and the integrative traits of the task. Few causes are pushing information handling institutions out of their standard premechanization steps, and move to the starting of intellectual transformation (Odini, 1999).

These causes include the fast advancement of all types of informational materials, the rise in the requirement for access to the accountable matter of these materials, and the complication of attaining people to do the work. Nonetheless, the causes of these differences in information handling activities has not been escalated enough to outgrow a comfortable body of practice and knowledge which African training institutions can use as the infrastructure for their curricula. Resolving educational directions is a challenging task, especially in an always-changing situation, as developments emerge too fast, it will tend to make no sense of meticulously laid plans (Odini, 1999).

### 4.2.1.2.3    Contribution of IS to HRM in University

HRM is one of the management practices that gives a huge impact on the overall results of the university. Other management could be the school library, department management, property management, student counseling center, etc. In this literature review, it will basically show how the university's HRM can be improved through IS and IT and how to implement to the university. When proposing an IS to support the function of the university, the selection of the procedures becomes very important. The advancement of this project is based on the "Information System Life Cycle," which will be shown and explained in the next few paragraphs. Information System Life Cycle is one of the cycles that is accepted in its vary modification. Along the side of that, we have enforced principles of structured outline and investigation (Rodriguez et al., 1995).

In this manner, the first step took part is an analytical amendment of those actions which, in the prospect of other consultants in the field, should be attempted by HRM, in order to differ them later with those executed by the university to test their implementation level. The amendment of the current system was not only done to focus those activities which either insufficiently execute or not at all but also take into account the different methods IT would make an improvement for each action like the latest trend of technologies such as database technologies, communication, and multimedia (Rodriguez et al., 1995).

The analysis of this research is shown in Table 4.1, which summarizes and combines all 14 activities regarding HRM in an organization. It basically demonstrates the IS and IT level used in the university and the proposal of progression which eventually arise.

In general, as can be seen, the activities that are already computerized are those that are closely connected to the administrative processes which are most powerfully controlled in state universities, for example, selection, recruitment of staffs and students, and salary administration. It can be said common motif in other universities too (Rodriguez et al., 1995).

In addition, five activities are mentioned in Table 4.1: socialization, organizational development, motivation, health and safety, and lastly HR auditing. In these activities, there are no important benefits from IS support compared to traditional methods. This may be the outcome either of the unorganized nature of the assigned tasks, which will create their method difficult if not impossible, or of the ineffectuality organized information in these cases (Rodriguez et al., 1995).

Lastly, in the four activities such as analysis and description of jobs, career development, performance evaluation, and training and development, important skills could be acquired with the support of competent IT (Rodriguez et al., 1995).

At this level, investigation of information requirement for every activity was executed based on a particularized study initiated in combination with senior management of the HR department. From this research, a number of systems necessities arise, serving as a foundation on which the layout of system was finally based. This includes the comprehensive investigation and details of the work to be accomplished in each activity in order to recognize information needs which are connected to capacity, chance, dependability, and styles of presentation. Interviews, job description, observation, etc. are few of the research tools and techniques which were employed in other universities as well (Rodriguez et al., 1995).

**TABLE 4.1** IS/IT Support to Activities Concerning Human Resource Management.

| Human resource area | Activity | Current state | Proposal |
|---|---|---|---|
| Provision | | AIS | |
| | | AIS | |
| | | NO AIS | |
| Application | Analysis and job description | NO AIS | DIS |
| | Planning and forecasting | AIS | DIS |
| | Career development | NO AIS | |
| | Performance evaluation | NO AIS | DIS |
| Development | Training and development | NO AIS | DIS |
| | Organizational development | NO AIS | |
| Maintenance | Motivation | NO AIS | |
| | Salary administration | AIS | |
| | Salary benefits plan | AIS | |
| | Health and safety | NO AIS | |
| Audit | Human resources audit | NO AIS | |

AIS, activity undertaken with IS/IT support; DIS, develop IS support; NO AIS, activity undertaken without IS/IT support.

### 4.2.1.2.4   Collaboration Partnership Between Academic Institution and Organization

As a part of an initiative which could be done by the higher academic institution to help the undergraduates to understand the complexity of IT technology is by cooperation and collaboration with the organization either government or private by providing them with the real practical experiences where through this activities, the undergraduates could explore the real situation and environment of work. Besides, through this collaborations, undergraduates could apply the theories that they had learned in their institution and also gain new knowledge and experiences as what has been stated by Yew (2008) in his journal which concluded that undergraduates will experience knowledge such as learning specific software packages, setting up computer networks, installing the software, understand teamwork, and building communications skills. As to ensure that this initiative is running smoothly and both sides are achieving the overall aims, goals, requirements, and expectations, the institution and the organization itself can analyze the performances of the graduates through an outline or create a formative assessment.

Supported by Malik and Goyal (2003), it is stated that in order to achieve the win–win situation and to maintain the market-facing organization, collaboration and trust between the organization levels who are involved in this activity are highly required to be allocated as the most suitable initiatives to satisfy the needs of all the parties which is to solve the problem by illustrating the differences rather than by lodging various point of views, it is a matter of decisiveness and teamwork. Moreover, not only to sustain the effectiveness of IS in the organization and to figure out the weaknesses of the organization but also to seek for guidance for future improvement, an Adaptive Communication Environment (ACE) model framework which stands for adapt–collaborate–evaluate model has been proposed to which it is an easy, simple, and flexible method to assess the organization in establishing an effective and healthy organizational environment.

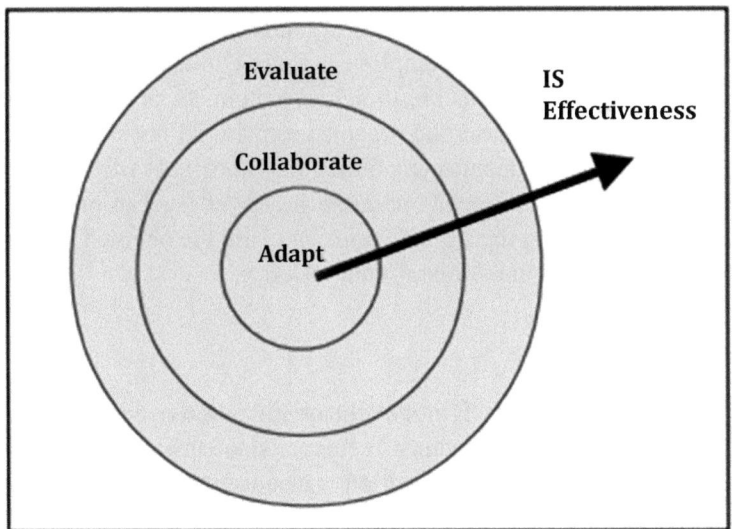

[Rings indicate continuous processes taking the visibility of IS effectiveness from inside-out.]

**FIGURE 4.5**   Structure of ACE models.

Figure 4.5 shows the structure of the ACE models to help in observing the effectiveness of the organizational environment. Presented in a combination of three rings which can be seen from inside to outer layer are: adapt, collaborate, and evaluate; each phase has different tasks to be completed before proceeding to another phase. Additional explanation is mentioned below for further understanding.

- **Adapt**

In order for an organization to be known to be adaptive is on how they react to the new developments in technology and response toward the risks and benefits which might appear and install the technology while modifying the possible problem arise. The organization might suffer a little amount of loss from the phase of adapting, but as a return, the organization will gain more experiences which can lead them to respond fast in the competitive environment. Additionally, the participation from managers and employees is an obligatory to remain to be adaptive.

- **Collaborate**

Similar to the adapt phase, known as organization practicing, the collaborate phase certifies that all the employees share common business goals and feel correspondingly responsible to achieve them. In order to achieve and pass the collaboration phase, the organization should not build any limit or boundaries toward its department, functions, and level of hierarchy (high and lower level) other than encouraging the lower level to participate with confidence, convincing the suitable training to aid the people to realized that IT and IS are the essential tools of their tasks.

- **Evaluate**

As the final phase in the ACE model, the organization needs to evaluate both the IT and IS performance against the exact standards. It acts as an instrument for constant feedback; evaluation is also used to make sure that all the blockages, problems, and failures are put on the list of priority. In order for this phase can be conducted, controlled, and lessened to other employees, organizations should undergo the adaptive and collaborative environment first or else, the environment evaluation will not work as expected.

### 4.2.1.2.5  Network Leadership/Employee Relationship

Hornstein (2008) indicates that to react to the implementation of technology and the skills changes, involvement between the leader, managers, and the staffs were highly required. Several methods should be undertaken which comprise the participative leadership, empowerment, system thinking, and the eight skill change process. This method will be elaborated further as follows:

- **Participative leadership**

As for the participative leadership, for employees to be more dedicated to their organization and the goals, leadership behaviors toward their employees and their full participation play very important role. It not only could improve the relationships between both employees and employers but also it helps in decreasing the problems that often occurs between them. Furthermore, it also could increase the performance of the organization as the employees are willing to step into the natural change in terms of skill requirement from the traditional and basic method of handling system to more advanced types of technology.

- **Empowerment**

Supporting the employees with the changes through involving the employees by giving them the perception control to make decisions, share power and responsibilities in some situations is a part of empowerment. It not only could contribute in making the organization improve their consistency but also could be satisfactory in terms of successfully adopting the changes.

- **System thinking**

Prevention from the intervention and being impacted by the environment which could influence the organization, leader should work together with the staffs to sort out ways through adopting the strategy of system thinking inside the organization. The strategy of configuration between the impacted area, the people and also an action taken to ease the other impacted people especially those who are responsible for IT initiative, everyone should participate in making effort through contribute in not only in the integrated but also in the collaborative and the significant ways of communication, thinking skills, and approaches. As for the long-term benefit, to maintain the innovation and transformation of the organization toward the changes, the leader must prepare the staffs with enough energy and resources for an adoption of the practices which have not been used by others before. The reason is to support the organization in implementing business transformation initiative, to change the mindset, beliefs, values, and attitudes of the staffs which is mostly reluctant to adapt with the changing skills requirement provided by the organization.

- ## The eight-step change process

Developed by John P. Kotter in the eight-step change process, it is suggested that processes, procedures, structures, and system which included activity of gathering data, identifying options, analyzing, choosing, planning, organizing, controlling, and problem-solving were also the main support to achieve the successful skill changes requirement; but above all, leadership is the most important tool to aid the staffs to ensure that the systems and changes will run smoothly.

In the eight-step change process, the first step was establishing a sense of urgency, as an initial step to start, all the possible opportunities must be the first thing to be recognized and discussed; the second step is creating the guiding coalition by selecting the group of staffs from different fields to form a group of sufficient power to work together in leading the change. The leader has to be virtuous in arranging the staffs to ensure the group will not be lacking power. It is followed by the third step which should be taken in order to achieve the vision, developing a vision and strategy through forming a vision in each group, and giving instructions to aid each group to react with the change effort.

Next, as for the fourth step, after been practicing the third step and using it as guideline to move the group to next stages, the team should grab every possible opportunity in voicing out the new vision and strategies which have been created earlier. Empowering people to effect change is the fifth step where it is covered by process of changing the systems or structure that is related with the change in the vision; it involves the employees in the discussion and indulges them not only as partners but also as risk takers in the activities, actions, and ideas. In addition, this step also helps in providing the employees with the opportunity to plan for the future and take action for every tasks or decision and on how to handle or lessen the obstacles, too.

The six steps have explained about how to generate short-term wins via forecasting and producing a tangible improvement in the performance and recognizing and rewarding the employees who made the improvement of the performance successful. Changing of the systems, structures, and policies which do not fit along with the vision of the team and organization by using the authority, and in addition to supporting the decision-making, hiring and promoting selected people whose can fit with the implemented change vision is the description of the step number seven which is also known as consolidating gains and producing more change. Finally, toward the end of the step of anchoring new approaches in the culture (step number eight), this process guarantees that the organization could develop a great leadership development and succession, better performance, and effective leadership through interaction between the customer and productivity—oriented behavior.

## 4.3 RESULT AND DISCUSSION

### 4.3.1 KNOWLEDGE/SKILLS REQUIREMENTS IN RAPIDLY CHANGING INDUSTRY

Summary of required knowledge/skills: ranking

|  | N | Mean | STD | Ranking |
|---|---|---|---|---|
| 1. IS technology knowledge/skills hardware | 90 | 2.600 | 1.015 | 19 |
| Packaged products | 90 | 3.622 | 1.214 | 9 |
| Operations systems | 90 | 2.933 | 1.149 | 16 |
| Networking/communication sw and lang | 89 | 2.506 | 1.179 | 20 |
| Application program | 90 | 3.622 | 1.186 | 10 |
| Programming lang | 90 | 2.867 | 1.470 | 17 |
| Systems development methodologies | 90 | 3.333 | 1.290 | 13 |
| Implementation, operation and maintenance | 90 | 3.700 | 1.267 | 7 |
| Visions about IS/IT for competitive adv. | 90 | 3.511 | 1.256 | 12 |
| IS technological trend | 89 | 3.528 | 1.226 | 11 |
| 2. Organizational and societal knowledge/skills |  |  |  |  |
| Knowledge spec bus functional area | 90 | 3.700 | 1.156 | 7 |
| Knowledge spec industry | 90 | 2.944 | 1.293 | 15 |
| Knowledge of spec organizations | 90 | 3.778 | 1.109 | 6 |
| General environment | 90 | 2.844 | 1.170 | 18 |
| 3. Interpersonal knowledge/skills |  |  |  |  |
| Teaching and training skills | 90 | 3.222 | 1.139 | 14 |
| Interpersonal behavior skills | 90 | 4.522 | 0.877 | 2 |
| Interpersonal communication skills | 90 | 4.589 | 0.685 | 1 |
| International communication ability | 89 | 2.067 | 1.204 | 21 |
| 4. Personal trait and/or knowledge/skills |  |  |  |  |
| Personal motivation and working indep | 90 | 4.500 | 1.723 | 3 |
| Creative thinking | 90 | 4.144 | 0.787 | 5 |
| Critical thinking | 90 | 4.489 | 0.753 | 4 |

FIGURE 4.6 Critical knowledge/skills sets required by industries: an empirical analysis. *Source*: Yen et al. (2001).

Among the four critical knowledge/skill stated, the result of the study shows the significant prerequisite of knowledge/skills that the present educational trends and education needs are those future employees who have strong interpersonal skills, better communication skills, be it in oral or written types of communication, creative and critical thinking, and an engaging as well as pleasing working attitude (Yen et al., 2001) (Fig. 4.6). Jamwal and Singh (2001) also mentioned that for the company to efficiently deliver the MIS administrations to end clients, appropriate business and interpersonal skills are required. Hence, there is no doubt that apart from technical skills, the company also needs employees who have interpersonal knowledge/skills and personal traits and/or knowledge/skill as it allows the company to perform effectively. In addition to that, result is then further proven by the study that is carried out by Lee et al. in 1995, where the participants works in IS-/IT-ranked interpersonal skills and business functional knowledge as the most influential skills needs in their job than technical-specialties knowledge or technology management knowledge (Lee, 1999).

### 4.3.2   STRATEGIC WORKFORCE PLANNING OF HR IN ORGANIZATION

Chartered Institute of Personnel and Development (CIPD) conducted an experiment through their website to investigate to what extend the workforce planning is important. The experiments used open-ended question and short answer question. A total of 135 randomly respondents from private and public organization participated in the experiment. About 79% of the respondents viewed that workforce planning is important for their organization. Some of the findings of the experiment showed that the respondent agreed that workforce planning helps in building up a better comprehension of resources needed and forecasting future demands, and ensures that the right people are in the ideal place at the ideal time. It enables the organization to anticipate and reshape their workforce to meet future necessities and provides valuable as well as precise information especially for the senior officers to make right decision-making and plan for future operational strategies (Baron et al., 2010).

Besides that, with proper strategic workforce planning enables the organization to recognize if there are any skill gaps or any potential issues that might arise before it turns into problems within the organization. This is because avoiding the problems will not cost that much of money than settling them after they already happen. Also, workforce planning strengthens the organization's ability to maintain the achievement of business productivity and lets the organization react rapidly and moves strategically as the organization have already perceived the emerging challenges.

### 4.3.3   STAFF TRAINING

A research on to what extend the training could help the librarians' job performance was carried out with 30 participants who were library assistants at Babcock University Library of Nigeria. All the 30 participants acknowledged the statement that they required training as to guide them on how to look or find information electronically; it improved their working skills hence affecting their job performances where they could deliver their work effectively and efficiently. In addition to that, the result further uncovers that training enables the librarians to adapt to the 21st-century challenges of library working. It is clear that if the librarian receives sufficient training, it does not impact on their execution only but also an overall efficiency of the library (Bamidele et al., 2013).

Yan (1996) concluded that the training should not be a one-time training only, the librarians should be given ongoing training so that they could provide excellent services to the users as well as to let them keep up with the constantly changeable internet environment, especially in these rapidly growing technological innovations where everyone uses internet as means of exchanging services. It is clear that if the librarian receives sufficient training.

### 4.3.1   COLLABORATION PARTNERSHIP BETWEEN ACADEMIC INSTITUTION AND ORGANIZATIONS

#### Collaborative Effectiveness

The result of the collaboration with regards to IS project is very moderate. Only 33% of the respondents sense that there is collaboration. The result shows that the roles are played by diversified groups in various IS planning activities. Leadership roles overshadow in the case of organization and department management. A big percent of responses illustrate a supportive role, illustrating a passive participation instead of active collaborative participation. A supportive trait will not give much if there is no powerful leadership. The absence of negative roles played by any group is a great indicator. Personnel ego was illustrated as the highest reason to failure, and also other reasons were lack of constant training, unavailability of consultancy from experts, and lastly, losing highly skilled staff. It was reported that the status of IS in overall organization hierarchy is not very strong. Users determined that there was lack of interaction with IS staff.

### Adaptive Effectiveness

The result have indicated that the adaptation of IS and IT is very weak. It is indisputable that the transaction processing is being used frequently. However, the information being used is not reliable for decision-making, control, future planning, and information analysis. In addition to that, more than 80% of the respondents sense that within a year, they will have a routine of reporting, structured decision-making as well as control. Other than that, even after 2 years, almost 40% and 54% of the respondents does not plan to use IS for future planning and information analysis. The deficiency illustrated that the system usage is declining as the level of sophistication increases, offering a surface use of technology for information storage and retrieval.

### Evaluative Effectiveness

Any investment can have losses without frequent evaluations. When the process is well represented and the participants are highly participating on constructive evaluation, the evaluation will then be successful. A well-represented plan of evaluation, initiating roles and rules of evaluation is a must for an organization. The result shows that the control on IS use will be discharged, due to a weak evaluation process, resulting in the debtors over-looked and unanswerable, sufferers neglected, and organizational invest-ments not profitable.

### 4.3.5   COLLABORATION BETWEEN UBD AND OTHER LOCAL/ ABROAD UNIVERSITIES AND ORGANIZATIONS

University of Brunei Darussalam (UBD) has given their students a good opportunity for Discovery year during their third year. Some students will have their internship, Community Outreach program, Students exchange program, or have an incubation program at the university for the two semes-ters which is based on the students' Cumulative grade point average (cGPA).

Cord et al., 2010 and Toncar and Cudmore, 2000 researched that many students that enters university to graduate with a degree actually do not have any idea what job they are going to do/have in the future. By having internship programs at other organizations either in local universities or abroad, students will have a good opportunity to know what job they're having in the future, gaining experience from the organization, and a feasibility of their final choice if they are on their right professional track or not.

Besides that, universities can get various benefits from letting their students to have internship programs in the organizations. The university will know the important set of ability and skills that the students gain in order to adequately compete with other students/competitors in the future. Students can achieve these abilities and skills learned during an internship. Universities also allowed their students to improve their course of study to fit the needs of the students and anticipated workers (Zegwaard, 2014).

**TABLE 4.2** Ranking of Benefit Supplied by the Host Organization from the Internship Program.

| Benefit | Overall (N = 108) | | Local and central government (N = 31) | | Parastatal (N = 19) | | Private company (N = 52) | | Other types of organizations (N = 6) | |
|---|---|---|---|---|---|---|---|---|---|---|
| | $\bar{x}$ | S.D. | $\bar{x}$ | S.D. | $\bar{x}$ | S.D. | $\bar{x}$ | S.D. | $\bar{x}$ | S.D. |
| Corporate social responsibility (CSR) | 4.2 | 0.56 | 4.1 | 0.59 | 4.1 | 0.56 | 4.2 | 0.54 | 4.6 | 0.41 |
| Enhancement of corporate image (ECI) | 4.0 | 0.74 | 4.0 | 0.57 | 4.0 | 0.59 | 3.9 | 0.87 | 4.5 | 0.63 |
| Gaining new prospective and technologies (NPT) | 3.8 | 0.68 | 3.8 | 0.62 | 3.8 | 0.85 | 3.9 | 0.66 | 4.1 | 0.51 |
| Cost savings (CS) | 3.4 | 0.88 | 3.4 | 0.93 | 3.2 | 1.0 | 3.4 | 0.83 | 3.9 | 0.79 |

Ranking of benefit supplied by the host organization from the internship program was extracted from a case study in Botswana through the questionnaire that has been given to the host organization. Bennett et al. (2008) and Morrow (1995) has stated that in order to perceive the benefits from the internship programs, several ways have to be identified which are consisted of the satisfaction of corporate social responsibility, enrichment of the corporate image, gaining of new ideas, and cost saving. The result from the respondents shows that contribution from the internship program did not only benefit the organization itself but also the nation, the institution (university), and the interns (Klodwig and Christian, 2014).

Some indicated that gaining new ideas from the interns is one of the major benefits while others proposed that they could benefit from the direct cost saving. This is supported by the organization itself throughout the internship program, where they found out that the interns are helpful especially when they are enthusiastic about doing the tasks and work for hands on or practically. In addition, changes in the environment and ability to work out aside from the classroom excite them. The interns possess qualities such as hardworking, able to share knowledge and information with the existing workers, eager to learn new things and job, and applying their skills.. Moreover, since interns and fresh graduates are more sensitive toward the current trends, especially in technologies, it is such an advantage for the organization as interns will offer new ideas and outlooks in the organization. Furthermore, interns have been exposed and accustomed to the challenge; this could help to motivate the existing workers to challenge themselves to work more efficiently to avoid from being looking down upon by the manager and others (Klodwig and Christian, 2014).

As for the cost saving, once the interns have completed their studies, the organization could hire the intern permanently since the interns already have the basic knowledge and information about the organization, familiar with the workers which could lead to the happy working environment, and cut the cost of training and teaching. Apart from that, the aids provided by the interns can help to reduce the workload on hired staff. In addition, interns have added value to the organization and have helped in growing the prospective employee database (Klodwig and Christian, 2014).

Therefore, it is proposed to the university and the organization to take the internship program more seriously; also an agreement has been advised from both sides to collaborate through expanding the internship program in a period of 3–6 months rather than only for 10 weeks as this could help the interns to gain more experiences, the organization to find out the skills and ability of the intern, and the university to see the potential strength and weaknesses to improve the internship programs (Klodwig and Christian, 2014).

## 4.4   CONCLUSION

Sloman's (2011) study has the following findings:

Professionals who deal with HRs, and who fail to realize the potential importance of HRIS system will not be able to fulfill their role in the organization. They will not be able to provide information which management needs for successfully managing operating costs and for development of their employees. Management of the HR department should be more ambitious in terms of their requirements and to unite with the IT sector, to enable better functioning of the system (p. 19).

It is simplified that, changes in skill requirements are a part of the process which could drive the organization to be more efficient and productive as the evolution of globalization and advancement of IT has conquered the changes in an organization. While venturing into new trends to remain relevant, a corporate development partner has to be adopted by the HRM and MIS to achieve the main goals of an organization. Tripathi (2011) also stated that the technologically advanced MIS model definitely helps the HRM to keep the control on working of the staff at various levels. Supported by the cited journals and through the results and discussion provided above, it is concluded that the changing skills requirement is a part of the important aspect for an organization to achieve their goals and improve the performance of the employees beyond the standardized expectation even though some of the points do not fit with the changes.

## KEYWORDS

- knowledge identification
- composition
- modernization
- procreation
- management practices
- affluence of information technology
- organizational pattern

## REFERENCES

Aeron, H.; Kumar, A.; Moorthy, J. Data Mining Framework for Customer Lifetime Value-based Segmentation. *J. Database Market. Customer Strategy Manag.* **2012,** *19* (1), 17–30.

Bamidele, I. A.; Omeluzor, U. S.; Imam, A.; Amadi, H. U. Training of Library Assistants in Academic Library. *A Study of Babcock University Library, Nigeria.* 2013. DOI/full/10.1177/2158244013503964.

Bénédicte B.; Pierre-Yves S. From Technical to Non-technical Skills Among Information Systems Suppliers: An Investigation in the Skills Domain. *J. Enterp. Inf. Manag.* **2017,** *30* (2), 320–334.

Cephas O. Training and Development of Skills in a Changing Information Environment. *Libr. Manag.* **1999,** *20* (2), 100–104.

Cord, B.; Bowrey, G.; Clements, M. Accounting Students' Reflections on a Regional Internship Programme. *Australas. Account. Bus. Finance J.* **2010,** *4* (3), 47–64.

DeLone, W. H.; McLean, E. R. The DeLone and McLean Model of Information Systems Success: A Ten-year Update. *J. Manag. Inf. Sys.* **2003,** *19,* (4), 9–30.

Dench, S. (1997). Changing Skill Needs: What Makes People Employable? *Ind. Commercial Train.* **1997,** *29* (6), 190–193.

Feng, K.; Chen, E. T.; Liou, W. Implementation of Knowledge Management Systems and Firm Performance: An Empirical Investigation. *J. Comput. Inf. Sys.* **2004,** *45,* 92–104.

Garrido-Moreno, A.; Padilla-Meléndez, A. Analyzing the Impact of Knowledge Management on CRM Success: The Mediating Effects of Organizational Factors. *Int. J. Inf. Manag.* **2011,** *31* (5), 437–444.

Hornstein, H. Using a Change Management Approach to Implement IT Programs. 2008.

Jamwal, D.; Singh, N. Management Information System Competencies and Strategies. *Int. J. Comput. Sci. Technol.* **2011,** *2* (1), 109–112.

Kim, C.; Lee, I.; Wang, T.; Mirusmonov, M. Evaluating Effects of Mobile CRM on Employees' Performance. *Ind. Manag. Data Sys.* 2015, *115* (4), 740–764.

Klodwig, M.; Christian, M. Benefits to Host Organization from Participating in Internships Programs in Botswana. *Asia-Pac. J. Cooperative Educ.* **2014,** *15* (2), 129–144.

Kumar, R. Role of MIS in HRM. 2012. Retrieved from https://www.slideshare.net/Rupesh-Kumar22/role-of-mis-in-hrm (accessed Nov 2018).

Lee, D. M. S. Knowledge/Skill Requirements and Professional Development of IS/IT Workers: A Summary of Empirical Findings from Two Studies. 1999.

Li, L.; Mao, J. Y. The Effect of CRM Use on Internal Sales Management Control: An Alternative Mechanism to Realize CRM Benefits. *Inf. Manag.* **2012,** *49* (6), 269–277.

Malik, K.; Goyal, D. P. Organizational Environment and Information Systems. **2003,** *28* (1).

Martin, B.; Healy, J. Changing Work Organisation and Skill Requirements. **2008.**

McKeen, J. D.; Zack, M. H.; Singh, S. Knowledge Management and Organizational Performance: An Exploratory Survey. *Proceedings of the 39th Annual Hawaii International Conference on System Sciences.* Computer Society Press, Hawaii, 2006, 9 pages. http://dx.doi.org/10.1109/HICSS.2006.242 (accessed Nov 2018).

Mirbagheri, S.; Hejazinia, M. Mobile Marketing Communication: Learning from 45 Popular Cases for Campaign Designing. *Int. J. Mob. Market.* **2010,** *5* (1), 175–192.

Noll, C. L.; Wilkins. M. Critical Skills of IS Professionals: A Model for Curriculum Development. *J. Inf. Technol. Educ.* **2002,** 1 (3), 143–152.

Rodriguez, J.; Osorio, J.; Berriel, R. Contribution of the Information Technology/Information Systems to Human Resources Management in the University. *Inf. Technol. Educ. Manag.* **1995.**

Sinisalo, J.; Salo, J.; Karjaluoto, H.; Leppäniemi, M. Mobile Customer Relationship Management: Underlying Issues and Challenges. *Bus. Process Manag. J.* **2007,** *13* (6), 771–787.

Syed, N.; Xiaoyan, L. The Linkage Between Knowledge Management Practices and Company Performance: Empirical Evidence. In *LISS 2012;* Springer: Berlin, Heidelberg, 2013; pp 763–769.

The Role of Information Systems in Human Resource Management. 2011, pp 1–20. Retrieved from https://mpra.ub.uni-muenchen.de/35286/1/Chapter_2_draft_The_Role_of_Information_Systems_in_Human_Resource_Management.pdf (accessed Nov 2018).

Teo, T. S. H.; King, W. R. Integrating Between Business Planning and Information Systems Planning: An Evolutionary-contingency Perspective. *J. Manag. Inf. Sys.* **1997,** *14* (1), 185–214.

Toncar, M. F.; Cudmore, B. V. The Overseas Internship Experience. *J. Market. Educ.* **2000,** *22* (1), 54–63.

Tripathi, K. A Study of Information Systems in Human Resource Management (HRM). *Int. J. Comput. Applications (0975–8887),* **2011,** *22* (8), 9–13.

Workforce Planning Right People, Right Time, Right Skills. (2010) (pp. 1–46). UK

Yan, L. Training Library Staff to Adapt to the Internet Environment. 1996.

Yew, B. K. A Perspectives on a Management Information Systems (MIS) Program Review. *J. Inf. Technol. Educ.* **2008,** 7.

Zegwaard, D. Benefits to Host Organizations from Participating in Internship Programs in Botswana. *Asia-Pac. J. Cooperative Educ.* **2014,** *15* (2), 129–144.

# CHAPTER 5

# HUMAN RESOURCE MANAGEMENT INFORMATION SYSTEM: THE CHALLENGES

**ABSTRACT**

Human resource information system (HRMIS) or e-human resource management (e-HRM) is a software to easily conduct the operations of administrative or other necessary HR functions. This includes storing employee information, managing payrolls, and also e-HRM offers ease in planning, recruitment, performance appraisal, communication, compensation, and training. With the use of MIS in the organization, it would help increase effectiveness and efficiency of operations. Although it is also important to note the drawbacks of using a vulnerable information system. This is mainly on the security. Hackers favor a challenge. Therefore, the more advanced and precious the organization is, the more hackers are likely to challenge themselves to try to breach the firewall. Important information will be very dangerous when it is leaked. Personal information of employee, finance accounts of the company, and many more can be extracted thus, making the company easier to be attacked. This study focus on security challenging to implement e-HRM within organization that may lead importance of having security awareness for possible threats and security breaches and avoid attacking of information by social engineer.

## 5.1 INTRODUCTION

Technology has been around from the Information Age where the rise of computers and internet has started to blossom. Both of the factors have contributed to the evolution of technology infused in an organization. Information and technology are a key factor for an organization to be successful because it impacts the most due to its ever-changing nature. Without both information and technology, businesses will be left behind from their competitors due to lack of information being passed around in the organization and how to better manage their employees mainly the human resource (HR) department. Technology helps to enhance organizational productivity in terms of better execution in delivering information faster and accurate.

Advances in technologies have changed the way a business operates. It has done wonders and, in instances of choosing the right way to implement technology, has made numerous businesses gain an advantage over their competitors. Bryant (2017), have suggested that with technology, businesses can be quick in decision-making and saves space in terms of storage keeping. In addition, knowledge can be shared swiftly with fewer resources. Therefore, technology will be the main topic in this report in its advancement in management information system (MIS), especially used in the human resources management (HRM).

The HR department in an organization is concerned with managing the employees. Any aspect that directly deals with people in the organization goes directly to the HR department. This includes compensation, hiring, performance, employee motivation, communication, training, etc. It is important for an organization to have a highly skilled labor and employees in the HR department to provide them with competitive advantage in the market. Competitive advantage in relation to HR department would be retaining employees that always strive to achieve the organization's goal and to make full use of the resources available.

HRM is one of vital departments in an organization. Hiring the employees, paying the employees, training for the employees, and project management that are related to employees is all under HRM department. HRM is a system that provides the information concerning all the employees in the company. It is the system that could improve the communication facilities between the people in an organization that relate to HRs. By using HRM system, all employees' information about employee general details, contact details, skill details, educational details, and project details will be recorded in the system. HR is also responsible for maintenance of the information that are connected to the employees details background including the employee payment criteria based on the skills and experiences, training, and performance. In HRM department, they have the authority to access the employee's private information.

HRM system or human resource information system (HRMIS) or e-human resource management (e-HRM), or any other similar usage of the term, is a form of software that uses systems and processes to easily conduct the operations of administrative or other necessary HR functions. This includes storing employee information, managing payrolls, and more. In a research of significant contribution of e-HRM to HRM effectiveness, Ruel et al. (2007) concluded that, there is in fact an increase of HRM effectiveness with the help of e-HRM (2006). In addition, Leda, Maria, and Eleanna have suggested that e-HRM offers ease in planning, recruitment, performance appraisal, communication, compensation, and training. For instance, with information system, electronic mail is used as a means for communication. Bontis et al. (2003) have presented that 75% of corporate uses electronic mail as a medium to communicate. Also, intranet and e-forums allows access for employees to all kinds of information (Leda et. al., 2007).

Therefore, with the use of MIS in the organization, it would help increase effectiveness and efficiency of operations. Although it is also important to note the drawbacks of using a vulnerable information system. This is mainly on the security. Hackers favor a challenge. Therefore, the more advanced and precious the organization is, the more hackers are likely to challenge themselves to try to breach the firewall. Important information will be very dangerous when it is leaked. Personal information of employee, finance accounts of the company, and many more can be extracted thus, making the company easier to be attacked. Hence, it is important to study the possible threats of security breaches so that measures can be taken to implement to avoid attacks.

## 5.2  LITERATURE REVIEW

The 21st century has brought new technology and systems to HR that can lead to new collaboration. Data and information are significantly important assets for today's organization. All the HRM data and information can become more simplified and analyzed with information technology. Information technology is able to store records and information of the employees as well as access it with extreme ease such as fast to gather information and data, collected and delivered to the employees. Besides, it can increase the expectancies of HR department as information system has become the core value of HR. It reduces the HR administrative workload focusing on the core tasks such as providing the top management the information that they needed in order to make decisions. It can affect the organizational performance as a whole in the long run.

The technology is impacting with the world we are living in, which included HR that is always viewed as the area that deals with employees, decisions of the salary is being made, people get training from, hired or fired, etc., and it is where people from different backgrounds and skills are organized in order to accomplish mission, vision, and goals of an organization. According to Ngai (2008), HR function is all organizational members who are responsible to carry out duties at any level in order to accomplish the corporate objectives. The technology plays the vital role in order to assist HR department to anticipate and manage the organization as a whole.

HRM takes all the decisions within the organization that could affect the relationship between the organization and the employees. Consequently, to manage and control HR is difficult as it abbreviate all the circumstances that are related to employees and organization. Broderick and Boudreau (1992) argued that to achieve success in business, managing the HR is very important as it is the assets of an organization. Hence, the data and information that is collected from the employees should be protected from others who could harm them. The data and information of the employees should be stored with accuracy, relevance, and appropriate with safeguards to maintain the security and confidentiality.

The technology has become cognizant and important to human existence and world. According to Mathis and Jackson (2010), the technology is bringing impact to the HR activities at the workplace. By using technology such as HRMIS, it is becoming much convenient for the HR administration to store the data and information, retrieve, update, classify, and analyze the resources that are related to the employees. HRMIS has become strength for an organization in general. The major objectives of HRMIS is to update the latest and recent information and progression about the people and the work also to secure the data and personal information.

With the information system that is adopted in HRM, securing and protecting the data information of the staff is crucial. Most of the organizations are using information system for being effective and efficient in order to safeguard the information assets of the company; but with the rapid changing of the technologies, all the data information can be unprotected from others. With the modern technology that have been invented, Herath and Rao (2009) claimed, it is difficult to ensure the safety of the data and information. Data and information of an organization is seen to be an important asset in today's world. Failure to protect and to prevent the unauthorized access to confidential data and information will bring an impact to the companies and an individual reputation as the HRM is a department that recruits employees which are the assets of an organization.

According to Viljoen (2008), an organization must protect and secure the assets for the company and individual reputations. HRM is an important department within an organization where the data and information of employees are administered. The information could be correlated with identifiable skills and experiences, living, personal information such as race, gender, status, medical or criminal history, personal details, contact details, the opinion of others regarding the person, salary, and others that could be related with human resources.

The technology such as information system will provide security for the data and information and assurance that allow the HRM can ensure and trust. Moreover, information system can ensure the technology services that are utilizable and can prevent failure or error which certifies the sensitive and privileged information is disclosed from anyone who have not given the authority to access it (Crum, 2000).

In order to protect the data and information within HRM, the combination of technology such as information system can be accomplished. The fundamental of data protection strategy that is using information system for back-up and recovery, security of the system storage, and lastly information lifecycle management (Bradley, 2013). Moreover, the purpose of data protection is locking the file and record, database shadowing, disk mirroring which assure of the availability of the data, etc.

However, all of these benefits can be attained fully if the HRMIS can be implemented properly. Armoni (2002) stated that implementation of IS or upgrading previous system into an organization tends to fail due to several factors such as lack of support from management and inability to cope with the new system.

## 5.3 FINDINGS

The computer security is referred as the application of system that is used in order to conserve the confidentiality and integrity of information system resources (Stallings and Brown, 2008). Private information is crucial in order to keep the information and data of the employees safe from unauthorized users. Still, it is difficult to maintain the privacy of the employee's information. Integrity is a principle or behavior that can generate the information resources to manage in an authorized manner. It is significant to ensure integrity of the data, information resources, and the system. According to Kang (1998), there are fundamental differences between information security and privacy. Kang (1998) also emphasized the three groupings of clustered privacy which are physical objects, ability to choose without any interference, and finally regarding with the employee's personal information.

HRM is concerned regarding the safety of confidential data and information of the staff as the company assets. It has become a great issue for all organizations that are using information system as their core component in their business. With rapid advancement of technology happening around the globe, a company needs to be as actively innovative as possible in order to be able to compete with other organizations. With the implementation of HRMIS, it would benefit an organization due to its efficiency and innovation of using a computer-based management system. This can help the company to make better choices in employee selection, evaluation, promotion, and making its administrative work more effectively in order to be ahead of its competitors.

According to Viljoen (2008), information system has become one of crucial systems to manage all the data and information but at the same time, it has been significantly being considered as a serious problem that can affect government, organization, and general computer use over recent years. According to the Vassey and Glass (2002), "Information security is a loosely defined term used to describe an incident where the confidentially of information has been compromised, typically as the result of unintentional insider action." Besides the information system, technology nowadays such as email attachments files, cloud computing, and portable drives could expose the organization data and information.

Herath and Rao (2009, p1) argued, it is not sufficient to use the technology system tools alone although most of the organizations nowadays are using information technologies to protect the information. Therefore, the end-user security attitude and integrity are the biggest challenges in order to keep the information secure. Hence, every member of the organizations plays an important role and responsibility for the confidential and restricted information to be protected. According to Viljoen (2008), once every member of the organization is aware of how to secure the information, it is possible for an organization to manage its own people within the organization and information security.

Doherty and Fulford (2008) argued, the information is able to retain the employee integrity and confidentiality if the employee is being protected from threats against them. The potential damage and security breaches that might affect the HRM and an organization as a whole when computer-based fraud, the information resources is access to unauthorized person also sabotage and user errors can cause and result confidence lost, sensitive data disclosed, and asset lost. Furthermore, the main role of an organization is to control and manage the information security as there will be a number of users within the organization who may be untrustworthy and traitorous to the organization's information.

Human error or user error is one of the factors that threat the security of the information which can cause information leakage. Human error or user error is due to lack of information security awareness, training programs, and inadequate use of technology knowledge can involve security issues. Besides, security protection for information is not enforced or fully implemented by HRM. HRM is not aware of the users who are not an information technology professional which the unit is expect them to know everything about security or protection issues.

There are several of challenges that will be encountered in order to ensure the safety and security of the data and information of the employees within HRM department, such as training the users to manage the system such as HRIS or HRMIS which is the latest technology to manage and operate the HR department. In order for the employees to be more motivated before the implementation of new system, there should be a support from top management. Support from the top management will provide motivation for the employees to make changes. Also, operating HRMIS should be handled by a specialist or training an employee to be more of a specialist. However, it can disrupt the employee's daily routine and hiring of specialist may incur more cost.

## 5.4  DISCUSSION

HRMIS provides many benefits to an organization as compared to traditional HR. When using HRMIS physical files and papers that are used in traditional HR are replaced with files that are stored electronically in a database. The benefits that HRMIS offers over traditional human resource include better organization, space saving, reduced risk from damages and losses, effortless sharing, and backup.

HR department are able to access employee files with HRMIS using desktop computers or laptops which are more efficient and faster as compared to traditional physical files and papers. Searching for physical files in a traditional HR consumes valuable time with the vast number of files stored on racks and shelves. HRMIS organizes all the necessary information regarding HR on a single system which can be accessed anywhere and anytime with a computer. Old files and papers that consume massive space are replaced with HRMIS thus creating more space for other uses for the organization.

Traditional human resource uses physical files and papers which are prone to damages from accidents occurring in office. In an event where the office catches fire (big or small), files and papers are susceptible from the damages that the fire can inflict. Sprinklers that would be used during fire accidents would also harm these files and deter them from being readable. Salvaging files and papers from a fire accident would be potentially difficult and tedious as most of the files are destroyed or missing. Fortunate files that are not affected by the fire could still be used but the people of the department would have to reorganize and store these files to a new temporary place which occupy a great deal of time and effort.

Backup files are important to have in any organization where backing up original files and papers are copied and archived onto another location which are accessible during an unfortunate event causing losses of the original files and papers to happen. Backup with HRMIS are done either online or storage on a server hard drives. Traditional HR however, would require employees to make physical copies of files and reports which requires more time as compared to using HRMIS.

Electronic files of HRMIS that are stored online on cloud storage stores files that can be accessed at any locations with internet access. Internet coverage nowadays are widely spread across the country where people can access cloud storage virtually anywhere, anytime. Cloud storage refers to a remote server that stores files which are accessed through the internet. Any mishaps that would occur and damage computers and laptops would not affect HRMIS with files stored on cloud storage. The cost of paying internet service provider (ISP) weighs less as compared to having an internet connection issues which would affect cloud storage. With the rapid emergence of technologies people and organizations have heavily relied on internet to do daily task and activities.

Backup on HRMIS could alternatively be done by copying files to external storage devices. External storages include portable hard disks, USB flash drive, memory cards, solid state drives, DVDs, and CDs. Files are transferrable from HRMIS to external storages and shared between them effortlessly which then could be brought to any place when needed. When internet is not available in any case, external storages would be able to give other computers access to HRMIS files offline from the internet. External storages are mostly affordable to purchase and are widely available ranging from consumer grade to enterprise grade storage models.

Cost of implementing HRMIS might be high due to several costs that needs to be addressed, however, in the long run it can benefit the organizations in terms of space saving as stated above and the reduce usage of paper as more works are being done and saved inside the computer system. Employees can do different tasks simultaneously with the help of new system because they can do their daily routines all in one system whether it is to store client's information and to make reservation for an employee's training courses at the same time. Another benefit of this would be the fast communication being done with all departments as they are interconnected with each other in order to be a community. With the community-building, the organization can strive to achieve its goal more easily as the employees are working together more closely and to pass information to different departments faster to get the necessary results.

Risk management can also benefit with the implementation of HRMIS because it monitors the employee's job scope in the organization. As defined by Whatis.com, risk management is the process of identifying and controlling threats of an organization's capital. This also includes the strategic management errors and accidents that may hit their employees. In order to prevent accidents and errors happening, HRMIS can help to make its employees have the right training and courses for their respective job scope. The HR department work will be lessen due to the effectiveness of HRMIS as the courses and training being booked for the respective employees can be done easily and swiftly with regards to the old system where the HR department need to submit a manually written documents to the respective training department. This will make the employees to be more motivated as the organization are taking certain measures to make sure that they are responsible for the employee's action and well-being.

## 5.5 MAINTENANCE CHALLENGES WITH HRMIS

The biggest challenge of HRMIS as mentioned in the literature review is its security strength against unauthorized access and unintended publication. Presently, aside from security, it is the cost and staffing of implementing HRMIS which includes acquisition and maintenance costs, and hiring expertise to manage the system.

Confidential information kept in HRMIS are vast which are allowed access to certain department and employees. Level of access depends on each user with different rankings where high-ranking people in an organization would have a higher level of access as compared to vice versa. The head of department might have a full list of employee details under its department including their contact information, reports, and performance. Such information are not disclosed from the system to a salesperson who worked on the front lines.

There are unethical employees that would secretly benefit from confidential information stored in HRMIS. These employees would go to any length to gain unauthorized access to the system by stealing user id's from their superiors or by hacking into the system themselves. Attendance and performance records, salary details, and other information could be altered from an unauthorized user to their own benefit. These actions would harm the organization indirectly burying problems and faults within the organization.

There are cases where high-level access users would leak confidential information from HRMIS to other people internally or outside the organization. Personal information of an employee are shared to another employee for personal advantage and may be paid illegally with interest to the information. Corruption may also come from outside the organization, for example, a direct competitor who would want to know confidential information that may profit them against the organization. Employees that have no integrity that are given access to the system would potentially harm the organization just to gain personal profit.

External threats such as hackers are prominently present as a threat to any organization that uses internet as its daily operations. They will simply exploit a computer system and control them for different purposes such as disrupting the system by inserting malicious code and to modify or steal the organization's data. By inserting malicious code, the hackers could steal important information such as the employee's and client's records and delete them or potentially leaked them elsewhere. This action could ultimately cost the organization millions of dollar and danger of future attacks if they are unable to stop these threats early on.

Other challenges that an organization needs to address would be the cost of acquiring and implementing the HRMIS into the current system being used in the organization. Without any research or proper financing being done by the organization, it would lead to a loss if the proposed HRMIS failed to meet its expectation. Understanding the organization's HR information is essential as improving them would lead to better productivity and efficiency from the employees while having a rather comprehensive system to work upon. Identifying areas where it can largely benefit from new system of HRMIS must be the organization's main issues.

Small company or organization with a limited budget would have a hard time when trying to implement a new HRMIS as the cost may be higher than the benefit they would achieve in return. This cost would include maintenance and acquisition costs as with frequent updates of the system are needed to be relevant for the employees to work on. However, it is important for them to update their current system if they were to compete with the rather competitive nature of the global market. The organization needs to hire expert technician to monitor and maintain the system to keep it up-to-date. Such technician would require ample experience and skills which are not easily apparent for a business organization to assess. Hiring these experts would cost a great deal of money to pay for their salary. The expertise would also be useful during an event of system crash or accidents where information losses need to be recovered. All the necessary costs of maintaining HRMIS depends on the organizations available budget for the HR department.

Employees may not benefit from the implementation of HRMIS and this can prove to be a vital challenge for the organization as they are reluctant to use them. This can be in the form of trying to adapt the new system and complexity of using it due to the lack of skill and expertise. Danger of having this kind of employees may lead to less productivity and efficiency from them as they are the one who will be using the new system and this would slow down the organization's productivity as a whole. The HR department needs to have a key person that would handle the HRMIS system as stated by David et al. (2015). This would generate more cost for the organization due to hiring a specialist to take the job or training an existing employee to become a specialist instead.

Traditional HRs are still preferred to some employees as compared to adapting HRMIS into the organization. Employees that are not fond of computers and who do not know how to use computers may struggle to learn and use such system for the company's benefit. Older generations' people are mostly not exposed to technological advancements since their young age hence they may not know how to use computers. Teaching them to learn how to use computers let alone the system may take considerable time hence that would demotivate them from using the system. In fact, such employees would be more efficient when using traditional method of HR than using HRMIS.

In order to do the job accurately, massive training should be given to employee of a company in certain areas in order to use the system very well. Unable to perform the information system accurately, could lead the data and information to be easily stolen by another company. Another challenge is time to adapt the system which means the people in the organization need to take times in order to adapt the system. Training and learning for the adaption of the system is time-consuming process. Inadequate knowledge for using this system could lead in easy leakage of information. In order to manage the HRMIS effectively, expertise in running the system is crucial but it is difficult to maintain the system once there is shortage of IT expert in this area.

Security awareness is pivotal to ensure integrity of each employee in the organization. Employees should know the seriousness of security threats to information from unauthorized people. Workshops could be offered for employees to learn how dangerous can information leak be from personal level to organizational information. With the awareness employees are motivated to prevent breach of security within the organization and take preventive measures strengthening the overall security of the systems in the organization.

## 5.6 RECOMMENDATIONS

User's security awareness is important in order to protect the information as users who are managing the information resources must follow procedures with the implemented controls which are related to awareness and provide continuous training and education to the users. Wilson and Hash (2003) highlighted, information security awareness should be fully implemented to focus on the attitude of HR personnel in order to protect the information of the employees within the organization. Security awareness is a process in order to educate the people within the organization of the importance to protect the information by implementing the security program awareness. The program is designed to focus more on user's security awareness.

Moreover, establishing security policies and procedures within organization are crucial in order to ensure the information resources is protected from unauthorized person. Without proper procedures, education, and training, the users such as internal users, external users may have a harmful effect toward an organization. To prevent fraud, education and training concerning the security should be executed to every HR officers in order for them to understand their responsibilities and roles for effectiveness of the security measures. The information security education and training could lead to quality and secure data in HR department.

As well, information technology standards is crucial in order to govern, support, and control on how organization could manage the system infrastructure in order to protect the data and information resources. The IT Governance Institute and Office of Government Commerce (2008, p 19) argued that, managing the information as a routine everyday could help to ensure the effectiveness in management policies and procedures. Practice by controlling objectives of information technology which is designed for users, auditors, and as comprehensive guidance for the company management which requires all the members in the organization to have total responsibility in all aspects.

Besides, all users in an organization should not be given ability to access the information and all the systems and access control should be strictly imposed. In order to manage the access control, establishing user accounts by identification, authentication methods could limit access of the users to resources. The identification includes who is the user and to show who is the user. The authentication procedure is the process to verify the authenticate identity of a user such as password, identification card or physical characteristic such as fingerprint. It is to limit the number of users from accessing the information resources.

Strengthen the security by giving a right to the authorized user when the users is asking for personal data and information. It is to prevent from unauthorized users to access the data and information such as file access control scheme should be supplied in order to generate the conjunction with hardware to create a barrier between a user and the files' data. Also, in order to protect the confidential, a strong security which is personnel safeguards such as background checks of the users, disciplinary operation is essential in order to prevent leakage of information.

Additionally, information system security policies are also necessary in order to prevent threats which need to be implemented to check leakage of information. It is designed to inform the people within an organization especially the users who are accessible with the information to protect the information. The policies also inform how these responsibilities should be carried out which includes baseline to acquire, compliance the policy by auditing information system security. With the written policies it could help to ensure the people within the organization who could behave in a responsible manner to manage the information. According to Wood (2009), information system security policies can assist the HRM department in many different ways in order to achieve the objectives of the organization. The objective of the security policies is to educate and train the people within the organization to control the information in responsible manner. The information system security policies should be easier to be understood and practical in order to be more effective by making the information system security policies to be accessible to the units and users within the organization.

Penalties and warnings should be implemented in events of breaches on the system security policies. This would motivate employees preventing themselves from making an unethical behavior of leaking information from the system to irrelevant and unauthorized parties. Harsh penalties may, for example, include fines, salary deduction, suspension, or even termination of contract from the organization. Employees who are caught breaching the policies would be given penalties which will serve as a warning to other employees to avoid them from doing the same. The heavier the penalties are the more likely that employees would obey to the organization's policies.

It is important to motivate the information resource owners and the users to control the management commitment in term of security. It is to ensure the users and people within the organization could support each other in order to provide visibility needed by the information security team. The security computer training should be implemented such as guidelines to protect the information to be accessible by unauthorized users. Management in commitment of information system security is the important factor in a successful securing of the information.

Employees of the organization needs to have the right mindset being infused to them when the new system are being implemented in the first place. This is to help them in transitioning from the old system toward the new system in a smooth way rather than instantaneously that may result in less productivity from them. Having the right mindset is important in every aspect of daily lives as it enables an employee to work under stress or crumble under stress. Support from the top management is also a key factor in helping the right mindset for the employees. It shows that the top management are more than willing to be a part of the upgrading process.

To ensure the implemented system are being used to its full potential, collective feedback could be done in a monthly basis in order to check if the system contributes to a higher productivity and efficiency for the employees. Also, with this monthly feedback, the security of the system could also be monitored if any updates or changes are needed in order for it to be more secure. Regular checking and outflow of confidential information can be checked by the IT department. Employees need to understand the danger of leaking confidential information as this can tarnish the organization's reputation.

Passing of information into the correct recipient is crucial to an organization and leakages of such action can be avoided through encryption. The term encryption refers to changing electronic information into information that cannot be read unless authorized. Important information from the HRMIS are shared within the organization or to relevant people when needed. The information sent outside the organization through an email would be encrypted and a key should be given to the recipient of the information to decrypt the encrypted information. No other people can access the information without the key, hence preventing from unauthorized access.

Backing up data and information is an important procedure to any information systems. Organization with HRMIS could implement RAID (Redundant Array of Inexpensive Disks or Drives, or Redundant Array of Independent Disks) configuration to constantly backup the entire information on the system in real time. Any changes made on the main storage for the HRMIS will be backed up onto other hard drives thus having multiple storages of the same content as the original system. Small organizations could afford to buy multiple hard drives as they are readily available and cheap to set up. Big organizations could implement multiple RAID setups to backup massive amount of files and data for HRMIS and other systems they use.

Lastly, vulnerability management or vulnerability scanning process is acquired in order to prevent the leakage of information. Enforcing vulnerability management could control and manage the information security risks. To prevent unauthorized users from stealing information, identifying and aggravating the vulnerabilities in information system technology is crucial. The vulnerability management could reduce risk, with presence of security vulnerabilities in information technologies system. To ensure the implementation of vulnerability is achievable the roles and responsibilities have to be clear.

## 5.7   CONCLUSION

Effective information system in HRM is crucial with technology assist in order to increase speed, lesser paperwork, and cost-effectiveness but to maintain the security and protection of the data and information within the organization is the biggest challenges and crucial to secure them. Generally, the information security management system is to ensure whether the users are aware of their responsibilities and integrity in handling any information that are related to confidential data and sensitivity in accessing and using the employee data and information.

Emerging business creates more competitive environment and drives it to be the best they can be in different types of businesses. This has enabled some organizations to be more innovative and to take certain risks than the others in order to have that competitive edge over their competitors. In order to be one of the leading businesses in a particular industry, they need to be efficient in all departments of its organization and to work closely together in order to achieve its goal. Obstacles such as implementation of HRMIS should be overcome as soon as possible in order to generate high productivity from its employee.

It is the biggest challenges to protect the data and information unless complete training and education is given to the people in the organization especially to HRM unit about the importance of security policies; moreover, to motivate and encourage them to engage actively in secure computing practices. Whether internal or external, an organization should be fully prepared as people in the organization must be committed and responsible. Manager should always monitor their employees in regards to their work but they should also be proactive in their daily routine so information being sent is in their job scope for instance, HR employees should only access information on the welfare of other employees and information other than that should be restricted.

In order to be successful in today's business environment, an organization should embrace it with open hands and allow it to be a part of the organization's culture. For an organization to be relevant and stay ahead of their competitors, it is important for them to take advantage of the ever-changing technology in the world with having the best employees in HR. Shiri (2012) mentioned that an organization would have an increase in effectiveness and productivity of its employees if the HR department is able to have up-to-date system (HRMIS) and technology.

## KEYWORDS

- social engineer
- security challenging
- e-human resource management
- recruitment
- performance appraisal
- planning
- communication
- compensation
- training

## REFERENCES

Armoni, A. Data Security Management in Distributed Computer Systems. *Inform. Sci. Int. J. Emerg. Transdisci.* **2002,** *5* (1), 19–27.
Bontis, N., Fearon, M.; Hishon, M. The E-flow Audit: An Evaluation of Knowledge Flow Within and Outside a High-tech Firm. *J. Knowl. Manag.* **2003,** *7* (1), 6–19.

Bradley, T. Data Protection and Information Lifestyle Management. Chapter 1: Introduction to Data Protection, 2013. Retrieved from http://netsecurity.about.com/od/chapterexcerpts/a/aaexc_datailm.html (accessed Aug 22, 2017).

Broderick, R.; Boudreau, J. W. HRM, IT and the Competitive Edge. *Acad. Manag. Execut.* **1992,** *6* (2), 7–17

Bryant , B. J. Benefits of Technology in Business. Chron, 2017. Retrieved from: http://smallbusiness.chron.com/benefits-technology-business-336.html (accessed Nov 2018).

Crume, J. Inside Internet Security: What Hackers Don't Want You to Know. Pearson Education Limited: New York City, New York, 2000.

David, S.; Shukla, S.; Gupta, S. Barriers in Implementing Human Resource Information System in Organization. *Int. J. Eng. Res. Manag.* **2015,** *2* (5). https://www.ijerm.com/download_data/IJERM0205021.pdf (accessed Aug 27, 2018).

Doherty, N. F.; Anastasakis, L.; Fulford, H. The Information Security Policy Unpacked: A Critical Study of the Content of University Policies. *Int. J. Inf. Manage.* **2009,** *29* (6), 449–457.

Herath, T.; Rao, H. R. Encouraging Information Security Behaviors in Organizations: Role of Penalties Pressures and Perceived Effectiveness. *Decis. Support Syst.* **2009,** *47* (2), 154–165.

IT Governance Institute and Office of Government Commerce. Aligning CoBiT®, ITIL® and ISO/IEC 27002 for Business Benefit. Management Summary. A Management Briefing from ITGI and OGC; 2008. Retrieved from http://www.onuva.com/wp-content/uploads/2013/04/Aligning_COBITITILV3ISO27002_Bus_Benefit_9Nov08_Re search.pdf (accessed Aug 22, 2017).

Kang, J. Information Privacy in Cyberspace Transactions. *Stan. L. Rev.* **1998,** *50* (4), 1193–1294.

Leda, P; Maria, V.; Eleanna, G. E-HR Adoption and the Role of HRM: Evidence from Greece. *Personnel Rev.* **2007,** *36,* (2), 277–294.

Mathis, R. L.; Jackson, J. H. *Human Resource Management,* 13th ed.; South-Western College Publishing: Ohio; 2010.

Ngai, E., et al. Importance of the Internet to Human Resource Functions in Hong Kong. *Pers. Rev.* **2008,** *37* (1), 66–84.

Shiri, S. Strategic Role of Human Resource Audit in Organizational Effectiveness. *J. Manag. Public Policy* **2012,** *3* (2), 39–45.

http://jmpp.in/wp-content/uploads/2016/01/Shammy-Shiri.pdf (accessed Aug 27, 2018).

Stallings, W.; Brown, L. *Computer Security: Principles and Practice*; Pearson Prentice Hall: Upper Saddle River; 2008.

Ruel, H. J.; Bondarouk, T. V.; Van der Velde, M. The Contribution of e-HRM to HRM Effectiveness: Results from a Quantitative Study in a Dutch Ministry. *Employee Relat.* **2007,** *29* (3), 280–291.

Vessey, I.; Glass, R. L. Research in Information Systems: An Empirical Study of Diversity in the Discipline and Its Journals. *J. Manag. Inf. Syst.* **2002,** *19* (2), 129–174.

Viljoen, M. A Framework Towards Effective Control in Information Security Governance. Unpublished MTech. IT Thesis, Nelson Mandela Metropolitan University, Port Elizabeth, South Africa, 2008.

Wilson, M.; Hash, J. Building an Information Technology Security Awareness and Training Program. *Recommendations of the National Institute of Standards and Technology*; US

Government Printing Office: NIST Special Publication: Washington, DC; 2003, pp 800–850.

Wood, C. C.; Dave, L. *Information Security Policies Made Easy Version 11;* Information Shield, Inc., 2009.

# CHAPTER 6

# THE SHIFTING PARADIGM OF HUMAN RESOURCE MANAGEMENT THROUGH A HUMAN RESOURCE MANAGEMENT INFORMATION SYSTEM

## ABSTRACT

Continuous improvement in an organization may ensure the organization to be more successful. There has been an increase in the number of challenges followed by a rapid increase in the number of business organizations when compared to the previous years. Due to this, numerous number of organizations started to introduce the Business Process Reengineering in order to manage their business efficiently and to enhance their performances. This studyfocus on how reengineering processes and with the use of IT that support and advance every department, especially Human resources department, Accounting department, Administration and logistic. The study also focuson factors can influence an organisation to use reengineering and why the organisation choose reengineering processes as an alternative solution to solve the current problems faced and therefore, can assist in improving productivity and meeting the organizational objectives and goals.

## 6.1 INTRODUCTION

In the current organizational environment, the competition level is rising up where businesses must compete the pressure to improve the performance and keep on reducing the operation cost of the business. Enterprises must familiarize the changing situation to increase the sales revenue and their market share. Hence, numerous business organization start implementing business process reengineering (BPR) as a strategy to achieve such radical changes. Hammer and Champy (1994) defines BPR as "the fundamental rethinking and radical redesign of business processes to achieve dramatic improvements in critical, contemporary measures of performance, such as cost, quality, service and speed" (p. 32). Business process re-engineering (BPR) is a strategy implemented for business where it will help them to restructure their organization's workflow, business processes and to find a way to improve their performance in order to compete with their rivals. It focuses on the objectives of the organization and the process in which how it will be achieved. The aim of re-engineering is not only to improve their operation but also the organization as a whole. BPR was started in early 1990s where all the management strategies available during that period of time were not able to sustain the organization. To make BPR can be practical; both Information Technology and Information System (IS) are practice in all business processes. Business processes could create astonishing improvements by using IT. It permit firm to reshape their company using the best tactics, and support in collecting and evaluating information to make task uncomplicated. Among other enables, IS accelerates the management of certain processes in the company. They also have possibility to manage the firm's processes with those of its business associates, for example, customers and suppliers. Re-engineering concept does bring IT and IS functional role. Organization's working procedure is created using IT and help to synchronize the processes and functions that are involved in operating the business. Moreover, IS as an important role in the achievement of action right through the BPR life cycle and also in the redesign of the processes and their supporting system. The application of both IT and IS in the process re-engineering may contribute to positive and major outcomes from BPR.

## 6.1.1   TECHNOLOGY

At the point in time where, Clayton Christensen was a recently appointed educator at Harvard Business School and started his well-known investigation of why organizations fail, he adopted an uncommon strategy (Satell, 2013). He would not like to take a gander at any organizations, however, at previously successful ones. The kind whose stocks were valuable in the market and whose CEOs graced the fronts of best business magazines. Not the washouts, but rather the once extraordinary champions who faltered and fell. What he found was remarkable. While he anticipated that he would see firms who had essentially lost their direction, what he found were firms that took after the greater part of the prescribed procedures instructed at business colleges like his, for example, putting efforts intensely in R&D, tuning in to clients and concentrating on benefits. In the long time since he started his exploration, it has turned out to be evident that innovation movements can drastically change time-regarded business standards.

Information technology (IT) has developed into a crucial and major part of marketable strategy. From big corporations who keep up centralized server systems and databases to private businesses that possess a single personal computer (PC), IT is involved (MacKechnie, 2017). The trends and patterns of innovation of PC within businesses and the purposes for using it can best be managed by observing how worldwide the organizations are utilizing it.

Firstly, for communication in any organization, e-mail is the fore-most method for compatibility between representatives, providers, and customers. Email among many others was one of the first pioneers of the internet, providing a straightforward and fair intention to communicate. Throughout the years, various communication tools have additionally developed, enabling officials to communicate using live chat system, web-based meeting devices, and video conferencing systems. Voice over internet protocol (VOIP) phones and smartphones offer significantly more highly advanced routes for representatives to communicate.

Secondly, with regards to managing inventory, associations are needed to have enough stocks to take care of demand without allocating resources more than are required. Inventory management systems track the entire organization's upkeep, while also triggering a request for replenishment of extra stocks should said stocks drop below a predetermined value set by the organization. These systems are best utilized when the inventory management system is associated with the purpose of offer (POS) system. The POS system is a system whereby it ensures that each time an item is sold, the said item would be cut from the stock tally which makes an internal data circle that goes around all offices.

Thirdly, with respect to data management, the times of expansive record rooms, lines of file organizers, and the mailing of reports while incomparable to the speed of light, are still considered lightning fast. Today, most organizations store advanced variants of reports on servers and capacity gadgets. These records turn out to be immediately within reach to everybody in the organization, paying little heed to their geographical area. Organizations have in their reserves and keep up a colossal measure of authentic information financially, and representatives take advantage from prompt access to the reports they require.

In addition, corporations are using IT to improve the way they trace and supervise customer relationships. Customer relationship management (CRM) systems gather each communication an organization dealt with a customer, which aims to apply all information into knowledge. Hence, every that a customer contacts a call center with an issue, the customer support assigned will have the procedure to be aware what the customer has acquired, see shipping information, ring the training manual for that item, and adequately solve the issue. The whole connection is stored away in the CRM system, prepared to be reviewed back in case if the customer calls once more. The customer has a superior, more engaged involvement and the organization profits by enhanced productivity.

Finally, management information system (MIS) is where storing of information is used as an advantage because that information can be utilized appropriately. Progressive corporations use the information as a feature of their vital arranging procedures and, additionally, the strategic execution of the system. MIS allows organizations to track sales, costs, and profits of any period. The data can be used to track productivity thus amplifying degree of profitability and detect parts of change. Managers can track sales consistently, enabling them to promptly respond to lower than anticipated numbers by boosting worker efficiency or diminishing the cost of an item.

### 6.1.2 MANAGEMENT INFORMATION SYSTEMS

Management is to get the things done with legitimate asset at remedy cost, place, and plan and to coordinate assets as far as human or material keeping in mind the end goal of accomplishment to certain objectives. The five components of management are planning, organizing, directing, coordinating, and controlling.

Information is actualities, occasions, and exchanges which have been recorded. They are essentially the crude data sources which additionally get prepared to end up as information. At the point when certainties are separated through at least one procedure either human or systems, and are prepared to give certain sort of subtle elements, they are the information. Prepared data when introduced in some helpful and important frame, it is really the information we are taking a gander at.

System is a utilitarian unit, which includes set of techniques or capacities to create certain yields by preparing information or data given as info.

MIS is a programming instrument which gives an all encompassing report of prepared data in view of which management can take certain critical choice on which methodology and strategies could be made sense of. MIS gives data that is expected to oversee associations productively and adequately. MIS is any sorted-out approach for acquiring important and opportune information on which managerial choices are based. MIS encourages the basic leadership process and empowers the organizational planning, control, and operational capacities to be completed adequately. MIS is an investigation of how people, gatherings, and associations assess, plan, actualize, oversee, and use systems to produce data to enhance productivity and viability of basic leadership, including systems named decision support systems, expert systems, and executive information systems. The idea may incorporate systems named decision support system, expert system, or executive information system. The term is regularly utilized as part of observations and investigations concerning organizations and has ties and links with different regions, for example, information systems (IS), IT, web-based business, and software engineering. Accordingly, the term is utilized conversely with some of these areas.

Advantages that can be achieved utilizing MIS when organizations are able to easily recognize their positive attributes along with their flaws due to incoming reports being very easily near their vicinity, employees' execution record. Recognizing these perspectives can enable an organization to enhance its business procedures and operations. In addition, the accessibility of client data and input can help the organization to adjust its business forms as per the necessities of its clients. The viable management of client data can help the organization to perform coordinate showcasing and advancement exercises. MIS can enable an organization to pick up a competitive advantage. Competitive advantage is a company's capability to level up, increase effectivity, reduce costs, and when compared and contrasted with other firms in the market, is in a situation of favorable business position.

The manager of the organization is constantly kept updated on the patterns and data that are formed and also likely to form within the differing parts of the business. This aids in determining and long haul point of view arranging. The supervisor's consideration is conveyed to a circumstance which is normal in nature, inciting him to make a move or a choice on the issue.

Here, through big data and data sciences approach, makes possibility of data detailing for structured database as function as population learning base to recognize their behavior. The information is readily available and within reach in such a shape and form that immediately mixes and matches and also investigates, sparing the manager's significant time.

In this day and age of consistently extending complexities of business and additionally business association, with a specific end game to reap the rewards and evolve, they must have a legitimately arranged, examined, outlined, and kept up MIS so it gives required data to engage the organization to take convenient and sensible decisions.

Furthermore, an IS that is well and constantly maintained as a form of resource for the company is equipped with the ability to equitably gauge the practicality of data and precisely develop a blueprint for upgrades that improve business returns.

### 6.1.3   HUMAN RESOURCES MANAGEMENT INFORMATION SYSTEM

Human resources management information system (HRMIS) is a platform which provides guidelines for human resource information system in helping with operations, through customer-based technology and internet technology (HRMIS, 2015). A few examples of HRMIS softwares are Abra Suite, Oracle, and Spectrum HR.

HRMIS contains two parts. The principal part enables human resources (HR) experts to get to the systems through a client-server technology to manage HR capabilities. This HR capabilities to be incorporates benefits projects and advantages claims, such as; vocation designs, accreditation, contemplation for advancement or exchange, dependents, teach, instruction, business value, grievances, well-being and well-being and mishap revealing, dialect abilities, leave and leave privileges, posting information, open dissensions, compensation data, security or dependability clearances, aptitudes, pensions, examination, and instructional classes. Secondly, the online part got to by means of internet technology by every workers of the organizations and gives online access to leave self-service, employee self-service, and manager self-service.

Leave self-service enables a worker to use computer to ask for leave exchanges, and for administration officers to process demands for leave with payment. All workers can utilize the self-service forms to enquire of workers functions and permissions, such as; scheduled leave, see their detailed leave exchanges for that monetary year, ask for leave, and correct a leave asked for as of now.

Employee self-service enables the workers to view and refresh individual data on the web. This incorporates addresses, conjugal status, telephone numbers, work data, business history, preparing data, achievements, respects and honors, licenses and endorsements, enrollments, profession inclinations, aptitudes and dialects, and the capacity to self-recognize. Giving the representative this service enables errors in data to be recognized and corrected in a convenient way.

Manager self-service empowers directors to see on every minute of everyday premise position, work, and preparing data of their employees. Work data consists of both present and past postings inside the association. Position data contains the subtle elements relevant to the making of a position and also the correspondence prerequisites and the etymological profile of it. The preparation data consists of both individual courses gained by the representative and courses given by the organizations which are significant to obligations performed. The goal of giving access to representative preparing data is to guarantee the security of the overall population and all organization's workers.

## 6.2 LITERATURE REVIEW

HRMIS is a system whereby it allows the organization to monitor all personnel within the organizations and personal information about them. This includes, but is not limited to, name, address, date of birth, date of joining the organization, etc. It is usually done in a database or, more often, in a series of interconnected databases. HRMIS is a system that assists an organization to acquire, store, manipulate, analyze, retrieve, and disperse data around an association's HR (Tannenbaum, 1990). HRMIS is able to execute numerous tasks from a simple act of capturing and keeping information along with exchange of information, to more complex transactions. With the technology advances in today's world, along with it, the functions of HRMIS also go up.

HRMIS is an IS used to find data relating to HR. There have been many differences as on how information of the employees was saved in the past compared to the current IS used in an organization. IS analyzes data into useful information such as details of workforce, cash flow, and performance efficiency. This type of technology led to a more advanced human resource management (HRM), data are now easily stored in computers instead of manually, using written books or ledgers. Due to importance of HR and increasing number of IS usage, organizations now see these resources as a strategic building platform, hence both HR and IS are implemented together for maximum proficiency and productivity within the organizations. HRMIS is a computerized system normally consisting of interconnected database that tracks staff and their employment information (Gill and Johnson, 2010). It can be defined as a coordinated system used to collect, store, and analyze information for an organization's HR (Hendrickson, 2003).

The increasing difficulty and intensiveness of salaries payment systems at this time require more flexibility in changing and access to IS. Modern-day computers have made HRMIS available and affordable for any sized organization. HRMIS has developed from basic bookkeeping to a more complicated operation to aid management in making the right decisions.

Based on the most recent technological changes, computerized HRMIS was developed to combine the activities linked with IT and HRM into one common location with the help of using enterprise resource planning (ERP) software. The aim of HRMIS is to combine the multiple sections in human resources department (HRD), including payroll, efficiency of workforce, and initiatives into a less capital-intensive system than the main platform used to handle past operations.

Few management researchers suggested that "It is not technology, but rather the specialty of human and human management" that is proceeding with challenges for officials in the 21st century (Drucker et al., 1997). Similarly, Smith and Kelly (1997) trusted that future financial and long-term investment is achieved by making full use of talented group of people to be productive and efficient. There are number of HR officials that are busy focusing on dealing with their everyday tasks that they pay no attention to more important issues that are occurring. This can be problematic for any department; however, it can be particularly motivating for HR, which must face many speculations about the office's capacity. Along these lines, today with an expansion in the quantity of associations, HR is currently seen as a wellspring of competitive advantage (Kavanagh et al., 2012).

It is essential for firms to have a very talented human capital as a need to achieve the goals set by the organization. Along these lines, a successful manager of HR in a firm is required to train and recruit talented youth who can be fast learners of the modern technology. Currently, as modern technology is improving faster, adaptability skills and fast learning is compulsory in order to meet with the organization requirements to maximized usability of IT emerging technology. HR managers should know that the changes in technology would not just build the nature of workers' data, but in addition, will strongly affect the general adequacy of the association (Shiri, 2012).

To lessen the normal exchange and customary HR practices and to manage the complicated issues, the officials decided to robotize a number of these processes by implementing HRMIS or HRMS. HRMIS refers to the systems and operations at the joint between HRM and data technology. It is a system used to obtain, store, manage, breakdown, recover, and circulate data with respect to the corporate's HR. HRMIS has a database which is shared by all HR capacities that give normal information and managing all HR services.

Studies show that having talented employees that can adapt and learn to make use of current and new technologies, can bring great impact toward the organization as it gives them a competitive advantage toward their competitors. For example, accounting standards have been set in a software system where the accountant can now just key in the numbers as the formats and calculations are done automatically by the software format.

In the era of IT emerging technology, when the IT and IS become main part of managing data and information, as driver of HR management that may support HR in creating business methodology, and in this manner improving association execution (Gueutal, 2003; Lawler et al., 2004; Lengnick-Hall and Moritz, 2003). HRMIS is fully used to gain, keep, control, examine, recover, and arrange correlated information with respect to an association's HR (Kavanagh et al., 1990). IS helps HR staff to guide their focus toward more into business basics and vital level of assignments, for instance, improving leadership skills, and increasing the ability of management. HRMIS gives an opportunity to HR to assume a more vital part, through their capability and willingness to produce measurements which can be used to help key basic leadership (Lawler and Mohrman, 2003).

Currently, IS plays a vital role toward HR as it can help HR managers to be more productive and efficient. Researchers have found that the use of HRMIS can help managers to improve their leadership in HR role toward the employees. Organizations that adopt HRMIS will have a great benefit over their competitors as the HR will be very productive as they will be efficient up to their highest capacities. Moreover, communication in spreading the information between all managers and employees is important as it allows them to work in a straight line to achieve the same target or goals. HR can assess information and progress of their employees as how they are performing and where they are currently at in achieving the organizational target and expectations.

## 6.3   RESULT AND DISCUSSION

### 6.3.1   HISTORY OF TECHNOLOGY IN HRM

In the early development of HRMIS, it was mostly utilized for administration purpose. Traditional HR practices were limited to staff bookkeeping and were provided as a utility in the organization. HRMIS was not yet introduced at that time. Information was managed manually, often using a system as simple as an index card file. The HRD was usually small, with a

few administration staff members and lack of planning to reach the organization's objectives (Hendrickson, 2003). IS was used to collect leave
requests, workers compensation, salary variation, and pension entitlements.
Earlier, all HR activities in an organization were managed manually. There
were challenges in managing employees' attitude and formal training and
development method in earlier organization due to not so much focus on
HR-related activities. When HR was seen as key to the organization, several
factors completely changed attitudes toward HRMIS, and both large and
small business organizations   started utilizing HRMIS. The term HRM
replaced personnel. Organizations heavily rely on HR for their business
mission and vision. Also at the same time, many large organizations use IS
to maintain organization databases and technology. HRMIS has provided
a better solution to help record and store data and information and future
reports to be implemented for regulation by the government. The HRD staff
has become one of the most important users in the organization's computer
system (Hendrickson, 2003).

Before modern technology was introduced into HRM, offices and
management have been doing their work manually in terms of data storage,
transactions, projects, and others. They are required to keep a lot of paperwork, files, and other sorts of methods for data storage. There will always
be a room of storage which contains bulk of files and documents in paper
form on the shelves and cabinet. It has been done by most of the HRM but
this practice has its own limitations and drawbacks. It is difficult for the
managers to find the specific papers or documents for work when there are
piles of files and document in the room. The files and papers are not reliable
as they can be fragile if they are not stored well inside files or protected with
plastic cover and also they can be easily lost.

The traditional way of using technology in HRM has changed the ways
of communication in the current era. Communications between the departments used to be slower and cumbersome for the employees as to transfer
the information. Information is often distributed among the departments by
giving out memorandums, notices, letters, and others. There was a lot of
paper usage and wastage as it is often used only once. Paper is considered to
be reliable as it can be used as a hardcopy proof as an evidence or as a formal
standard of filing.

There will be a delivery guy in every organization to deliver any information inside or outside the organization. For example, when the organization wanted to do a business meeting with another company, they will hire
the delivery guy to send the letter and inform the other company manually.

Then, the company will also do the same as to approve the meeting. It is time consuming as well as less productive in achieving its organizational goals. The organization can make use of the time wasted doing something productive and it might as well be cost effective.

The HRD will be assessing, managing, and motivating the employees. The traditional way of assessing the employees would be by writing down all of the performances and achievements done by the employees on a piece of paper. It will be assessed and differentiated with other employees as how productive they are and getting closer toward the organizational goals. The information will then be kept in the specific room in the organization for future purpose. It can be reliable if it is kept in a secure place but then there will also be risk as it can be hard to collect in the future and might as well easily be lost. Assessing the key performance indicator (KPI) of each of the employees in the department used to take too much time in the past as it was done manually by the managers.

### 6.3.2   IMPORTANCE OF MODERN TECHNOLOGY INTO THE HRD

Technology has made major effects toward the function and has helped HRM to further extend of their performances. The people working under HR function in HRD are being to improve performance in any method or strategy even by using IS to improvement business processes more efficient and responsive decision of HRM that may positive impacted to the clients such as shareholders and investors, workers, and managers. Web-based technologies are greatly used for all HR key activities. Some of them are: employment application creation, employee benefits enrollments, and training on using technology-based resources (Mathis and Jackson, 2011). As some of the organizations are using IS and IT for registrations in recruiting new employees such as online registrations or submission of curriculum vitae and online video interview.

One of the major advantages of using IS into HRM of the organization is to enable some of the HR staff from their work momentarily so they can actually help and focus more on strategic objectives in HR organization and growth (Pinsonneault and Kraemer, 1993). Correspondingly, Broderick and Boudreau (1992) found that using IS can make the HR staff's everyday task much more easier and automatic such as payment and financial processing, administration works, and transaction activity such as employee payment. This will help the HR staff to focus on other matters such as improving company activity and production. Use of human resources information system (HRIS) allows the staff to improve the company performance level and increase motivation for the employee to work better (Bussler and Davis, 2001).

Modern technologies have helped the HR work efficiently and be more effective. The HR professionals are taking a long period of time with the use of the traditional method which is with papers kept at the shelves. Using the modern technology to assist the business organizations in practice will improve the time management of their HRD and drastically improve the organization to work efficiently and more effectively. It has also brought a great impact toward the company by the use of modern technologies to operating cost effectively. Modern technologies have simplified the work into the simplest form as, for example, the task of 20 employees can be done by one employee with the use of modern technologies. It has also reduced the time consumption of operation as it helps the company to be productive and be more efficient. This will lead the company to focus and reach its organizational goals.

### 6.3.3  IMPACT OF TECHNOLOGY ON HR PERSONALS

As technology already exists in this era and is being commonly used around the world in every organization especially in the HRD, professionals have experienced more efficiency, adapt to changes, and comprehensiveness of information employed (Gardner et al., 2003).

In additional to information responsiveness, the HRM also helps professionals workers to be aware of the latest public trends in policy decision-making and employment management practices in the business organization. To be more effective the HRM needs to make sure the current trends and data are related back to the business laws and regulation, so this would help the HRM  to be able to adapt the changes with the current trend and able to solve any unexpected issue or drastic changes that might occur outside of the company, for example, riot or wars. Furthermore, this type of practices need to be encouraged since this practices will able increase the value and contribution of the HRD making positive impact back to the business organization.

In additional, also keep up with the constant changes of laws and regulation. The information of the law must be well organized and updated for the organization. Using technology, the business organizations access the internet to communicate with other managers and employees in the organization within the industry. With this, technology has helped the HR employees in gathering data and information. This information is needed to be constantly updated to make sure it meets the company ability. This use of technology not only allows the HR to receive and distribute the information to other departments in the organization but also has motivated the organizational employees to meet the expected level of productivity.

It has also brought an impact toward the employees to be able to capture and adapt the new technology in the current era. Everyday, there will always be changes as it is developing faster than before. Organizations must adapt to the current changes as these can lead to a competitive advantage against the competitors. Adaptation toward the current modern technology is a vital role as it will bring a great benefit toward the organization. These improvements and updates allow the organization to operate their projects to be cost effective and more efficient.

As technologies are changing and getting more improvements, it affects the internal and external communications of the organization. The modern technologies play a vital role toward the communication in the modern era, as it has improved ways of communications than before. Currently, there are social media, mobile communication gadgets, e-mail, and more. Modern technologies have brought great impact toward communication as it can be delivered instantly and has less cost than in the past.

### 6.3.4  BENEFITS OF USING HRMIS

Firstly, using HRMIS can help reduce paperwork usage in the business company. Every year companies spend billions of dollars to manage paper.

Using the HRMIS can help the HR offices to reduce and save wastage of using piles of paper by adapting and encouraging the organizations' employees to complete and fill the forms via online or through e-mail. The traditional method of recruiting is receiving resume from the applicants which will be a bulk that the manager will be storing in the office. This method consumes a lot of space for piles of resumes from all of the applicants as all of it is either thrown away or kept into the store as for future purpose. By using IS toward the HRD allows the management to keep the information from the resume inside the system by keying in all of the data up to the clouds. This information can be used later in the future if they need to find more employees for the company. In the system, the management can also sort the information as which attributes criteria based on the priorities and automatically stored. For example, the managers are seeking for a Master's Holder of Business Administration, then anyone who have listed that they are a Master's Holder in Business Administration will be sorted in one page as it is easier than finding them from all of the resume. Therefore, HR can find specific information directly from the system or the cloud than manually reading the resume from the applicants seeking for what they need.

It has become a trend where interviews can now be done by online live-stream video. It allows the applicant from overseas to apply any company around the world. Interviews have become easier as applicants can be in any location to perform the interview and all they need are only the equipment such as camera, computer, and also internet connection. It has brought a great benefit toward the country that have high unemployment rate as there are no jobs available there. The talented youths and unemployed can just apply overseas as it is easier for them to apply through online registration and also do online interview as it would not be expensive than before.

With the proper managing all the paperwork data can be stored in the cloud system where all data and information are stored in the internet digitally.

Secondly, is improving decision-making skills and increasing productivity in the HRD. With the aid of technology or HRMIS, the document can be fully stored and completed instantly to be forwarded to the respective department and organizations. Furthermore, using this HRMIS can help provide the information for the government and public to read or access to gain some information for themselves, for example, government can use the organizational data to make audits and shareholders can see the productivity of the organization's current operation, and other more uses. HRMIS can be used to store and retrieve file document more quickly and efficiently. This method can prevent spending time on paper and finding spaces to store file in the file cabinet which can easily go missing or get damaged.

Third is improving decision-making process. The purpose of HR is hiring, training, and evaluating the employees in the workplace but this method takes too long and has a lot of paperwork. The organization executive and business managers are able to save time by using the HRMIS to handle all of HR activities such as managing the new application and managing the interview process, the process of setting up goals and keeping track of employees training, and scheduling reviews. The HRMIS can be used to stall the hiring, firing, and training process if there is a possibility of insufficient paperwork and no submissions. This system can track the record of employee's mistake or lack of requirement or productivity rate and will instantly alert the employees of their conditions at workplace and HR administrator will also be notified with employees. As the HRD is collecting all of the information of the new recruitment in a system, it allows the management to decide in a single platform as they can differentiate all of the information and choose the attributes that the organization wants that can lead them to achieve their organizational goals.

A properly designed HRMIS system can continuously monitor the productivity of the company according to the company's daily routine while at same time gets supervisors help to deal any case with the company. The routines can be monitored from a recorded video which will be stored inside the company's cloud or storage as they can see the performance of their employees in the future. These data and videos can be used for future reference on identifying the employees KPI in which it must be in a straight line in achieving the organizational goals. It can also be recorded as an evidence for future purpose in recapping the incident. The HRMIS system can also generate reports to help highlight of any issue and problem where manager of the company is needed to take action to solve the issue. The main organizational aim is to achieve high productivity for every employee as it leads to high efficiency of the workplace as to meet its organizational goals.

Here, the services provided by the organization dealing with well-trained of employees and workers. HRMIS is a tools to increase the employee's capabilities on their performance of services. HRMIS is used to implement and manage organizational strategy and training process in order to give the new employees a chance to develop their skills thus enabling them to reach their potential. Employees who are productive, active, adaptable, and engaged are the best examples of employees who will be able to provide better services to the customer, which creates a good image of the company. The HRMIS can be used to monitor the employees training activities about any new change of procedures or equipment used, so this will help the employees to maintain or even increase their productivity of the company rules or requirement.

It can also improve the communication and process of transferring information within the organization and also external information transfer. The use of modern technology toward the HRD allows them to share information or updates of the company's progress toward the organizational goals to all of the employees instantly. One of the uses of the modern technology is e-mail, social media, and others. This system allows the management to send information toward all of the employees in an instant for any urgent matters, updates, or tasks given. It is also efficient than the traditional transferring of information back in the past as they usually sent memorandum, letters, and also mouth-to-mouth. This traditional system of transferring information among the employees used to consume a lot of energy and time as it could have been used for something productive that could lead them to achieve the organizational goals. As it gives a great impact toward the HRM to share information among the organization, therefore it improves the knowledge management of the organization. Access to information is easier when modern technologies are implemented in HRM as the organizations will be storing their information to the cloud or storage. The employees or managers can directly get the information or files that they need than using the traditional way of keeping the information. In traditional way, information was stored using papers and files, and all of it was then kept in a secure room. It used to take time just to find the specific files if it was not in order. While, by using the cloud or storage, the manager can directly get the files just by using a few simple steps.

Currently, organizations are developing a specific software for HR. It has brought great impact toward the department as to be more efficient and as well improve its productivity as to achieve the organizational goals. This software allows the organizations to reduce their recruitment of HR managers which leads to less man power needed. The software has made and accomplished all HR managers' work and make it easier and more simple as consequences the organization only need fewer worker on certain jobs. This will also lead to less cost for the organization in hiring HR as they are using the software.

Lastly, HRMIS is used for a rapid increase in the company's rate of productivity. New workers are usually offered the job to fill vacancies left by the previous well trained and more experienced staff members. This will increase the employees' morale as they get more work done but with raise or promotion. The HRMIS can help with the recruiting, hiring, and also can provide training to new employees as soon as possible so they can fill the new roles, with the best continuous training. This will help the organization to increase the employee productivity rate, efficiency, and reduce the risk of employees quitting. And these are some of the possible benefits on using HRMIS into the organization.

### 6.3.5   RISK OF USING HRMIS

One of the risks is the failure to select project management when trying to adopt uses of technologies in HR. Many business companies rely on other employees such as vendors. Sometimes selecting employees who are not expert in project management, this will cause a delay or unsynchronization with the company priorities and would lead to more conflicts and delay the working process.

When selecting the suitable project management, the person must be at least an experienced project leader and have a chosen team consisting of at least two people that are flexible at work. If there are a lot of people working the project, it will be more difficult and lead to a horrible consensus to the business. The team incharge should be strictly focusing on HRMIS implementation, while other regular job duties should be taken over by another employee and managers in the business.

Second is poor and incomplete planning procedures. According to statistics, 68% of all IT failures are caused by poor planning according to the 2009 Standish CHAOS report. Without proper planning, each step of implementing HRMIS facing with possibility of fault. However, by taking of stakeholder requirement, the processes of HRMIS adoption will more efficient, that may lead to increase the success rate. Organizing the offices so the HR process can be more automated and quick to make faster decisions and even save budgets. Using the HRIS can be beneficial and quick rather than wasting time taking a long time to find the employee data and information through all the thick and messy files.

Next issue is the lack of consideration for employee opinion, usually consider the other employees thoughts and suggestion whether the employees have better ideas and maybe specialty in technologies. Excluding the human factor during the implementation of HRMIS is considered to be a second reason for HRMIS implementation project to be likely to fail. This issue will lead the employees to catch up with the new working system on a consistent basis; this approach might be difficulty and possible decrease in productivity since the lack of communication between managers and employees. This process of adapting the new system can take a long period of time to make all the employees to feel included in the new system.

More issues such as entering the wrong data into the system. Usually after the new system project has been started, most of the companies make a grave mistake of rushing of transferring the information and data records from paper-based system into the computer system. The business processes of HRMIS may lead to the rushing of transferring information enhancement, and any technical issue of HRMIS implementation, caused of malfunction of the system. Here, the HRMIS may get affected become and unreliable for further functions as supports organization.

It has come to an issue when HRD are doing online live video stream for interview. However it might be unreliable as the interviewer cannot really see applicants physically, since interviewer only see them in the monitor as a visual of how unique and style of the applicant behavior. It can also be a major loss for talented applicants if they are not equipped with standard cameras which may have caused them a rejection which does not have any relevance with their job interview. For example, the interview can be affected when the video is blurry or is of low quality as it can interfere with the communications and might as well cause the applicant to lose the chance of getting the job. This has come to an issue toward the applicants who are applying for a video job interview overseas.

Another issue is insufficient data security. As we all know that HRMIS data is considerably highly sensitive, as it contains the employee's personal information and proprietary of the organizations information is stored in the company files. As the organization will spend more if they want a higher level of security protecting their files and information that are kept in the system or cloud. The most common problems that the organizations are currently facing are unauthorized access, hacks, glitches, spam, scam, and other cyber crimes. All of the information from the company's cloud or storage can be easily leaked or destroyed by an unauthorized access which can be done by a group of professional cyber crimes. For example, when the organization's HRs are relying too much on the modern technologies such as the clouds or storages, all of the employees' information can be leaked out or can also be used as a tracker for crime use. It has happened not long ago that an innocent employee was suspected of doing the crime as his information was left behind such as a copy of his identification card or others.

When an organization is using IS toward its HR, there should be organization's access codes. These access codes are only known by the employees or specific departments that are handling the system. The access codes can also be leaked if the employee accidentally shared/or shared them with someone else. With the access code, the person can access all the information of the organization and can also make use of it for bad purposes.

Current problems in the modern era are the spread of viruses and malwares. Viruses can spread into the system from opening attachments or links that are from scam emails, online pop-up advertisements (click baits), downloading or installing softwares from malicious or unrecognized sites, outdated software, etc. Viruses and malware softwares can bring a great deficit toward the organization's IS and storages. Viruses and malware software can infect the information kept and can also destroy and corrupt the data and softwares. It can also decelerate the performance of the system which causes the organization to work slower and a specific task becomes time consuming to perform. The other forms of virus that can also affect the organization's software are worms. This software will fill up the storage or cloud of the system with unnecessary files or corrupted softwares without notifying the organization. This problem can lead the organization to upgrade to purchase a bigger size of storage or cloud as it is full but then it is the worm piling up the files in the storage with blank software that only consumes the space and size.

Another risk which can affect the organization to be less effective in achieving its goals is relying the security on another organization. When an organization relies on the use of IS, it needs to hire an organization that specializes in security for the storage or cloud. The organization needs to plan very well prior in choosing the right security to protect its softwares and confidential data. When there is a breakout or cyber crimes are happening toward the security company, it can affect the information that they are protecting as it can be leaked and there also can be an unauthorized usage for their own purpose. The security organizations should be reliable and very trusted as they are entitled to handle all of the information of the organization and also all of the confidentialities that are being kept inside the IS storage and cloud.

These are few more issues regarding of the use of HRMIS into the HRD and the organization activity and other companies could also face different issues depending on the company production and position.

### 6.3.6 SOLUTION OF THE RISK

For every problem that might occur in HRMIS there will always be a solution to the issue, such as to solve the problem in the paragraph regarding HRMIS having security issues. This can be avoided by having security before and after implementing the HRMIS data. Before transferring any information into the computer system and set up, proper security measures are needed to protect the data. After the data has been transferred into the system, then security measures by using security clearance level is conducted. Here, the system make a leveling authority of accessed, such as username, password, authority; user, admin, superadmin.

Secondly increase security and reduce the risk by installing a backup hardware or backup server and backup hard disk to protect the data. This will keep record of the data being keyed in by the office everyday such as transaction, information detail of the company employee, and others important information needed for the organization to work. So in case of emergency, if the main data or main computer is damaged or lost, the HR employee can rely on two backup data servers to restore their data and it can be used again. For example, the organization office computer server has suddenly closed down and all data has not been saved in time, so the only way to continue their work and fix their data is by logging into the backup server to recover back the data needed to finish their work and any important documents.

In other hand, two approaches of data invalid or rushing key-in are implemented. First approach is parallel testing and other security measures. This approach is to ensure the information being transferred and keyed-in valid into the system, which are properly arranged and accurate according to the files. Second approach is double check that the data being transferred in the system are all accurate and check for any error. Here, so HRMIS can spot the mistake before fully implementing to avoid project fail.

An organization should do an intensive planning and decisions toward improving its security of keeping the data in the system and also the security handling it. They should do an intensive research and choose reliable and trusted organization of security as it should be kept in a secure place. It is an investment that the organizations must focus on as more expensive it is the more secure their information are being kept. The high standard of security level can reduce the risk of being hacked or accessed by any unauthorized personnel. The security should also be able to detect the unauthorized personnel for future purpose and also an evidence for investigation purpose.

While using modern technology toward operation of HRM, there should be maintenance of at least once in 6 months in a year. Maintenance required to service, clean, repair, or update the current software so as to prevent unwanted softwares inside the system. The system's software should be updated so it can recognize unwanted softwares, viruses, and malwares. The system needs to be cleaned and serviced as to clear all unwanted softwares and broken softwares that use and consume the space in the cloud or storages.

### 6.3.7   RESULTS

To be an effective HRM, the continuing update and implementation is required. Whit this improvement, organization has a unique selling point compare to the competitors. Here, the information system has a main function as an enabler of reengineering processes of HRM adoptions. IS keeps on upgrading itself which has led to an improved technique of collecting data or information through the development of HRMIS systems (Kavanagh and Mohan, 2009). Most of the Enterprises often involve in partaking for IS after successful implementation of ERP and CRM solutions and aim to improvise the processes in making decision about selecting the suitable employees, background check, and others. IT has provided a lot of methods to help the HRMIS applications and a lot of associations to improve their productivity by increasing the performance of HRM (Parsa, 2007, p 70). Use of technology in HRM helps not just reducing the complexity of administrative procedures such as recruitment, organizing the workers, but also setting up the rules and regulation.

The information and data that has been collected in the HRMIS provides a guideline or method for management decision support. With the proper HRM, companies will be able to provide the calculation of the possible outcomes to the whole business of the company. The calculations consider the parameters such as the health cost per worker, salaries rate, working position, investment rate, costs, and increase in per employees or human cost. This causes a sudden increase in the use of HRMIS systems implemented into most of business companies around the world today.

Using HRMIS can help an organization to manage the employee, transaction, spending, and other important documents. Also, using HRMIS can provide security and competitive advantage in using IS in the organizations compared to their competitors who might not afford or are unable to adapt with HRMIS system. HRMIS can boost the organization's productivity as to work in an efficient manner so as to achieving its organizational goals.

## 6.4  CONCLUSION

The importance of using MIS into an organization activity is similar to a human being, whereas the organization is considered to be the entire human body and the use of MIS or HRIS is the heart of the company. And the function of the heart is to pump in and out blood around the human body which is similar to the function of MIS which is transferring data and information in and out to the whole organization and also providing information to head or manager which is the mind of the organization. Assume that organization is a human being, with complex of functions of each parts of body. Its sound similar as working in the organization, as MIS is important to keep the information more efficiently and avoid delay especially during busy or emergency situation. So, this will fully satisfy any circumstances which may occur.

MIS is considered to be a part of the organizational family and it keeps the organization properly running. The use of MIS will provide a certainty of well-organized and reliable information that is gathered and is being used at domestic organization and the information provided can be from other organizations, research studies, variety sources, articles, and information from other companies who have reached their goals and have become successful. The system has the ability to fully use of the data of a single employee, group of peoples, the management head such as the organization supervisor in each department and the organization managers.

The function of HRMIS is to provide data and setting up the rules and regulation needed for the organization and also help with the HR functions according to their capabilities, HRM consider to be a guardian of representative records, in any case, the existence of a HRMIS so this will make the data to be more organized and accessible for other utilize and useful or aid for the admin office of basic leadership. This IS can help produce and provide more viable and faster results than using a paper-based form of information. HRMIS are able to recover and keep track and be aware of any kind of information.

The redesign of HRMIS will not be able to bring about new information, and it will not be able to provide any new or recent information interfaces to outsiders; it will also not be able to provide variety of new services and information to organization employees.

The main aims and objectives of an organization are to be able to provide benefit of conveyance to the representatives that are quicker, increasingly proficient, and also improving the validity of the person or employee that has applied. Using a system that will bolster an online engineering will enhance the productivity of the application. This will profit the association and offer a tenable technique for managing HR in the association.

By adapting the new system of HRMIS, there will always be some problems caused by the IS and even the HR staff and employees. All of these issues cannot be avoided unless the users have done their research and are well prepared when adapting the HRMIS into the organization. After all the issues have been fixed, the system should be fine and the organization can fully utilize the benefits of using HRMIS system into handling and managing the organization properly and efficiently, thus able to reach the organizational goal.

These studies have shown that some of the systems sometimes can be confusing and be difficult to properly function, and somehow this will be able to change or improve the HRD rehearses with some hierarchical system, recognize change territories, and stay up-to-date with the current trend and method used in organizing the business. Using the HRIS system can enable an organization to do research or survey and predict any crevices or potential issues and increment of the dedication of HR experts to continuous changes. Some part of the report can help the HR employee to check whether more organizations have embraced the HRMIS. The possible uses of HRMIS selection is more important to see if the HRMIS are being fully utilized for more than their normal uses or everyday activity.

## KEYWORDS

- shifting paradigm
- human resource management
- business
- organizations
- purpose of offer
- customer relationship management
- management information system

## REFERENCES

Chanda, A. Strategic Human Resource Technologies: Keys to Managing People; SAGE Publications: India, 2017.

Drucker, P. F.; Dyson, E.; Handy, C.; Saffo, P.; Senge, P. M. Looking Ahead: Implications of the Present; Harvard Business Review, 1997; Vol. 75 (5), pp 18–24.

Gardner, S. D.; Lepak, D. P.; Bartol, K. M. Virtual HR: The Impact of Information Technology on the Human Resource Professional. *J. Vocat. Behav.* **2003,** *63* (2), 159–179.

Gueutal, H. G. The Brave New World of E-HR. In Advances in Human Performance and Cognitive Engineering Research; Emerald Group Publishing Limited: 2003; Vol. 3, pp 13–36.

Human Resources Management Information System (HRMIS). 2015. http://www.rcmp-grc.gc.ca/en/human-resources-management-information-system-hrmis (accessed Feb 2019).

Kavanagh, M. J.; Gueutal, H. G.; Tannenbaum, S. I. Human Resource Information Systems: Development and Application; PWS-KENT Publishing Company: Boston, MA, 1990.

Lawler, E. E.; Mohrman, S. A. HR as a Strategic Partner: What does it take to Make It Happen? *Hum. Resour. Plan.* **2003,** *26* (3), 15–29.

Lawler, E.; Levenson, A.; Boudreau, J. W. R Metrics and Analytics: Use and Impact. *Hum. Resour. Plan.* **2004,** *27* (4), 27–35.

Lengnick-Hall, M. L.; Moritz, S. The Impact of e-HR on the Human Resource Management Function. *J. Labor Res.* **2003,** *24* (3), 365–379.

MacKechnie, C. Information Technology & Its Role in the Modern Organization; 2017. http://smallbusiness.chron.com/information-technology-its-role-modern-organization-1800.html (accessed Feb 2019).

Parsa, H. G. Critical Factors in Implementing HRIS in Restaurant Chains. In Advances in Hospitality and Leisure; Chen, J. S., Ed.; Elsevier: UK, 2007; Vol. 3, pp 69–86.

Satell, G. 4 Ways in Which Technology is Transforming Business; 2013. https://www.forbes.com/sites/gregsatell/2013/04/02/4-ways-in-which-technology-is-trans-forming-business/#e9e3538d9c74 (accessed Feb 2019).

Shiri, S. Effectiveness of Human Resource Information System on HR Functions of the Organization: A Cross Sectional Study. *US-China Educ. Rev.* **2012,** *9,* 830–839.

Smith, A. F.; Kelly, T. Human Capital in Digital Economy. In The Organization of the Future; Hesselbein, F., Goldsmith, M., Beckhard, R., Eds.; Jossey-Bass: San Francisco, 1997; pp 199–212.

# CHAPTER 7

# REENGINEERING WORK PROCESSES FOR IMPROVED PRODUCTIVITY: A MANAGEMENT INFORMATION SYSTEM PERSPECTIVE

## ABSTRACT

Reengineering is changes the process and structure of an organization by using information technology (IT) to attain a greater advantage in achieving an organization's goals. Reengineering involves relearning current business processes, identifying processes that are less efficient, coming up with an action plan in eliminating wastage and unnecessary efforts, and designing new processes for an organization to gain competitive advantage. Few factors that lead to competition are demand, product quality, price, service given to customers, an organization's position in the market, suppliers and buyers bargaining power, and threat of new entrance. While applying reengineering to an organization, it is important to empower people by giving them authority and capability to make full use of their skills and ability, providing proper working environment as well as up-to-date equipment, and enhancing their skills through training. To ensure that reengineering process flow smoothly, it is important to have a close relationship with employees to provide them with information for keeping them on track and improving their performance. In other hand, human errors can also be avoided by providing employees with the right tools based on IT emerging technology.

## 7.1 INTRODUCTION

According to Chen (2001), reengineering is changes made in the process and structure of an organization by using information technology (IT) to attain a greater advantage in achieving an organization's goals. In general, reengineering involves relearning current business processes, identifying processes that are less efficient, coming up with an action plan in eliminating wastage and unnecessary efforts, and designing new processes for an organization to gain competitive advantage. Uncertainty and comprehensive competition occur due to globalization which has changed the business environment in the market place. Few factors that lead to competition are demand, product quality, price, service given to customers, an organization's position in the market, suppliers and buyers bargaining power, and threat of new entrance. Continuous improvement programs sometimes are not the best way to overcome the problems. Hence, organizations use reengineering which leads them to rethink about their current processes to identify business weaknesses that direct an organization to completely redesign the structure, technological aspects, human resource, and other organization's dimensions to be able to cope with changes and improve the performance.

While applying reengineering to an organization, it is important to empower people by giving them authority and capability to make full use of their skills and ability, providing proper working environment as well as up-to-date equipment, and enhancing their skills through training. To ensure that reengineering process flow smoothly, it is important to have a close relationship with workers to provide them with information for keeping them on track and improving their performance. Other than that, human errors can also be avoided by providing workers with the right tools. All department or divisions in an organization share information to keep everyone up-to-date and to ensure smooth flow of the reengineering processes. It is not suitable to transfer numerical information using paper as plans are usually being changed following negotiation within the process. When plan changes, respective people need to do amendments in the calculations. If the right tools are not being provided, this will affect the efficiency and output of the process. Providing right tools will make the numerical information of the plan to reach every division consistently and will also improve productivity.

Objectives of reengineering are mainly to improve customer service by eliminating customer complaints, to speed up processes in completing task by cutting out unnecessary processes, be transparent throughout an organization processes to optimize decision-making and to cut operational cost, be flexible and able to adapt changes in market conditions, improve quality of products, be innovative through leadership, and to improve organization's efficiency and effectiveness. Information technology plays an important role as it eases every stage in reengineering such as analyzing process, designing process, implementation process, and evaluation process. Here, project management tools can help to facilitate the reengineering design process, tracking information on customer requirements through their complaints and feedback for organization further improvements. Customer feedback can be gathered by using database aimed at improving the customer satisfaction. Also, with the use of telecommunication technologies, it improves organization collaboration among different functional units, exchanges information, and accomplishes common business process.

There are many benefits of reengineering when implemented successfully; thus, it is important for an organization to outline the factors that cause them to reengineer to get the company's overview of the things that should be done in order to achieve the maximum result. There are many factors that cause an organization to reengineer. Two major factors that might influence the decisions are external and internal factors. The external factors are factors influenced from outside the organization. This may include the customers, competitors, government regulations, and changing market conditions. The internal factors are existence problems that the companies may be facing within the organization itself. This may include the need for technological advancement, to increase the efficiency of the organization, and to reduce the overall cost.

Customers play an important role in the business–customer relationship. Depending on the product and the availability of the product, it can define whether the customer have more control over it or otherwise. With a product that is widely available everywhere, according to Hammer and Champy, the customer has the ability to dictate the quality of the product and the price that he is willing to pay. Organizations that are not fast in responding the demand may be required to reevaluate their strategies to gain competitive advantage in the industry.

Another example of external factor is the changing market conditions. Some organizations are able to adapt to the rapid change while others fail to adapt. In today's competitive market, there are growing number of competitors with greater product innovations and technological advancements, and thus, an organization should adapt quickly to the changes in order to keep up with the current trends. As the competition is growing not only locally but also globally, it has inevitably put an organization to more competitive rivalry. The government regulation and political pressures may force an organization to revamp the entire organization's process.

The internal factors that drive an organization to reengineer may include the need to improve their technological advancements. In this 21st-century era, advancements in technology are common and needed in order to succeed or to be better than competitors. Two of the major uses of the technology are to keep and analyze information and to cut down the decision-making process time. Better execution of the technology enables an organization to have a competitive advantage over others. The need to increase efficiency may also be a driven cause for the organization to reengineer. With the economic slowdown and increase of competitive rivalry, organizations may seek for other means to increase efficiency and productivity. This may include improving the accuracy of a system, avoiding repetitive error, as well as speeding up the development of the product.

It is important for an organization to outline their strategic focus as a guideline in making future changes. Depending on what the focus is, whether it aims to reorganize the organizational structure or to diversify an entirely new product, an organization may reengineer to redefine strategic focus.

## 7.2 LITERATURE REVIEW

There are several definitions interpreted by many authors of a business process reengineering (BPR). According to Hammer and Champy (1993), business process engineering can be defined as "the fundamental rethinking and radical redesign of the business processes to achieve dramatic improvements in critical, contemporary measures of performance, such as cost, quality, service and speed." While other authors such as Davenport and Short (1990) emphasized BPR as the work flows and procedures within the organization itself.

BPR plays an important role in achieving long-term corporate objectives by associating work processes in an interactive way with customers. Senge (1990) advocates a system concerning the suppliers, customers, and the future. The system has designing to reshape the performance for further organization improvement processes (Gulden and Reck, 1991).

Market uncertainties, increase in globalization, and business conditions are the factors that influence the changes in the organization. According to Milan et al. (2014), BPR is one of the most important management practices because it assists the organization on how to improve the competitiveness and sustainability in the time where there are increasing number of competitors, economic fluctuations, the changes caused by globalization and internationalization.

Reengineering often denotes in the changes of the organizational structures as a whole such as changes in work design, changes in roles and responsibilities of the employees, manager, supervisors, and executives, and changes in system of values and incorporates culture. This is because reengineering is considered as radical redesign of the organization where the changes are from creating the new process instead of modifying the existing process. The originators of reengineering, Hammer and Champy, explained reengineering the business process is from radical redesign and fundamental rethinking in order to succeed dramatic results of improvement in critical contemporary standards of success including costs, quality, speed, and services as cited in Milan et al. (2014).

Over the last decades, one of the most popular management tools to improve the business processes to be more efficient and effective is BPR. This tool will help to satisfy the customers in a way that it will meet the needs and add value to the goods and services that are being provided, as well as this business process will help the organization to meet their strategic goals and objectives. Reengineering approach toward successful implementation in the organization such as improving management information system (IS) will involve risks. Hence, it is crucial for the organization to properly manage the processes so that they will be able to meet their aims and objectives upon the implementation.

It is necessary for an organization to increase their effectiveness and efficiency of the management in order to improve the productivity and functions of their operation. This is supported by Milan, et al. that through continuous improvement, workforce will lead to increase in the effectiveness and efficiency of the management functions.

By using Lean Six Sigma in the supply chain management, it can help in improving the service care for the patients. According to Spagnol et al. (2013), the first step that the organization needs to do toward waste reduction is identifying the nonvalue-added activities in every process which can be disregarded but will not affect the whole procedures. There are seven lean principles and tools that the organization can use to recognize the waste. By eliminating all the wastes in the supply chains, it enables and helps the organization to be efficient. Six Sigma program focuses on controlling the quality of the production of the goods until it becomes end product that add value to the customers; it also ensures to reduce errors and avoid defects.

As stated by Cui, et al. (2017), inventory inaccuracies and errors are one of the main problems that an organization undergoes. Loss of items, misplacement, and undetected items can lead to shrinkages and inconsistency of records between the number of physical goods and the data or information in the system. This is one of the hindrances in achieving sustainable supply chain. Due to these problems, researches have come out with an idea to use radio-frequency identification (RFID) tag that uses electromagnetic fields that can transfer data. It can be scanned up to few meters away which means that it increases the visibility of assets as it can be tracked. RFID can also enhance and speedup the process of extracting information about the products, which will result in the improvement of inventory management (Tsao et al., 2016).

For the success of reengineering, leaders are responsible to lead the change by understanding the working environment, planning the system and processes, organizational culture, structure of the organization and communicating continuously with workers to keep them informed (Guo, 2004). He also states that leaders are supposed to inspire, give motivation, and provide authority, monitor performance, and establish excellent communication with workers. This statement is supported by Bertolini et al. (2011) where they claim that a few reasons for lack of success in reengineering are the unwillingness of accepting changes due to no leadership and lack of supervision from higher management that causes lack of motivation.

In the dynamic environment, businesses should be able to stay competitive to enhance their performance and capabilities. Thus organizations require continuous improvement. The key to survival of an organization in such circumstances to achieve their success is the application of total quality management (TQM). According to Topalovic (2015), TQM is one of the most important techniques to improve the quality and is mainly applied in the organizations around the world.

Optimizing business processes such as cutting cost and reducing management can help in improving return on capital employed and profitability which are the major aims of reengineering. It involves vital changes that are not only focusing on the business processes but also to how the organization is being operated and managed (Zhao, 2004). He also believes that IT platform will bring greater potential on substantial benefits to the organization, by which business processes will extensively, radically, and fundamentally implemented within the organization through IT emerging technology as enabler of reengineering processes.

## 7.3   RESULT AND DISCUSSION

### 7.3.1   CASE STUDY ON CONTINUOUS IMPROVEMENT PROGRAM

#### 7.3.1.1   CASE STUDY 1: IMPLEMENTATION OF TQM IN HOSPITALS

For the past two decades TQM initially started in Japan in the beginning of 1980s and spread to Australia and the Western countries. TQM becomes popular and very important as many organizations were applying TQM to improve and develop their managements. Al-Shdaifat (2015) defines TQM as a customer-oriented management as it concerned between the work processes and people to maximize the customer's satisfaction and to improve organizational performance.

The foundation of TQM system is the orientation of customers. The companies produce the products and services to meet customer's satisfaction; hence, the companies must be highly sensitive in understanding customers' needs and satisfaction to strengthen the company's performance and capabilities to provide maximum quality to the customers. Moreover, Topalovic (2015) mentioned that implementing TQM has proven that it helps to strengthen customer satisfaction; it also showed improvement in the financial performance as cited in Agus et al.

In the present time, most of the health organizations face many challenges, which require high demand in quality services, including high cost of health services, rapidly growing technological dependence, pressure to reduce the health costs, ability to cope with the quality offered in the international organization such as standards and licenses, and ability to meet patient's needs. Hence, the health organizations are forced to implement the system that will help them to centrally manage their healthcare processes at high-quality services and in measurable ways. Such a system to handle these challenges is TQM. Key principles of TQM include customer focus, quality obsessive, teamwork, scientific approach, continuing commitment and continual improvement system, training and education, freedom of control, and employee involvement and empowerment.

The case study on implementation of TQM in hospitals evaluate the implementation of TQM principles in Jordanian hospital from the frame of perspective of the nurses, differences of the implementation of TQM from different sectors, and sociodemographic factors including age, gender, education level, experience, departments, daily-served patient that affects the implementation of TQM in those hospitals. Since majority of 80% nurses are working in any of the health sectors including government, university, military, and private, the survey was conducted on them as they play an important role toward the implementation and success of TQM program in Jordanian hospitals.

Factors analysis had been used to identify TQM principles which are being used in Jordanian hospitals. It is derived using varimax rotation method from the principal component which includes continuous improvement, training, teamwork, customer focus, and top-management commitment; these factors reached 70% out of 40 variances included in the questionnaire. The five principles ranging from 41.6% to 53.9% of 40 variances, and the finding shows that the least implemented is continuous improvement factor, whereas the most implemented principle is customer focus factor. Moreover, the relationship between the TQM availability and the extent of implementation of TQM principles in other sectors and Middle East countries is being discussed in the findings of the case study.

However, result shows that the principles of TQM were not fully employed, and it was concluded that the implementation of TQM in the Jordanian hospitals is not effective. The ineffective implementation of TQM may be because of deficiency of knowledge regarding the importance of TQM in an organization, inadequate training program, and poor financial support to improve healthcare services and satisfaction to the patient. However, private company tends to successfully implement TQM more than other sectors, this is because of the essence of competitiveness; they put more efforts to give greater and efficient satisfaction to the customers as well as thus, will be able to increase profits in private hospitals. In addition, study shows that there is no relationship between TQM implementation and sociodemographic variables present in poor-quality department in the hospitals. The degree of TQM implementation will improve by the availability of quality department. To implement TQM successfully in the organization, an effort should be taken into consideration to set up a foundation and examine the credibility of the healthcare system.

### 7.3.1.2   CASE STUDY 2: LEAN HEALTHCARE SUPPLY CHAIN MANAGEMENT: MINIMIZING WASTE AND COSTS

Lean philosophy focuses on the speed and efficiency of a process or an operation by eliminating a nonvalue-added activities as well as eliminating waste with the aim to produce products or operate services at the lowest cost and as fast as possible. Lean program can help to eliminate the seven types of waste within the organization. The seven types of waste are activities from the elements of production that does not add value to the products which includes overproduction, transportation, overprocessing, rework or defects, motion, waiting, and inventory (Catia et al., 2014).

Meanwhile, Six Sigma is a quality control approach that focuses on accurate and precise data collection which will be used for enhancing process performance, eliminating defects, and reducing errors. As stated by Raval and Kant (2017), lean and Six Sigma are usually used together as a methodology that focuses on improving customer satisfaction, reducing errors and defect, and improving the organizational performance as a whole.

Supply chain is a system where a sequence of activities are involved from the producing of products till the delivery of products to the final customers. By managing the supply chain activities, organization can maximize customer value and ensure that they satisfy the end customers, as well as to achieve sustainable competitive advantage by running the flow of the supply chain in the most effective and efficient ways.

There are several lean tools that can help identify and combat waste found to provide customers exactly what they want. Some of the lean tools are, total productive maintenance where it restructures the organization and improves the people and equipment to accomplish the goals of zero defects; just-in-time is a tool that uses minimal inventories of raw materials where it will only produce finished goods when it is necessary. Poka-Yoke is another tool that minimize the errors or defects caused by human failures that will generate loss later on. Layout tool is whereby the organization use the provided space for production to produce highest quality products in a short time.

Using Lean Six Sigma on the supply chain can help increase efficiency of the operations of the organization. By decreasing order fulfillment time, Lean Six Sigma will identify area with the most waste and variance so that it can be corrected. Lean Six Sigma can also optimize order fulfillment where it makes sure that the system information are not outdated nor have waste that will lead to inefficiency.

As a result, the application of Lean Six Sigma techniques helps the healthcare sector in minimizing the waste generated in the supply chain, ensuring that it reduces cost but produces high-quality products and services. Lean techniques will help the organization in identifying the waste that should be eliminated that enables in the improvement of quality of products and services and ensures productivity, and added value to customers. Thus, Lean Six Sigma is an important tool for the healthcare supply chain that seeks sustainability and aims for customer satisfaction.

### 7.3.1.3   CASE STUDY 3: IMPROVING HEALTHCARE QUALITY—A TECHNOLOGICAL AND MANAGERIAL INNOVATION PERSPECTIVE

It is important to have continuous development of innovative technology in a healthcare industry in order to get the best quality healthcare services. According to Kim et al. (2016), although healthcare is one of the important key sources of business innovation, the quality performance of the existing healthcare on both developed and developing countries is still insufficient. Thus, having an innovative healthcare system is important to every subsector.

Healthcare industry consists of many sectors, for example, the provider (doctors and hospitals), the regulatory agencies, the pharmaceutical companies, and lastly, the healthcare's buyers and payers (Burns et al., 2011). In order to have a successful implementation of innovation in a healthcare organization, several initiatives are required, for example, introducing an entirely new different model for the industry, combining knowledge on different tactics to avoid social vulnerability, committing to the use of innovative technology, and many more (Kim et al., 2016). The study also stated that there are two ways in which the healthcare industry can innovate: one is by having knowledge protection according to laws and second is knowledge sharing for within or outside the organization.

Both theoretical and empirical studies are needed as a guidance in innovative engagement from each sector. It is common that a healthcare service provider's aim is to enhance quality of life of patients or to reduce number of patients getting sick. However, it is important for the healthcare provider not to have overarching goals. Cleven et al. (2016) made an investigation in Swiss general hospitals and found that their data in structural equation modeling (SEM) signified dependable provision between processes, cost efficiency, and clinical quality, and these were all linked with hospital's performance and the satisfaction received from their patients. Jang et al. (2016) on the other hand studied U-healthcare service which is a type of service provided by healthcare providers that uses mobile to perform their services. The study emphasizes positive view of U-healthcare services on delivering better communication and sharing of knowledge amongst doctors without disrupting the quality of healthcare to their patients.

An example of efficiency of innovative leadership and supply chain on healthcare was examined by Yoon et al. (2016). The study used SEM and hypothesis testing of analysis in several Korean hospitals. The study uncovers the importance of supply chain innovation in gaining efficiency in delivering the quality and operational improvement of healthcare. It highlighted that operational efficiency can be achieved using innovative supply chain and leadership. In the study of "the effects of healthcare supply chain in managing healthcare costs" made by Kwon et al. (2016), three strategic ideas were identified. One is maximizing revenue to the service provider, second is understanding the supply chain principle, and lastly, improving its processes. It suggested that both efficiency and effectiveness existed to produce a surplus in supply chain when the additional resources are being put into sections that are beneficial to customers.

A study made by Xing and Oyama (2016) outlined the problems of supply and demand in the local public hospital's system in Japan. The study discovered several problems in the Japanese healthcare systems. Many of which include high operating costs, debts, and unorganized management. The local public hospital was introduced to give equal accessibility and the development of financial performance. The study suggested that the local public hospitals play vital part in promoting innovative technology to get a higher quality of healthcare in the entire system.

To conclude from the case studies, it is crucial to have a continuous development of technological innovation in managing and improving the healthcare. With better execution and ideas that are suitable for each organization, the organization can have better quality of healthcare.

### 7.3.2    BUSINESS PROCESS REENGINEERING

BPR includes changing the organization's structures and processes which compromises skimming through the entire technological, human, and organizational dimensions. The IS plays an important role as it allows efficiency of running the BPR in terms of providing office automation, flexibility in manufacturing, faster delivery to customers, as well as providing paperless transactions. With the combination of information system in conducting the BPR, it provides a better efficiency and effectiveness in the organization's work performed.

There are several steps in composing the BPR in an organization. The BPR is characterized in three major elements: the inputs, the data processing, and the outcome. BPR primarily focuses in reducing the time and money consumed by an organization. An important factor in conducting a successful business process engineering is commitment and teamwork from the top management to the lower management. Although implementation starts from the higher management, it is important not to reduce the empowerment of the lower rank as it may lead to demotivation. Just as important as the teamwork from each involved parties, the use of information system (IS) is crucial in conducting the BPR more smoothly. IS enabled an organization to be more organized amongst each department. For example, there are some cases where there are physical barriers in communication just because not every department is in the same location. Thus, this is where IS can be used to gain better efficiency in communication. Workers can operate as a team using the technology intranets/extranets, disregarding distances barriers.

BPR is widely used by organizations as it provides various improvements over a short period of time. There are several key actions needed to take place in order to apply BPR to an organization: firstly, selection of strategic focuses, for example, taking into account the main focus of change that needs to be redesigned; secondly, simplifying new processes by eliminating unnecessary steps and optimizing efficiency of the process; thirdly, organizing a team of employees that will be involved in the process by giving a coordinator responsible for each process as to direct the team to be in line with the strategic focus; fourthly, organizing the workflow of the process; and lastly, assigning responsibility of each team members' in the process and introducing the redesigned process to the organization's business structure.

In order to select the best possible methods to apply BPR, an organization should always have a clear vision and objective. To have a clear vision, an organization should have the right team in charge of delivering the BPR with the right attitudes, motivation, and a clear well-defined objective that the team can easily understand. The vision acts as a guideline for the BPR. Based on the finding of Zigiaris (2000), all methodologies should comprise six stages: the envision, initiation, diagnosis, redesign, reconstruction, and lastly, evaluation stage. The first stage is the envision stage, this is, where the company define their strategic focus and the IS opportunities are recognized. The initiation stage is when the organization assigns the team and goals the organization wishes to achieve. The next stage, diagnosis stage is when documentation of processes and subprocesses such as resources, IT, and costs is carried out. It is then followed by the redesign stage; this is when the new process is developed. The second-last stage is the reconstruction stage; this is when the techniques of the management converts to make sure that the relocation of the newest process will run smoothly. Finally, the last one is the evaluation stage; this is when the newest process is being thoroughly administered in order to conclude whether the organizations have met their expected goals.

### 7.3.3  ROLE OF IT IN BPR

The requirements in many aspects of an organization are being changed due to reengineering where IT plays an important role. This statement is supported by Zhao, as stated in the literature review, where he believes that reengineering heavily relies on IT to undertake project planning, project management, organizational analysis process modeling and mapping, etc. Organizations experience changes in office environment when technology in the personal computer is getting advanced which makes workers to be more independent. Focusing on manufacturing companies, when reengineering, IT helps to improve the performance drastically by improving the accuracy in sharing information regarding company's strategies and goals. Also, IT minimizes human errors in repetitive and complex tasks and reduces the time taken to accomplish tasks. IT also helps company to improve its position in the market by being the first to market a new product in gaining competitive advantage. In addition, Gunasekaran (1997) states that for a consistent flow of information along functional areas, an electronic data interchange (EDI) helps to eliminate the barriers between functional areas. The use of IT helps manufacturing industries to compete by accessing global suppliers on databases.

Looking at an organization as a whole, internal communication is a crucial part to achieve a desired outcome. To achieve that, installing electronic-mail system allows employees to receive and sent messages. Information Technology makes possibility to retrieve previous data by using other big data, data mining, and data sciences, to successfully compete in this era of digital business, with highly accuracy and reliability to make organization final decision (Gunasekaran, 1997). Reengineering affects major business processes such as order flow, product, marketing and sales department, service processes, accounting and personnel. Analyzing the flow of information and materials is crucial in reengineering depending on the objectives of the organization as it improves the overall performance of the system. Research proves that organizations improve their productivity dramatically and are able to save more costs. The objectives of reengineering focusing on products are to reduce lead time for designing by using IT as well as to cut down unnecessary process. IT also plays an important role in marketing and sales department as it helps company to meet customer satisfaction by providing quality products and services. Information gained from market research and feedbacks can be analyzed by using specific software to get more accurate results. By using IS, organizations are able to improve their customer service by improving distribution and logistics operations. Delivering goods to customer on time will improve customer relationship which benefits the organizations in a long run. With the help of IT, cost of product at various stages of operation can be collected and stored in databases using computerized IS. Also, to benefit company in a long run, IT helps to incorporate innovations, flexibility, and productivity into the accounting performance measures. It has been known that reengineering causes stress as employees are expected to take more active role in every process. This problem is dealt with the use of IT as it improves employee's cooperation by having a more reliable communication system. Also, to ensure continuous improvement of reengineering process, organization should establish a monitoring system which will improve working practices. In addition, organization should analyze the importance of training, leadership, and reward system that can be used to increase motivation level.

### 7.3.4   BENEFITS AND LIMITATIONS OF REENGINEERING

An organization can benefit greatly from reengineering process, especially in reducing costs and improving productivity by implementing efficient processes. For an organization that had undergone multiple redesign processes but still had not managed to achieve the highest level of efficiency, reengineering could be the method to achieve maximum efficiency. By developing an efficient system to help the organization in their operations, it does not only reduce costs while sustaining the quality of the outputs and services but it also satisfies the consumers' demands in receiving a satisfactory service and to purchase quality products that are worth the consumers' money.

Reengineering can also help the organization to work in a maximum speed and avoid waste of time by eliminating unnecessary activities that were used to be difficult by the help of technology. Enterprise resource planning (ERP) is a system or process under BPR that can help the organizations in managing their daily operations. ERP is one of the cost-effective methods that makes the task simpler and shorter by integrating various functions into one main system to streamline information across the whole organization.

Employees' job satisfaction can also be improved when there are faster and simpler ways to complete their given tasks. Reengineering process will also ensure that the employees will be assigned their task on time so that the employees will have enough time to finish their given tasks effectively with a better outcome.

In healthcare sector, reengineering can help to clarify the aim of the organization in focusing more on the patients. By allowing the employees and staff in discussing the reengineering process, it can make the employees feel valuable to the organization. The empowering of staff through listening to the employees' voice and including them in the decision-making will lead to a positive reaction in organizational development of the healthcare organization. Reengineering will help healthcare sector in many ways such as eliminating the traditional way of keeping records into a more efficient way by the aid of technology and internet. With this, instead of wasting time in looking for hardcopy files in the room full of cabinets, physicians, clinicians, or nurses can just access the information of the patients with a click on their computers. This will also give more time for the doctors in attending more patients and simultaneously will reduce the patient's waiting time. Hence, it reduces costs and improves service quality.

On the other hand, reengineering is a very high-risk project that involves a big amount of money and thoughts into the idea. According to Beer and Nohria (2000), 70% of all industries that implement reengineering process could not achieve their targets and have failed. This means that there are limitations that will lead to the failure of the process such as employee's resistant to change because they does not understand the importance of reengineering and how it will affect the organization. Employees might also be against the change as they are afraid on how the process will affect their job as downsizing could be one of the effects of reengineering. Even though reengineering is a top-down approach, employees will also be demotivated when they are not involved in the planning and change process, especially for healthcare sector where the clinicians and physicians who are the key players and play an important role in the organization, and hence the employees will not comply and will not cooperate with the process. Additionally, lack of support from the top management in leading and being committed about the change is another cause that can lead to failure as the aim of reengineering does not deliver properly.

Another main limitation is the organization's lack of readiness for change that includes structural and cultural change. Every organization is different and has complex nature of works that needs to be taken into consideration before proceeding with the reengineering process. As a result, there needs to be a lot of efforts from the management level to explain further in details until the employees in all levels can absorb the idea of reengineering which will also involve the changing of structure and culture of the entire organization. Therefore, a lot of time will be wasted in the preparing process before the organization can work efficiently and effectively.

### 7.3.5   *CASE STUDY ON REENGINEERING*

#### 7.3.5.1   *CASE STUDY 1: IMPROVING KNOWLEDGE MANAGEMENT IN THE HEALTH SERVICE—REENGINEERING APPROACH TOWARD SUCCESSFUL IMPLEMENTATION*

One of the examples of reengineering organization is National Health Service (NHS) in the United Kingdom that adopts the use of technology in the healthcare system to improve the ISs in general. This includes restructuring of the roles and functions of IS as an element and its association to health service. The objective is to reduce costs, enhanced productivity, and service improvement. Thus, processes will benefit the service users, stakeholders, partners, and employees to meet their mutual expectation from the use of IT to support their preferences that gives value to them.

According to Gyampoh-Vidogah and Moreton (2009), implementation of BPR concept will be advantageous to an organization if it is systematically and carefully redesigned because simply deploying IS without properly examining specific functions and problems will lead to the failure. Therefore, NHS had created a group to steer the planning and implementation of the reengineering process. Initially, the group reviewed the proposed aims, roles, and functions, redesigned work, and approved further development of the project. Their main concern was to avoid past failures in IT/IS project implementation in order to achieve implementation objectives.

In addition, several problems had been identified with the reengineering process such as the current knowledge portal is not delivering an appropriate information regarding the program and its activities; an editorial group was saturated to edit contents centrally; to support the information needs, the internal communication was not set up properly; gathering the new information relating to contents was not very well structured; search for all relevant national mental health information is not a priority and there are no links to some of the relevant remote locations; arise of technical challenges of enterprise or professional software to develop other areas; there was lack of financial support to support implementation; and the resistance to change in using new technology into an organization (Gyampoh-Vidogah and Moreton, 2009).

Therefore, in order to improve the issues regarding the IS in the organization, they structured their key activities and stages to successfully implement the MIS using BPR. Such key activities and stages include setting up their goals and objectives; this is where the organization had to set their objectives in order to achieve their measurable goals. When goals are established, the activities of reviewing the program and selecting of communication strategy can be carried out easily.

Next key activities are reviewing the program and its activities; this will examine the added values of the final product. By doing so effectively, the organization needs to eliminate nonvalue-adding processes or activities to be more specific on the areas that need to improve in order to increase the efficiency and the productivity of the healthcare services and also be able to identify the communication strategy and the information exchanged during the process of implementation.

After reviewing the process and identifying the communication strategy the organization needs to estimate the system costs to enable cost justification process and to determine the cost–benefit appraisal to evaluate the cost of proposed system that will lead to an improvement and expected savings. In addition, it is also important to examine the possible challenges those will arise during the project cycle such as technical challenges; hence, effective technological platforms should be hired to support the organization because the outcomes of the processes can lead to achievement of the goals and objectives of the system.

Then, re-engineering the process of healthcare system, it requires a clear understanding of the information running in the organization to gives an overview of the processes, and then the organization has to evaluate the change management regarding information systematically. A proactive approach should be taken to deal with the change management which is being classified into three aspects. The first approach is adapting, next is controlling, and followed by effecting change; this is to track the system details that is running in defining the procedures that will be implemented as well as the technologies that deal with the changes. The change management will help to lower the risk and costs of change and enables a focus strategy to maximize the positive impacts of change. Thus, the implementation of the ISs in NHS was successfully carried out.

As a result, method of reengineering can successfully be used in healthcare services through various stages when considering the factors that indirectly influence the healthcare environments by obtaining more empirical details on information management system within the organization, more discussion toward prevalent technologies and theories related to health services, and healthcare development of change. Moreover, perspective from both management and IS should be able to identify the tools and methodologies that are efficiently needed to diagnose the success cases experience and also the impact to the IS in health service.

### 7.3.5.2   CASE STUDY 2: THE ELECTRONIC HOSPITAL INFORMATION SYSTEM IMPLEMENTED AT THE DISTRICT GENERAL HOSPITAL TRINCOMALEE—AN EXPERIENCE OF BPR

In managing organization efficiently to achieve the goals for health industry, it is necessary to have a good health management information system (HMIS). An effective decision-making process is one of the results from having HMIS. According to Jayawardena (2014), most country's health information system (HIS) is not up to the standards and is experiencing many shortcomings. Hence, electronic hospital information system (EHIS) is used in minimizing errors and ensuring accuracy on public health information as well as to generate daily activities efficiently. It has been proven that the EHIS is successfully operating at the District General Hospital (DGH), Trincomalee. The short-term objectives of the project were to enhance the efficiency of hospital's IS, reduce cost, minimize the usage of paper in the hospital IS, and to maintain health statistics of the hospital in a proper way. In the long run, the objective is to build and sustain patient's databases when analyzing data and to help with decision-making process by referring to evidence stated in the system.

Jayawardena (2014) explains that the use of EHIS that has changed the clinical practices and linked all the sections. For admission, patient receives a card with patient identification (PID) number. This PID number will then be used for patient's second visit to track the details and medical history of the patient in the system. During consultation on the second visit, the medical officer will check from complaints list database to look back at the history of the patient during their previous visit. For medications, doctor will prescribe the medicine from the medicine list database which include the dose and frequency of specific medicine. Patient will only need to bring their card to the pharmacy to take their medicine. Use of the database will prevent patient from receiving wrong medicine; the system will automatically check the stock of every medicine. For laboratory test results, doctor could select the relevant department of which the test has recently being carried out. Then, the doctor will be able to see the result of the test and the information such as test date, test name, test type, and doctor's name who ordered the test. This system helps to prevent duplication of laboratory report of the same patient.

The EHIS creates greater advantage such as providing longer time for interaction between doctors and nurses with patients , improving quality of patient care, improving communications, easy access to information that leads to improve in quality of documentation, reducing errors, enhancing the ability in tracking patient's record, improving hospital image and reputation, avoiding misuse of prescriptions to the pharmacy, and preventing staff from misusing the laboratory reports as they has limited access to the system (Jayawardena, 2014). The outcomes of reengineering are simply to improve the quality in all dimensions beyond customer expectation. McAdam and Corrigan (2001) claim that reengineering meets customer expectation by ensuring the service given is what the patients want, guaranteeing that personnel has correctly carry out procedures that are necessary to a specific patient, and using the resources in an efficient and productive way following instructions given by higher authorities to avoid wastage. However, there are a few challenges faced by the management. One of the major challenges is that some workers are too comfortable with the traditional system. From their point of view, the new computerization system implemented gives workers extra work burden. In overcoming the problem, training was organized to familiarize workers with the system. As stated by Guo in the literature review, leader also plays an important role in reengineering. The training conducted and leadership can successfully change worker's mindset where they finally accept to work with computers.

To conclude, DGH, Trincomalee's gain greater benefit from computerizing their sections and functions. By issuing PID number to patients, it is easy to track patient's record during their next visit which also includes patient's history. Improving medical officer's productivity can be done by minimizing the use of sheets when requesting laboratory investigations. The result of laboratory test was being stored and will only be printed when necessary. This is to avoid duplication of test on the same patient, minimize wastage of stationary, and reduce patient waiting time. For easy retrieval of reports, radiological images are being stored in the computer. Lastly, patient can do online booking for surgery's date. The approach toward reengineering proves that the system used minimize errors and provide security and confidentiality to patient's information. Hence, it is important to maintain the system so that the hospital will be able to meet future demands and challenges. In addition, one of the most important factors that leads to the successful computerization of the health record system is the commitment given by the staff. To sum up, key success factors to reengineering are communication, organization's strategy and policy, processes, IT, and organization's structure, culture, and people.

### 7.3.5.3  CASE STUDY 3: RFID-ENABLED PROCESS REENGINEERING OF CLOSED-LOOP SUPPLY CHAINS IN THE HEALTHCARE INDUSTRY OF SINGAPORE

RFID is a technology that organizations have been adopting rapidly in their business; it uses tiny computer chip that is implanted in a tag similar to bar code identification. RFID uses small radio frequency devices that transmit radio waves that enable identification and tracking of products as well as retrieving information related with the products such as the serial numbers, date of production, expiry date, price or purchase records, and so on. RFID tag includes a read device that can be scanned or read up from several feet away, as well as a host system application that can store data, process data, and transmit information.

According to Kumar and Rahman (2014), healthcare sector is the most likely candidate for implementing BPR. This is because the operations are usually repetitive and deal with high-volume items that are tangible such as letters, bills, medical appliances, soft goods, and so on. In this case, implementing RFID in inventory system and management can help hospitals in improving efficiency and effectiveness of doing work. RFID in inventory management can help in dealing with inventory inaccuracy such as theft, errors in transaction, errors in supplies, scanning errors, and so on. RFID can also help reduce the information distortion which is caused by bullwhip effect that gives forecasts supply chain inefficiencies. Furthermore, RFID technology can speed up the operational flows as the tracking, shipping, and counting, and also, check out process that can be improved.

As mentioned by the manager from linens department, there is a loss of approximately 12% of the total linens for every 3 months. Twelve percent is quite a huge number of linens' loss and should be a concerned to the department as to avoid from losing any more linen and keep on ordering more linen for every 3 months. Before using RFID, the main issues of the existing procedures related with linens department include: first, the operational inefficiencies such as waste of time where the storage room clerks and nurses will go to every department to search for linens, service delays due to insufficiency of linens, and so on. The second issue is the inability for the linens department to track where the linens are, especially the used linens called soiled linens, which were packed in bags before it was transferred through pressured pipes which connects every ward in the hospital to the linens department. Once the soiled linens were packed into bags, it was hard for the staff to count the number of individual linens in the bags. Hence, staff will just estimate the number of linens before sending it to the subcontractor for the cleaning process.

The above issues mentioned can be rectified with the implementation of RFID system. It can track the inventory movement as the tracking system in the devices or tags that will be able to locate all the linens as well the data of each item. However, the linens department needs to replace the existing conveyor for making it working effectively once the antenna is mounted for an easy scanning of linens to know the true quantity of soiled linens before loading it into the delivery truck. On the other hand, in order to keep track of the number of linens used and to avoid unused linens that later the staff and nurse need to gather manually, the hospital can deploy RFID post that allows nurses to easily pick-up clean linens once they need and the amount they need it to avoid unnecessary retrieval of unused linens later. Furthermore, there are antennas and readers that can track the whereabouts of the linens and product's information such as the lifespan, expiry date, usage status, and the loss of linens can be recorded.

The case study shows that installing RFID system in the linens department does affect the workload of nurses, for example, it reduces the searching time for linens at every department and ward, nurses do not need to make unnecessary phone calls to every department asking about the whereabouts of the linens, and so on. Additionally, it also improves the operational workflow as now the linens can be tracked from the dashboard in the linens department and eliminates the lack of visibility problem. As well as there are no more problems of shortage of linens due to inaccuracy of inventory because the new RFID-enabled system will provide all the necessary information on the linens availability rate on the shelf, utilization rate of each ward, and so on. RFID helps to improve the efficiency and productivity of staffs in managing the linens, which hence also eliminate the need to buy or replace the lost linens. Finally, RFID-enabled system is very useful in repetitive operations where each of the goods needs to be tracked down to avoid misplaced or missing item.

### 7.3.5.4   CASE STUDY 4: BPR APPLICATION IN HEALTHCARE IN A RELATION TO HIS

The use of information system in health care is also known as Health Information System (HIS). There are several benefits of using IS in the healthcare organizations; many of which, can lessen errors, can execute the healthcare organizations functions such as order entry and decision support system according to Khodambashi (2013). Despite the benefits, there are also some drawbacks and failure reported for the use of HIS but healthcare organizations must always attempt to get the best possible outcome; thus, the evaluation of HIS is important in ensuring the benefits can be fully utilized. The objective of using IS in a healthcare organization is to reduce cost but still get the maximum benefits possible with a proper design of technology without jeopardizing the quality of the services provided. A healthcare organization is able to achieve this by applying the BPR in relation to HIS properly.

As explained by Khomabashi (2013), business process is defined as "as set of logically related tasks performed to achieve a defined business outcome." Clinical process is almost similar to the definition of business process with the only difference is, it comprises clinical activities. It is crucial to examine the overall involved parties in the clinical workflow as to execute the best possible method of BPR that can improve the overall HIS. The case study focuses on the techniques used in healthcare, putting more emphases on the effectiveness of using information in a health organization as a whole. Management techniques have also been applied to further improve the HIS processes and the selection depends on factors such as resources and objectives.

There are several reasons that cause a healthcare organization to apply BPR but the most significant reason is to reduce the overall cost. According to Khodambashi (2013), in order to utilize the maximum benefits of using this technology, the healthcare IS needs to be reviewed and evaluated to find the redundant parts. In order to cut down the cost, the redundant parts can be eliminated by simplifying the process and eliminating the unnecessary stages in the process. In addition, despite reducing the cost, BPR can also improve the empowerment and employees' satisfaction.

Based on the case study, the impact of BPR can be categorized into the cycle time, the cost of executing the BPR, and flexibility of the organization to respond to the result of the implementation. Organizations that use IT in line with their BPR were able to improve the redesign process. It enabled the organization's team to gather, analyzed, accumulate, and deliver information more efficiently, and it also boosts up the team's communication and teamwork amongst each other.

As mentioned by Khodambashi (2013), there are some vital points that should be considered through the reengineering process. The higher management team is vital in leading through the process. The commitment and willingness of the higher management team influenced the team at the lower levels. BPR allowed the employees to evaluate the process and to give their opinion regarding the selected process in order to boost up the efficiency. The case study emphasized the need to be involved with the stakeholders to get their feedback and put it into consideration. The stakeholders in a healthcare organization are patients, nurses, doctors, and family.

Although using BPR in healthcare organizations has its benefits, there are also some limitations to it. As involvement of the employees from the higher management to lower management is crucial, there are some cases where some employees are resistant to change. Because not everyone is cooperating well with the changes, it makes it harder to carry the BPR smoothly and as planned. When all employees are not involved in the planning and changes made by management, the process may not be well accepted and may cause employees to be demotivated.

To conclude, the benefits and limitation need to be weighted in before making decisions. With better delivery and organized plan, IT can be fully utilized by the healthcare organization.

## 7.3.6  IMPACT OF REENGINEERING

From the case study of reengineering above, in trying to implement the use of technology in the healthcare system, there are several issues that need to be considered before the implementation process. In the early stage after the implementation, there are usually problems faced by the employees and staff regarding the new change. Since reengineering involves the whole organization, and healthcare is quite a big organization, the change process is usually discussed among the top management and rarely involves the employees and staff that will be the one using the new technology system in the work. Without proper explanation and giving readiness for the employees to digest the idea that will affect their work in the future, it will lead to problems such as employees do not really understand how to do their work because of the new system, which makes harder for the organization to achieve efficiency level. However, in the long run, after reengineering, when all the staffs and employees are already aware and used to their work, slowly, the organization operations can work effectively and efficiently.

Even though reengineering is a very high-risk project that usually involves a lot of money in carrying out the change, when implemented successfully, it will reduce costs of the overall operations in the healthcare sector while still achieving maximum quality of products and services. With the use of IS that can store data and share the information in the intranet throughout the whole healthcare organization, it can minimize human errors and eliminate waste of time. Thus, staffs and employees can focus more on their core work which is focusing on attending the final customers which are the patients.

Reengineering in healthcare sectors does not only affect the human resources but it also affects all the stakeholders. When employing a new system that involves technology, a non-IT literate person will be at disadvantage. Due to this, some employees and also patients will be against the new system and prefer the traditional way of doing work.

## 7.3.7   RECOMMENDATION

The initial formation of reengineering is glorified; it delivers revolutionary improvement process and is being conducted around the world. Generally, in today's trends, reengineering is one of the tools that raise the interest of the organization as it provides opportunities to improve the performance and increase the productivity as well as efficiency of the business as a whole. In addition, it allows the organization to be more competitive and remain sustainable in uncertainty market due to globalization and continuously changing conditions. As supported by Milan et al. (2014), in his study of implementation of BPR in human resource management stated that in order to improve the functions of the operation within the organization, it is necessary to increase management's effectiveness and efficiency through a process of reengineering. However, reengineering process should design and manage critically of all its activities and task should be in parallel to its objectives, thus will contribute the organization to achieve its strategic goals. Simultaneously, reengineering process can lead to great success and also a great failure.

The purpose of reengineeringis to revamp the current processes to make it more efficient, creating an effective way of organizing tasks to improve employee's productivity, and introduce useful IT system to improve business processes to realize organizations' goals. Some organizations might find reengineering success very critical; this is because the organization uses unnecessary actions or initiatives to run the reengineering activities which will not give values to the output of the implementation process. Hence, there are other ways to support the reengineering processes in order to prevent failure such as hiring IT expert to aid in the implementation of reengineering because IT is a key component that enables running of the new processes smoothly and essentially; apart from that, IT tools help to simplify the processes, reduce inefficiencies, and access data and key reports easily. Dr. Robinson of IBM UK mentions two main reasons the organizations need to consider reengineering are rapid IT innovation and intensive increase in global competition.

Prior to the new implementation, the organization must make certain a financial planning. This is because, implementing new system requires a large financial contribution in order to prevent failure and be able to increase their competitiveness and productivity. Some organization faced reengineering failure because of the financial constraint. Therefore, the organization must consider the cost of investing a new system that will give a great benefit to the company after the change.

Moreover, reengineering requires proper planning. Some organization experience failure in reengineering processes is because they neglect of human factors in BPR such as human, organizational, cultural, and political issues. Many researchers highlighted that reengineering typically focuses on business processes instead of the welfare of the employee. In fact, during the reengineering process, it is necessary to give more attention to human dimension such as employees to advocate the new system implemented. Resistance to change is one of the major factors that leads to reengineering failure. Thus, the organization must always communicate with the employees and provide them information regarding the changes of the implementation of the new processes. By providing them information, it will help them to uphold to the new changes; hence, it is crucial for the top management to clearly understand the impact caused by the new changes to the employees' roles and responsibility. Therefore, the management should give them training to develop their knowledge, skills, and ability toward the new responsibilities so that they can easily adapt to the new implementation system.

An organization with high-quality human resource management can contribute to significant improvement of the performance of the operations as well as efficiency of the system; this creates a great competitive advantage to the organization from the other competitors. In addition, to support the improvement, employees should be given autonomy of those new processes because it will give positive impact for both individuals and also the organization. Autonomy is an ability of the employees to control their own work situation. As for the employees, when they are given an opportunity to approach their work according to their skills and personalities, the organization will develop as long as all the bases are covered because they feel more responsibility and ownership to their own work; hence, they can invest their energy in performing the tasks. Moreover, with this autonomy given to the employees, it will improve their morale and motivation toward the new system, increase employee engagement from top management, and thus, lead the employee to contribute new innovation that can make the organization to be more competitive and overall successful.

## 7.4  CONCLUSION

In conclusion, it proves that continuous improvement programs are not always the best decision for an organization to gain competitive advantage due to the constant change in the market place. Organizations believe that reengineering will give a greater benefit and opportunity for business to be able to step forward of their competitors. To achieve that, organizations need to concentrate on the crucial part of business processes that give greater effect on the competitive factors. IT is a crucial part in the success of reengineering. IT helps organizations to meet their objectives as it provides information across the departments or divisions in the organization by establishing good communication, improving process performance, and optimizing, modeling, and assessing the consequences of reengineering. However, without adopting the change in mindset concerning the importance and function of IT in organizations process, the process will be ineffective. Other than that, it is also necessary to take into consideration the importance of people. In making sure the implementation process of reengineering a success, proper approach needs to be taken while dealing with employees. Nature of reengineering is that it changes all aspects of an organization which include changing the task of an individual following changes in the process. Therefore, all employees are required to learn new skills in improving their knowledge and capabilities. In analyzing the skills needed and identifying the changes that have to take place, organizations need to undertake need analysis and job analysis. Coping with employee's reactions is always the biggest challenge in implementing reengineering as it causes massive change in the organization. Hence, higher management needs to explain the situation properly to avoid employees' resistance to change and throwing out their fear of changing daily routine or tasks. To do that, increasing employee's moral can be done by making employees feel how important they are in achieving the success. This will also help to eliminate their negative feelings.

## KEYWORDS

- IT emerging technology
- reengineering work process
- improvement
- productivity
- Lean Six Sigma
- total quality management

## REFERENCES

Agus., A.; Krishnan, S.; Kadir, S. The Structural Impact of Total Quality Management in Financial Performance Relative to Competitors Through Customer Satisfaction: A Study of Malaysian Manufacturing Companies. *Total Qual. Manage.* **2000,** *11* (4–6), 808–819.

Al-Shdaifat, E. A. Implementation of Total Quality Management in Hospitals. *J. Taibah Univ. Med. Sci.* **2015,** *10* (4), 461–466. http://www.sciencedirect.com/science/article/pii/S1658361215000761 (accessed Aug 24, 2017).

Attaran, M. Exploring the Relationship Between Information Technology and Business Process Reengineering. *Inf. Manage.* **2005,** *41* (5). https://pdfs.semanticscholar.org/0c76/4edb9a2c75a6650f009d580a4490803b070e.pdf (accessed Aug 22, 2017).

Bertolini, M.; Bevilacqua, M.; Ciarapica, F.; Giacchetta, G. Business Process Re-engineering in Healthcare Management: A Case Study. *Bus. Process Manage. J.* **2011,** *17* (1), 42–66. http://www.emeraldinsight.com.ezproxy.ubd.edu.bn/doi/pdfplus/10.1108/14637151111105571 (accessed Aug 24, 2017).

Catia, M. L.; Scavarda, A.; Vaccaro, G. Lean Healthcare Supply Chain Management: Minimizing Waste and Costs. *Indep. J. Manage. Prod.* **2014,** *5* (4). http://www.ijmp.jor.br/index.php/ijmp/article/view/245/427 (accessed Aug 26, 2017).

Chen, Y. *Business Process Reengineering.* The University of Warwick. 2001. http://www2.warwick.ac.uk/fac/sci/dcs/research/em/publications/phd/ychen/files/chap-3.pdf (accessed Aug 22, 2017).

Cleven, A.; Mettler, T.; Winter, T. R. R. Healthcare Quality Innovation and Performance Through Process Orientation: Evidence from General Hospital in Switzerland. *Technol. Forecast. Soc. Change* **2016,** 113, 386–395. http://www.sciencedirect.com.ezproxy.ubd.edu.bn/science/article/pii/S0040162516301512 (accessed Aug 25, 2017).

Gunasekaran, A. The Role of Information Technology in Business Process Reengineering. *Int. J. Prod. Econ.* **1997.** https://www.researchgate.net/publication/222493466_The_role_of_information_technology_in_business_process_reengineering (accessed Aug 22, 2017).

Guo, K. L. Leadership Processes for Re-engineering Changes to the Healthcare Industry. *J. Health Organ. Manage.* **2004,** *18* (6). http://www.emeraldinsight.com.ezproxy.ubd.edu.bn/doi/pdfplus/10.1108/14777260410569993 (accessed Aug 23, 2017).

Gyampoh-Vidogah, R.; Moreton, R. Improving Knowledge Management in the Health Service: Re-engineering Approach to Successful Implementation. *Springer Link* **2009.** https://link.springer.com/content/pdf/10.1007%2Fb137171_4.pdf (accessed Aug 16, 2017).

Jang, S. -H.; Kim, R. H.; Lee, C. W. Effect of U-Healthcare Service Quality on Usage Intention in a Healthcare Service. *Technol. Forecast. Soc. Change* **2016,** *113,* 396–403. http://www.sciencedirect.com.ezproxy.ubd.edu.bn/science/article/pii/S0040162516301871 (accessed Aug 25, 2017).

Jayawardena, A. S. The Electronic Hospital Information System Implemented at the General Hospital Trincomalee: An Experience of Business Process Re-engineering. *Community Med. Health Educ.* 2014. https://www.omicsonline.org/open-access/electronic-hospital-information-system-implemented-trincomaleean-experience-reengineering-2161-0711.S2-001.pdf (accessed Aug 23, 2017).

Kumar, A.; Rahman, S. RFID-enabled Process Reengineering of Closed-loop Supply Chains in the Healthcare Industry of Singapore. *J. Clean. Prod.* **2014,** *85* (1), 382–394. http://www.sciencedirect.com/science/article/pii/S0959652614003928 (accessed Aug 23, 2017).

Kwon, I. -W.; Kim, S. -H.; Martin, D. Healthcare Supply Chain Management; Strategic Areas for Quality and Financial Improvement. *Technol. Forecast. Soc. Change* **2016,** *113,* 422–428. http://www.sciencedirect.com.ezproxy.ubd.edu.bn/science/article/pii/S0040162516301585 (accessed Aug 25, 2017).

McAdam, R.; Corrigan, M. Re-engineering in Public Sector Health Care: A Telecomunication Case Study. *Int. J. Health Care Qual. Assur.* **2001,** *14* (5), 218–227. http://www.emeraldinsight.com.ezproxy.ubd.edu.bn/doi/pdfplus/10.1108/09526860110401340 (accessed Aug 24, 2017).

Milan, R.; Milan, B.; Marko, C.; Jovanovic, V.; Dalibor, B.; Bojic, Z.; Avramovic, N. Implementation of Business Process Reengineering in Human Resource Management. *Inzinerine Ekonmika-Eng. Econ.* **2014,** *25* (2), 211–222.

Patwardhan, A. Patwardhan, D. Business Process Re-engineering: Savior or Just Another Fad? *Int. J. Health Care Qual. Assur.* **2008,** *21* (3), 289–296. www.emeraldinsight.com/doi/full/10.1108/09526860810868229?fullSc=1 (accessed Aug 23, 2017).

Rachel H. K.; Gary M. G.; Chang W. L. Improving Healthcare Quality: A Technological and Managerial Innovation Perspective. *Technol. Forecast. Soc. Change* **2016,** *113*, 373–378. http://www.sciencedirect.com.ezproxy.ubd.edu.bn/science/article/pii/S0040162516302943 (accessed Aug 25, 2017).

Raval, S. J.; Kant, R. Study on Lean Six Sigma Frameworks: A Critical Literature Review. *Int. J. Lean Six Sigma* **2017,** *8* (3), 275–334. https://doi.org/10.1108/IJLSS-02-2016-0003 (accessed Aug 27, 2017).

Spagnol, G. S.; Min, L. L.; Newbold, D. Lean Principles in Healthcare: An Overview of Challenges and Improvements. *IFAC Proc. Vol.* **2013,** *46* (24), 229–234. http://www.sciencedirect.com.ezproxy.ubd.edu.bn/science/article/pii/S1474667016321930 (accessed Aug 28, 2017).

Topalovic, S. The Implementation of Total Quality Management to Improve Production Performance and Enhance Customer Satisfaction. *Proced. Technol.* **2015,** *19*, 1016–1022. http://www.sciencedirect.com/science/article/pii/S2212017315001462 (accessed Aug 24, 2017).

Tsao, Y. -C.; Linh, V. -T.; Lu, J. -C. Closed-loop Supply Network Designs Considering RFID Adoption. *Comput. Ind. Eng.* 2016. http://www.sciencedirect.com/science/article/pii/S0360835216303564 (accessed Aug 26, 2017).

Xing, Z.; Oyama, T. Measuring the Impact of Japanese Local Public Hospital Reformon National Medical Expenditure via Panel Data Regression. *Technol. Forecast. Soc. Change* **2016,** 113, 460–467. http://www.sciencedirect.com.ezproxy.ubd.edu.bn/science/article/pii/S0040162516301834 (accessed Aug 25, 2017).

Yoon, S. N.; Lee, D. H.; Schniederjans, M. Effects of Innovation Leadership and Supply Chain Innovation on Supply Chain Efficiency: Focusing on Hospital Size. *Technol. Forecast. Soc. Change* **2016,** *113*, 412–421. http://www.sciencedirect.com.ezproxy.ubd.edu.bn/science/article/pii/S0040162516301597 (accessed Aug 23, 2017).

Zigiaris, S. Business Process Re-engineering. Dissemination of Innovation and Knowledge Management Techniques. 2000. http://www.adi.pt/docs/innoregio_BPR-en.pdf (accessed Aug 25, 2017).

# CHAPTER 8

# WORKFORCE DIVERSITY: INFORMATION COMMUNICATION TECHNOLOGY INNOVATION AND CREATIVITY

## ABSTRACT

Diversity classified into two dimensions: primary and secondary dimension. The primary dimension consists of visible elements, which include age, gender, and sexual orientation, whereas differentiating elements such as religion, education, income, and geographical location are part of the secondary dimension. Managing workforce diversity very challenging for most organizations. Workforce diversity is basically concerned with the similarities and differences of employees in the organization and has become an essential element for businesses or organizations to expand their business globally and strive to achieve competitive advantage. Nowadays organizations have begun to realize that diversity is as important as any positive factor that will lead to a greater outcome for the organization as well as building up the organization to be a stronger competitive enterprise, globally. Organization must put a great effort on developing a better strategic plan of managing the workforce diversity as there is rapid growing of diversity in which resulting an increase in women and people with disabilities entering the work nature.

## 8.1 INTRODUCTION

Diversity can be defined as understanding, recognizing, and accepting individual similarities and differences between an individual and a group of people. It can be classified into two dimensions: primary and secondary dimension. The primary dimension consists of visible elements, which include age, gender, and sexual orientation, whereas differentiating elements such as religion, education, income, and geographical location are part of the secondary dimension. According to Ashton (2010), these elements are not noticeable in the first encounter until occurrence of some interaction that involves both individuals.

Managing workforce diversity in the workplace can be very challenging for most organizations. Workforce diversity is basically concerned with the similarities and differences of employees in the organization in terms of their age, gender, ethnicity, lifestyle, religion, race, and sexual orientation which distinguishes them from one another. In recent time, workforce diversity has become an essential element for businesses or organizations that are aiming to expand their business globally and strive to achieve competitive advantage. Therefore, now organizations have begun to realize that diversity is as important as any positive factor that will lead to a greater outcome for the organization as well as building up the organization to be a stronger competitive enterprise. In other hand, facing with digital business and digital ecosystem as globalization challenging, organization must put a great effort on developing a better strategic plan of managing the workforce diversity as there are rapid growing of diversity. This Growing of diversity may lead to resulting of an increase in women and disabilities people entering the work nature.

Managing workforce diversity can create old staffing-related problems such as employment leave, which are the main focus for human resource departments to fulfill the position when there is an increased demand of productivity but less staffing resources. Hence, the implementation of workforce diversity has become a major concern in many human resource management departments where there are several essential elements they need to consider in association with workforce diversity such as problems, challenges, and the management of workforce diversity. Organization must think critically what diversity actually means, how can they develop and manage it in effective ways as well as valuing every concepts of workforce diversity management in a way that it has to go beyond the traditional strategies. This is to cope with the demands of society that are living in an era of modern and advanced technology.

Human resource management allows human resource departments to deliberately make use of every working personnel, especially when there are diverse workforce in a way that it will improve the organization's performance and productivity. However, human resources need to be aware of their problems and elements that may affect their role as they are the one who will manage the diversity in the organization. One of the main tasks of managing workplace diversity is to identify what are the differences that exist among individuals and to provide opportunities and equality for the individuals to contribute their skills and talents for the organization (Foma, 2014). This will generally produce an environment where all employees feel included and valued to the organization; thus, it can help to increase employees' motivation and commitment. Some challenges and problems of having a diverse workforce in the workplace are discrimination issues, prejudice, harassment, and unfair treatment. These problems might lead to job dissatisfaction as well as reduce the rate of employees' productivity.

One of the main elements under managing workforce diversity is equal employment opportunity (EEO). EEO is a regulation including legislation and policies that enforce impartial and fair treatment toward every employee. In other words, EEO aims to treat every employee fair despite differences of age, race, beliefs, values, and physical ability. Agencies that are responsible for enforcing this legislation and policies are The Equal Employment Opportunity Commissioner and Office of Federal Contract Compliance Programs (Foma, 2014). Apart from that, discrimination law and affirmative action are the other elements that can be used to manage and execute the workforce diversity in the organization.

Workplace diversity has numerous benefits where it should be one of the main concerns of every business or organization to compete effectively in today's world. One of the benefits of having diverse workforce is that since most of them are particularly from different background and personality, they will be able to come up with unique solutions to the problems in a short span of time. On the other hand, there are several disadvantages of adopting diversity in the organization. One of these is regarding the cost and expenses of the workforce diversity management such as hiring interpreters and conducting training program. Therefore, businesses and organizations should prioritize hiring diverse employees because it can greatly impact the progress and productivity of one's organization. Additionally, eliminating discrimination against individuals is the main goal of workforce diversity management, which would give them an opportunity to be treated fairly with other coworkers.

In this study, it will outline the review of literature that is related to workforce diversity as well as its benefits and drawbacks to the organization. It will also be exposing the challenges and problems faced by the organization, examining the solutions to the problems, and finally, exploring the managing tools of diversity management.

## 8.2  LITERATURE REVIEW

Bedi et al. (2014), in their research, stated that workforce diversity is the center of the similarities and dissimilarities among employees of the organization, for example, age, gender, heritage, ethnicity, physical abilities and disabilities, race, and sexual orientation. They also referred workforce diversity as policies and practices that seek to include people within an organization who are different from other employees.

## 8.2.1 MANAGING TOOLS

Dike (2013), in her study on impact of workplace diversity on organizations, explains the importance of leaders and managers to understand discrimination and its consequences, recognize their own cultural preferences, and see diversity as the differences among individuals to create an effective and successful diverse workforce. Creating it depends on the manager's ability in figuring out what suits the best for the organization based on the dynamics and teamwork of the workforce. Managers must focus on personal awareness when creating an effective and successful diverse workforce. Employers and employees need to outline their personal prejudices first, before a session of training, to change their mind and behaviors. However, one session is not enough; organizations need to constantly develop and maintain ongoing training.

Kowal et al. (2013) stated that diversity training is defined as training that has the main goal to increase awareness of racial, ethnic, and cultural differences and build skills to promote more diversity and less racism. The training can range from web-based programs to lectures and workshops to field trips and outing to cultural engagement activities. An approach stated in their article is reflexive antiracism. Reflexive antiracism approach is a diversity-training course for White people working with Indigenous Australians, which was from the experience of the authors. To be effective antiracists, specific aspects of reflexivity are needed. Particularly, avoid labeling minorities as "good" and White people as "bad."

Madera (2013) conducted a survey to investigate a firm's diversity management practices as found on the firm's websites. The result of the survey shows that out of the 14 organizations that state the availability of diversity training through their corporate website, only 10 of the organizations have an actual training to manage workforce diversity. Diversity training, diversity speaker series, cross-cultural communications, and 1-day session training are some of the trainings conducted.

## 8.2.2 BENEFITS TO THE ORGANIZATION

Dike (2013) identified that workplace diversity can be beneficial to an organization. Diversity can reduce turnover and lessen absenteeism. In a more diverse workforce, organizations can best serve different customer groups in different markets. Organizations can also create products, which can attract diverse consumers or customers. The study also identifies that workforce diversity is a critical measure to organizations who seeks to establish globally.

In Kim's study (Madera, 2013), a diverse workforce can improve marketing strategy of an organization; this is due to diverse ideas and perspective from the diverse workforce. Having a diverse workforce can improve firm growth because diversity management programs promote employees' organizational attitudes that affect individual performance. A diverse workforce can also enhance firm image; this is because organization with ideal diversity management practices performs better than organization without diversity management practices. Lastly, a diverse workforce can make valuable human resources.

According to Rykers (2016), an increase in female workforce in an organization can have greater outcomes in the organization, for example, increase in innovation and productivity and increase in economy benefit. Many studies have linked that the higher the representation of women in an organization, the larger the outcome.

Braunstein et al. (2014) mentioned that organization can benefit from workforce diversity by innovation and creativity, and it can also raise legitimacy and strategic capacity. When diversity is perfectly handled, it can revamp organizational performance. Within politically centered organizations, a diverse management can be beneficial and can also have greater political authority.

Ali (2015) identified that a gender-diverse workforce may contribute a sustained competitive advantage in an organization. With gender diversity, problem solving can be improved and also creativity and innovation in the organization. This is due to different experiences that male and female employees have. Gender diversity may also help in understanding the needs of female and male customers. In developing a profitable product and giving best and effective needs, understanding the customers' needs is important. In an organization, employees' creativity and innovation can be improved by gender diversity because of their various abilities, point of views, and upbringing. A workforce with gender diversity can also create high-quality decisions because of their outlook, which lead to different possible choices.

### 8.2.3   CHALLENGES

According to Jonsen et al. (2013), organizations faced challenges if diversity effort in the organization only taken by employees to understand it but without being openly expressive about it. An increase in number of people not working and marginalized and their skills will not be used properly, if organizations manage diversity management without the existence of proper context. Another challenge found by Jonsen et al. (2013) is if the upper level in the organization is conquer by technical rationality and instrumental reasoning, then diversity is permanently difficult in the organization. Voluntarism, an action without being forced or threaten, is also one of the challenges that Jonsen et al. (2013) gained from their research. Voluntarism leads to challenges of workforce diversity because workforce diversity becomes a noncompulsory choice for the organization. One main reason discovered by Jonsen et al. (2013) behind the challenges of workforce diversity is the exaggeration on the voluntary participation of organizations in managing diversity without any consultation from employees in the organization.

According to Rykers (2016), unconscious and implicit bias can also be one of the challenges in workforce diversity. It can happen when unseen behaviors affect a decision by judging and assessing the people by their background, culture, or personal experiences. Implicit gender biases and unconscious sexist bias are also some of the biases.

According to Rykers (2016), stereotype threat refers to catch all the different assumption and sense that stereotypes inflict on a person, and stereotypes can have negative or positive impact on employee's performance. The challenge is to recognize the effect and minimize its influence in an organization. Stereotyping in an organization can directly affect the performance of the employees that are stereotyped. It also can lead to decreased engagement among employees, minimized career desires, increase in stress, bad performance, and increase in failure rates.

### 8.2.4   SOLUTION

In Jonsen et al.'s (2013) research, a solution for the challenges in the management of workforce diversity involves all the actors or employees beyond the organizations; this is due to the challenges of diversity that take place in an environment which have various access to power, voice, and legitimacy. In finding solutions of the challenges of workforce diversity, involvement of the stakeholders of the organization, which include shareholders, consumers, employees, and the general public, might be important. For instance, the combination of customers and shareholders may strongly impact an organization's decision about diversity and equality. This is due to the power the shareholder and consumers have to pressure the organization.

The state of involvement in handling the challenges of workforce diversity is also one of the solutions. According to Jonsen et al. (2013), the government has a crucial role in tackling the challenges because the government is the lawmaker, the makers of the markets, and supporter of public services. However, the success in solving the challenges depends on the government power and influence.

Solution to reduce biases is by being alert of one's own biases and let ample time to evaluations based on the person (Rykers, 2016). One of the examples of reducing gender bias is amend personal details from job applications or resume. Another example is by training and develops strategies such as help employees act when a bias situation arises, to address employees' biases.

According to Rykers (2016), recognizing and identifying the elements of stereotype in organization can increase the success of the system in the organization. Some interventions that organizations can do are by growing personnel diversity, eagerly identify similarities among employees, and actively promote various problem solving.

## 8.3   DISCUSSION

### 8.3.1   BENEFITS AND DRAWBACKS OF WORKFORCE DIVERSITY TOWARD AN ORGANIZATION

An organization that practices and has diverse working personnel can bring various advantages as well as disadvantages toward the development of an organization. Hence, it depends on how well the organization executes and embraces the workforce diversity.

#### 8.3.1.1   BENEFITS

##### 8.3.1.1.1   Diverse Talents, Skills, and Experiences

A diverse workforce can develop numerous talents, skills, and experiences as they come from different age group, culture, and languages, which may be beneficial to the organization and their performances. Having diverse employees with different capabilities will be able to create work plan that optimizes productivity where it allows managers and employees to reach their goals within the organization (Foma, 2014).

Furthermore, the differences in skills and talents of each working personnel may increase the tendency of great teamwork in organization as they are able to assist and learn from each other. As a result of having diverse workforce with diverse capabilities, organization will be able to operate more cost efficiently as they do not require any trainings if they want to implement a new system in the company. It is because their workforce has experience on that specific management system where they can share their experiences with the other.

Hiring people from a diverse background and personality can greatly impact the progress of the company as they can help organizations with entire view of the market, identify new opportunities and unmet needs for the organization to continue growing, and achieve success. Other than that, having diverse workforces that are basically owned, a diverse experience can create a better understanding in providing a better customer service as they know how to serve them as if they hold a common ground.

### 8.3.1.1.2    Creates Innovation

In recent times, team working is commonly used in most businesses and organizations as it ensures that the results of each performed task and the delivery of goods and services are being done effectively (Dike, 2013). By working together with people from different backgrounds and experts, each employee will be able to evolve creative thinking through bouncing off ideas of each other and offering feedback as well as suggestions. Through the exchange of ideas and strategy, unique solutions can be developed to solve any problems within the shortest possible time. Therefore, workers from many diverse backgrounds could generate fresh ideas, whereas a homogeneous workforce is not likely to arise any innovative solutions to any problems faced in both national and global market.

Most people have a unique way of thinking, learning, and processing; thus, composing a group that consists of diverse employees with different minds may develop several solutions to a problem. Groups like these might have less tendency to have conflicts as they are able to see things in different perspectives and manage to detect any errors that may be presented. Any organization that has these types of groups has a lower turnover rate, has less chances of having flaws in their products, and tends to be able to come up with many creative techniques as opposed to their competitors (Foma, 2014).

### 8.3.1.1.3 Globalization Opportunities

Language and cultural differences tend to be a barrier for a company that wish to expand their business globally. However, workplace diversity can intensely build up the company's relationship with broader range of customers and clients as hiring employees who are able to speak different language can impact the effectiveness of communication.

Thus, it provides great opportunities for the company to work on global basis and interact with customer who speaks different language or come from overseas. For example, some companies in the southwestern part of the United States of America prefer to employ customer service personnel who are capable to speak in more than one language to serve customers who speaks Spanish as their native language (Bedi et al., 2014). This is to enhance the image of the organization as the customers might feel comfortable when dealing with customer service personnel. Another example is a hotel, where if they employ people with different races, they will have confidence to accommodate and facilitate customers from different races as well. This will produce a great competitive advantage for the organization, especially when operating abroad.

When an organization expanded their operation to another country, problems such as culture shock may be encountered by the working personnel as the business becomes international and operated in another country. This is where the importance of cultural diversity of the employees can be derived as globalization in recent time has triggered more interaction amongst people from different culture and background (Dike, 2013). Moreover, other problems may arise from globalization, which includes gathering information about local customs and laws in the foreign country, assessing risk, and designing strategies to overcome these risks (Martin, 2014). Therefore, when a company has culturally diverse workforce knowledge, the ability from individual organization personnel can be fully utilized to achieve objectives and solve the problems. For instance, a local company trying to expand its business in Singapore, where employees can draw information about the federal and provincial laws that apply to the business, insurances required, and applicable tax regulations when outside company wants to operate in their country.

### 8.3.1.1.4 Enhance Workforce Performance

Every organization has its own unique company structure and objectives; thus, different strategies and techniques may be used to increase the company's productivity. Maintaining and increasing the company's productivity can be one of the major challenges for most leaders and managers as well as to the company itself. Therefore, through adopting diverse workforce strategy and managing it very well can easily enhance the productivity of the company. For example, when management considers the welfare of its workers by offering them proper compensation, health care, and employee appraisal, it enables workers to feel the sense of belongingness to the company irrespective of their cultural background by remaining loyal and hardworking which helps to increase the company's productivity and profit (Bedi et al., 2014).

In addition, having culturally diverse workforce can be very beneficial to an organization as it is able to harmonize the working environment within the organization. In other words, diversity unites all of the employees together despite their differences. Thus, it has positive impact toward the employees where they will be able to work without stress condition and no one will feel left out or in doubt with their existence in the company (Foma, 2014). Therefore, it generally increases the working productivity of each employee. Moreover, every diverse employee possesses a unique strengths and weaknesses that are derived from their culture. This occurs when each worker's unique trait is managed properly and effectively in the organization where it can leverage the strengths and complement its weaknesses to highly impact the workforce (Dike, 2013).

### 8.3.1.2   DRAWBACKS

#### 8.3.1.2.1   High Cost of Diversity Management

In managing diversity in an organization, mandatory training should be given by the managers, supervisors, and employees for them to be able to learn the better ways to communicate with diverse workforce and clients. This training is needed for a company that aims to manage the diverse working personnel in an effective way as well as gaining job satisfaction.

Presently, there are a quite numbers of diversity management program available regarding the size of the company and its workers. For example, diversity awareness programs, where it provides accurate information about diversity, develop an atmosphere in which people are free to voice out their different perspective and to reveal personal biases and stereotypes. However, some training programs are very costly as it might require high amount of participating as well as traveling cost.

Furthermore, to have an effective diversity management, an organization must undergo few steps, which include diversity assessment, development and implementation of diversity workplace plans, and various approaches to diversity training (Kokemuller, n.d.). All of these are associated with direct and indirect costs such as the use of consultant and outsourced professionals, materials involved, period of time on which the employees providing and receiving the training as well as implementing the diversity management.

### 8.3.1.2.2 Communication Issues

Workplace diversity can be an obstacle to the company, especially the way the employees communicate with each other. In fact, when an organization forms a group that consist of workers from different religion, culture and perception might reduce the level of conflict among the group member compared to teaming up of all homogeneous workforce. This is because since they are diverse in experience and background, they will be able to learn and accept each other's opinion and suggestion.

However, workplace diversity can place an obstacle in the way of effective communication, which can lower the cohesiveness among workers as well as reduce the productivity of each employee. Even though spending time with employees by getting to know them helps to reduce and in some instances eliminate the communication barriers during a long term, it is difficult to control an individual's first impression and coworker's orientation period when there is a culture clash (Dike, 2013). Apart from that, diversity workforce from different culture and countries might increase the number of communication filters and language barriers that aggressively impact the internal and external communication process. To handle this issue, some organization hires interpreters and diversity trainers to help the employees to face the communication challenges when working with diverse coworkers.

### 8.3.1.2.3 Discrimination

One of the significant disadvantages of working with a diverse workforce is discrimination in both managers and employees. Discrimination that is related to workforce diversity consists of different aspect; one of it is a race-based discrimination. Race-based discrimination refers to processes of discrimination founded upon ethnicity, perceived "racial" distinctions, culture, religion, or language. This type of discrimination has negative outcomes for individuals and society, which results in bad health, social issues, and economic costs.

In the workplace, race-based discrimination might lead to poor job quality, reduced organizational productivity, commitment, trust, satisfaction, and morale as well as increase in absenteeism and staff turnover. Therefore, if the company that has human resources processes in place to manage diversity, then discrimination toward diverse workforce can be prevented. However, companies that are slowly developing diversity without a strategic plan will be able to discover more discrimination between managers and subordinates and between employees. Consider that discrimination is an unfair treatment of someone because of distinguishing traits. Naturally, if you have a diverse workforce, there is more opportunity for discrimination since diversity is based on distinguishing traits among workers (Kokemuller, n.d.).

### 8.3.1.2.4   *Myriad Accommodations*

Managing accommodation, within diverse groups, such as; providing translating materials into various languages, having interpreters when meeting employees for different country, adjusting business operation hours to match with the employees preferences, especially with different working style, accommodating employee absences due to religious practices; is quite important to make a unique selling point of the organization compare with the competitors. These demands seem to be a challenging since company face difficulty on making the diversity management. Here, a special treatment of diversity management through IT emerging technology will make easier to be implemented in the organization.

Besides that, it requires more time for the human resource personnel of an organization to keep track on the needs of diverse groups in the workplace. This is due to their request and work constraints, which based on religion, nationality, gender, and race might change and grow. As a result, the organization might take a long period of time to acclimate the heterogeneous workforce to the current changes and trends; thus, it can extremely affect the effectiveness as well as the productivity of the employees and definitely the organization as a whole.

### 8.3.2   *CHALLENGES OF WORKFORCE DIVERSITY*

#### 8.3.2.1   *COMMUNICATION BREAKDOWN*

According to Boyaci (1996), communication plays a vital role for achieving the organization's objectives by the process of sharing information with one employee to another in social context. It is important for an organization to have a smooth flow of communication so that the objectives can be met and problems can be minimized. Communication breakdown can occur between employees in an organization. People from the same race or who speak the same language can even face the problem of miscommunication with each other. This shows that there is a higher chance for people who speak different languages that working together may lead to the problem of communication breakdown. This communication breakdown problem is mostly caused by language barriers. It can delay communication process and can create misunderstanding of what should be done during working hour. This is the reason why it is important for managers to train their international employees of the organization country's mother tongue. Performing communication training for the international employees is beneficial for both the organizations and the employees themselves. Due to language barrier, international employees are reluctant to share their suggestion on how to improve the organization as they are not confident to speak with their managers. Employees may provide useful information for the manager, and thus, employee's participation is encouraged.

## 8.3.2.2  CONFLICT

Robbins (2003) stated that everyone had experienced conflict in their life and it is a natural process due to life uncertainty that can occur every day. Conflict mostly occurs when an individual works together with a large group of people. It commonly occurs because group members have lack of trust and lack of confidence when working with each other. Different people have different ways of thinking and it is normal for them to judge one another either from their physical presence or attitude wise. Having doubts toward coworker's ability in handling their work can result in having conflict. This attitude of employees can slow down the organization's growth. Therefore, trusting fellow coworkers is very important while working. Senior employees who have worked for an organization for so many years mostly will not rely on the new employees to get their job done alone. Managers are the one who will mostly be responsible to assign training for the new employees. Senior employees will be an additional help for the new employees if they are having problem with their work. The challenge for the senior employees is that they need to supervise and teach the new employees on how to perform their job correctly. International employees will mostly face the problem of lack of confidence, which is a problem for an organization. The reason why they mostly have lack of confidence is due to language barrier and they are afraid to express their opinion as well as to help contribute in the decision-making process because they will probably feel that their opinion will not be interesting to the locals. All employees should be treated equally and managers have to make sure that the employees are ready and reliable enough to take the responsibility to do their work. Some employees prefer to do their job individually and they are not open to communicate with others. This type of employee would be difficult to work with because they will only trust themselves to get the work done.

## 8.3.2.3   DISCRIMINATION

Discrimination in workplace is a major issue that is happening all around the world. It is illegal to discriminate people based on their race, gender, age, and religion when hiring in the workplace. When hiring an employee, the necessary criteria are to look at their skills, abilities, and qualifications. An example for discrimination during hiring is when a Caucasian manager does not consider a qualified African to work at their companies, but instead automatically the Caucasian gets the job because they are of the same race. This action is considered as discrimination and it is called racism. Nowadays, racism can be seen everywhere around the world, and in some countries, it is considered as serious offence and the punishment is severe. Another example would be judging people based on their appearance. A manager hires male employees because they feel man can do the job better than women. This is unfair because male and female should be treated equally and managers should not look down on female's abilities. In most developed countries nowadays, gender equality is increasing and improving better than previous years. In some countries especially in middle-eastern countries, gender inequality is still a big problem that needs further assistance. Therefore, managers should be honest and evaluation should be fair for everyone; thus, it will be a healthy organization.

The amount of workload should be distributed equally among employees. No employee of the same position should have more workload than the other. If it is based on salary, the job should be equally distributed. If it is based on commissions and employee get paid for doing extra job, managers should not be bias and make sure everyone gets the chance to attend the extra customer or do overtime. One example of a bias manager would be when local employees are given benefit such as overtime but international employees is not given any overtime. This is considered as race discrimination.

Gender discrimination problems exist in organizations. Big roles and responsibilities are mostly handled by male, whereas the not-so-challenging jobs are for women. The possible reason for this is because previously men were mostly the breadwinners for their family. Women are believed to have bigger responsibility at home for their family and spend their time as a housewife. Therefore, male is seen as a dominating figure in the workforce compared to female. Currently in most developed countries, female participation rate in workforce is increasing. According to Heron (2016), the labor participation rate for women is increasing significantly and the gap between men and women participation rates is reduced. Gender discrimination can also occur when manager is paying both equally qualified employees but both of them receive different salaries. The salary of men and women that have the same qualification and responsibility should be paid equally. Another gender discrimination is related to female pregnancy that should be given special considerations. Employers should treat pregnancy as temporary illness and they should understand that the burden of carrying a baby is not an easy job. Therefore, managers should not push pregnant employees to work hard as this can cause stress and it is very bad for both the mom and the baby's health.

Religion discrimination is a very sensitive issue that is happening and should be controlled. Employers should not discriminate other religious rules and customs. In fact, the employers should respect and accommodate their employee's religious beliefs as long as it does not cause any harm to the organizations. An example of this would be for Muslim women's attires. In some countries, nurses wear short skirt at the hospital. Muslim women must cover certain part of their body and wear hijab. Although the short skirt is the formal uniform of the hospital, the manager should also respect and allow Muslim women to wear long sleeve uniform, long skirt, and wear hijab. This is considered acceptable because it does not affect the organization badly and the reasoning is very strong as they need to follow their religious beliefs. Another religious discrimination would be when Muslim employee cannot have their holiday to celebrate their religious events. Employer should allow the employees to have some holidays so that they can celebrate their event just like other people who celebrate their religious events.

## 8.3.2.4   HARASSMENT

According to Doyle (2016), there are different types of harassments that occur in workplace. Harassment in workplace does not only involve sexual harassment but it may also include offensive jokes, physical assaults, blackmail, threats, and intimidation. Anyone can be the harasser including the owner, manager, supervisor, or coworkers while the victim does not necessarily be the one that is harassed but also it involves the people who are affected by the harasser behavior. Harassment can occur during interview and interviewer should not ask personal question such as marital status because it is not related to the candidate's abilities, knowledge, and qualifications to perform the job. Fortunately, law is strongly against harassment nowadays and the harasser will be punished accordingly, but sadly, this antiharassment law is still not powerful enough to stop harassment problems entirely. Comaford (2016) states that workplace bullying is a common problem exist whereby an employee is bullying another employee with bad intentions. Mostly the bully would target weaker people as their victim because they feel that they are more superior in the organization. The bully wants to make a good impression for themselves and will do anything to make sure the victim stays bad in front of other people. Another example for workplace bullying would be against the LGBT members as some people do not accept this group of people. They would be bullied just because some people find out that they are different. The next common harassment that can occur in the workplace would be sexual harassment. More women than men are the victims of sexual harassment in the workplace. An example of sexual harassment would be a manager who offers to increase salaries of their female employees for sex. This behavior is cruel and shows clearly that the manager has no respect for women whatsoever. Therefore, managers should make sure that their employees' satisfaction level is high, because without a happy and motivated workforce, an organization will not grow quickly.

Harassment can also occur due to race and skin colors. Africans are usually the victims of insults and bullied in the United States of America. This feud always happened between the Black and White people. Insulting people based on their skin color is offensive and it should be eradicated. Employers should keep their eyes open and make sure that no such harassment exists in the organization. This type of harassment can cause high turnover rate and absenteeism rate, which can slow down the organization's performance.

## 8.3.2.5   ABSENTEEISM AND HIGH TURNOVER RATE

The number of people absent and leaving is increasing due to discrimination, harassment, and other problems that happen in workplace. Employers should make sure that their employees are free from discrimination and this will lead them to a healthy organization. Discrimination can cause people to lose their jobs and suffer from depression. It is very important to fight against discriminations. The challenge for managers is to make sure that their employees are happy and they need to keep on looking for any discrimination activity inside the organizations. Absenteeism is still considered bad as this shows that employees are unwilling to go to work. The possible reason for this could be because they were bullied at their previous workplace, so they are unwilling to go to work. High turnover rate is worse than absenteeism because high turnover indicates that many of the employees are leaving the company. This is a very bad sign for the organization as this can reduce their performance.

### 8.3.2.6  RESISTANCE TO CHANGE

Changes are usually implemented in an organization mostly for positive reason such as to solve problems and to help them go forward. However, not all employees favor changes as it is troublesome for them to change the way they work that they already familiarize with. Technology advancement is one of the reasons why most organizations wanted to make a change. Older generations are mostly the one that is unwilling to change. The reason is because nowadays organizations are mostly using computers and technology to perform their day-to-day operations. Not all people especially the older age-group people are literate in using a computer. Therefore, they do not accept changes because they have not got the skills to use a computer. They feel that the changes made by the organization have made their work life complicated. If an organization wants to make changes, they should prepare training sessions for their staff so that they will be ready once the change is being implemented. Change can produce anxiety and uncertainty for employees because it can be troublesome and the employees require another way of thinking and doing so that they can adapt to the new change.

### 8.3.3  SOLUTIONS

Managers may be challenged by the reduction in work productivity of its workforce or even worse by losing them due to diversity or differences occurring among individuals or group of people in the organization. In particular, lots of problems and challenges may occur at the workplace caused by this diversity among workforce.

Less cooperation due to diversity between the employees in the workplace may also give impact in the organization in achieving its goal. This is because each member in the workplace must cooperate together to ensure that each organization's goals are successfully implemented. Thus, managers should create a fair and safer environment so that everyone in the organization, without any exception regardless of the differences that might occur between each of them, can effectively work together and everyone has the access to both opportunities and challenges occurring in the workplace.

Managers who failed to do so may experience further conflict in the workforce and continued decrease in their productivity. Therefore, it is more than just recognizing differences between people in the workplace; it requires actions and strategies to resolve this workforce diversity challenges. The following steps are the solutions that can be taken and used by companies, especially by the human resource manager to fix the problems and challenges due to workforce diversity in a workplace and to prevent it happening again in the future.

### 8.3.3.1  UNDERSTAND DISCRIMINATION AND ITS CONSEQUENCES

First, both managers and workforce should understand discrimination and its consequences to employees and organization. According to Rospenda et al. (2009), workplace discrimination refers to actions of organizations as well as people inside them that set biased terms and conditions that will affect the capabilities of members of a group. Discrimination laws as one of the organization's laws will charge penalties to anyone who violates them. For example in Australia, their national and state laws enclose antidiscrimination and EEO in the workplace to obtain workforce that are free from discrimination and harassment (Business, 2017).

Instead, change the mindset of the employees to see the differences as an advantage and not underestimate others because of the differences they have. Doing so will prevent employees to discriminate each other, especially in a diverse group of employees. In addition, having the knowledge on ways to manage discrimination also can avoid managers from making biased decision, whether harsh or favorable, based on their religion, skin color, age, race, or even gender.

### 8.3.3.2  AVOID CONFLICT

Another important step that is important to be taken is to avoid the causes of conflict. To avoid conflict, it should actually start from the beginning that is to be taken by the manager, where the manager itself should first identify any misconceptions and biases toward individual before reaching it to their employees or coworkers. They must first manage their self-attitude and behavior. For example, during job interviews, examine your own behavior as manager when a certain gender, ethnicity, or age group comes in, will there be first assumptions on yourself as a manager or interviewer on whether to prove or disprove the candidate before the interview is even started. Here, self-awareness from the manager itself is the key to develop a safer, fairer, and less conflicted workplace, especially in a diverse group of employees.

By letting and encouraging the employees to be themselves, it will not only comfort them, but it may also make them feel accepted in the organization. One way that manager especially human resource manager can do is to highlight the differences or uniqueness among the employees and list out its benefits and good impact it can give to the organization. One way that human resource manager may highlight the differences of uniqueness among the employees, list out benefits and good impact to the organization, such as preferences of organization culture, that may lead to beneficial of the organization. Here, dealing with people that have vary culture within the organization may advantages and give more benefits to the organization as unique selling point to serve the customers.

Another thing that managers should to do to avoid conflict in the organization is to act fairly instead of acting uniformly. Acting uniformly means treating everyone in the organization in exactly the same way. Instead, treat everyone fairly by respecting their differences. For example, when scheduling a mandatory meeting, make sure the meeting does not fall on a specific religious holiday date. It is to prevent an employee in the organization from that specific religion to felt left out to anyone in the organization. Moreover, it shows an insensitivity attitude, which may breed resentment in any employee that will cause more conflicts and problems if they have to choose either to skip an important date for their religious to attend the meeting or to just skip the meeting and attend their religious festivals or activities.

### 8.3.3.3   IMPROVE COMMUNICATION

One way to improve employee communication skills that can be done by the human resource manager or any manager is by encouraging the employees to interact or communicate with others of different generations and backgrounds. For example, it includes everyone ranging from younger to elder generations, also diverse in terms of their gender, culture, religion, and so on in a department to work together completing the same task and to achieve the same goals. By doing this, it will encourage employee to study more about communication styles of different group of people and to slowly understand and appreciate each other despite differences between them.

Another way that can be done by the human resource management is to arrange team-building activity for the employees. Some activities that can be done during team building are any strategic games, role-playing, sharing life experiences and stories, and much more. This will open the door for all the employees to not just share something unique but also to speak out. By this, it will help the employees to improve their public-speaking skills besides to help them to ease their nervousness, and most importantly, it will improve their communication skills.

### 8.3.3.4  REDUCE RESISTANCE TO CHANGE

To change the mindset of "we have always done it this way" requires awareness from the employees to first identify what are the changes all about. It is the job of managers to make employees understand the changes by getting them clear of what advantages or benefits will be obtained from the changes. This will make it easier for employees to accept the changes made. According to Heathfield (2017), benefits to an individual employee are the most important thing for them since nothing is as important as knowing that it will give the positive impact to their own career on the organization besides from benefits of changes it will give to the group, department, or organization.

When change was introduced, it is important to understand the employee's point of view that whether it is an agreement or disagreement, their emotions, and thoughts. Never overlook employee's responses. For example, majority of the employees agree with the change and had no problem about it, but this one specific employee disagrees as it touches on his or her religious sensitivity; then as a manager, something needs to be done like to make some adjustment from the newly introduced change or to look for new ideas without touching any sensitive issues, especially in a more diverse group of employees. Respecting and letting employees to express their point of view with a nonjudgmental environment, or in other words by making them involve or participate in creating the change, will eventually reduce their resistance to change.

## 8.3.3.5   AVOID WORKPLACE HARASSMENT

The first thing that can be done to avoid workplace harassment is to prevent it from happening. Prevention can be made through the manager itself. That is, the manager must play a role to keep a respectful working environment so that harassment will unlikely to occur in such environment, in which every individual regardless of their positions in the organization treated with not only respect but courtesy, fairness, and dignity. Indeed, to avoid harassment by cultivating a culture of respect is much easier than managing harassment complaints. Therefore, continuous effort by every individual in the organization to demonstrate respectful work environment can prevent workplace harassment.

In the case where harassment has happened or any particular behavior that can be qualified as harassment in the workplace or organization, it is a human resource manager's role to make an immediate action. Thus, as a human resource manager, it is very important to keep themselves up to date on ways to deal with harassment. Other than that, they also need to educate other employees in the organization from all levels ranging from frontline workers up to executive leadership level. One of the efforts that can be done to make a workplace free from harassment is by developing a training session that specifically focuses on harassment in the workplace.

## 8.3.3.6   REDUCE ABSENTEEISM AND HIGH TURNOVER RATE

Providing employee support through programs offered by an organization can reduce absenteeism and high turnover rate. Human resource management can arrange programs such as inviting outside speaker to give motivational talk to the employees. Besides from making themselves more motivated and boost their enthusiasm in doing work, this can also help them to deal with their work and personal issues. These issues might be due to discrimination, harassment, or other problems caused by workforce diversity. So, it is best to have the motivational talk by professional speaker regarding this matter and how they should react to these issues without giving a negative impact to their work.

Reinforcing an attendance policy can also reduce absenteeism and high turnover rate among employees. For employees' absenteeism, warning should be given either verbal or written during the early stages. Then if the problem still goes on, managers can make more severe actions such as salary will be cut to certain amount, temporary suspension from work, or even worse termination of work as the last stage if the problem still persists. On the other hand, promise to give reward to any employees with full or higher attendance rate will give them motivations. Rewards can be given in terms of salary increment, presents, or to offer promotion.

### 8.3.4    MANAGING TOOLS

Human resource management helps to manage workforce diversity. The human resource management systems serve many purposes such as recruiting, selecting, placement, benefits analysis, requirement projections, and other services (Sousa and Oz, 2015). Human resource can be a problem-solver by finding funding sources and making sure that people in the organization are present at meetings for diversity task force.

Human resource manager or any manager plays an important role in understanding or knowing what is best for the organization. Managers must be able to design and carry out a diversity plan. To design it, managers are needed to understand the concept of diversity, be aware that there is diversity in every area of management, must have self-awareness, and have willingness to alter practices that leave out different groups. Managers must recognize that people have differences and people or employees who have this differences need to be encouraged to let their individualities show. Managers should also fairly treat their employees and respect their differences.

### 8.3.4.1    STRATEGY

Human resource management can help diversity in a workplace by planning a diversity program or strategy. One of the examples of a management diversity program is Pillsbury's managing diversity strategy, which consists of six stages (Lim et al., 2013). The first two stages are used for establishing framework on which the change will take place. Awareness session with senior management to understand their role and how to manage diversity is one of the activities included. Stages three to five give more detailed knowledge about how diversity can benefit an organization. These sessions explain the fundamental belief that leads to racist and sexist attitudes. The last stage, stage six, involves merging the cultural strategy into the business plan of the organization. Changes in human resource management policies, which include strengthening the recruitment, hiring, and retention of a diverse workforce, that will be necessary for achieving the long-term strategy are one of the strategies that will be included in the cultural strategy. The goals of Pillsbury's managing diversity is to make everyone in the workforce feel a need to change, to widen participant's understanding of cultural differences, respect each person places on cultural differences, and the way that including a diverse workforce will improve the organization.

Human resource management can also create strategies to increase awareness about workforce diversity in the organization. There are many strategies that human resource management can create for the workforce. Three of the strategies are by decreasing prejudices and application of stereotypes, by reducing unclear or inadequate communication with diverse others, and by building relationships with other workforce.

Ways for employees to decrease prejudices and application of stereotypes are by learning that diversity does exist and value and regard the basic differences, telling their own biases and prejudices and learn to reduce them, and stop believing in fictitious story about others when meeting friends or colleagues. To reduce unclear or inadequate information of indigenous employees, other employees can educate themselves about diverse people with someone who has experience with workforce diversity, listen carefully and ask when one does not understand, refrain from using terms that refer to certain groups and indicate the person is an exception because of diversity, and refrain from judging a person based on their fashion, manner, accent, or eye contact. Lastly, creating connection with other employees regardless of their differences, and receiving feedback from employees about how people treat them, can help build relationships in the workforce.

## 8.3.4.2   TRAINING PROGRAMS

Identifying the particular organizational needs and culture is what an effective diversity training program begins with. There are many ways to create effective diversity training. Some of them are by merging the diversity training with organization's education and training systems, by fusing diversity training with other diversity action within the organization, by delivering the training to all employees in the organization including managers, and by including responsibility.

Example of an effort to help diversity in a workplace is by conducting a cross-cultural education and training programs. This is for organizations that want to be successful in the global marketplace. To enter the global market, most organizations are providing cross-cultural training. The training deal with at least four elements, which are raising awareness of cultural differences, focusing on ways attitudes are shaped, providing accurate information about each culture, and building skills in the areas of language, nonverbal communication, cultural stress management, and adjustment adaptation skills (Lim et al., 2013).

Reflexive antiracism can also be one of the training programs for diversity in a workplace. The key aim for this program is to develop understanding of the interaction between identity formation and antiracist practice (Kowal et al., 2013). For example, in the reflexive antiracism session, participants are grouped together and were asked to brainstorm all the reasons that native Australians suffer from particular health and social problems at higher rates than non-native Australians. After brainstorming all the reasons, the reasons are arranged to categories. Then, managers or facilitators asked follow-up questions to clarify their position and frame the participant's views as opinions. Kulik and Roberson (Madera, 2013) stated that all diversity training programs shared one common set of goals, which is to increase knowledge about diversity, to improve attitudes about diversity, and to develop diversity skills.

Example of e-learning software is Lessonly, a software to get employees started on the rules or customs in diversity and ethics. One of the diversity training examples in Lessonly is "About Me" lesson creation. In this lesson, employee can be the creator and create their own lesson about themselves. They can create quizzes or short-answered questions to help other employees discover a common ground. This helps employees introduce themselves while still being productive in the organization.

### 8.3.4.3   MENTORING AND NETWORKING

Mentoring and networking are also one of the efforts to help workforce diversity. The importance of mentoring and networking are emphasized in the diversity management literature (Madera, 2013). Most diversity management programs use mentoring and providing networks for minority employees. Two major programs from mentoring and networking are creating programs where managers mentor minority employees and employee networking groups where employees take part in informal activities, discussions, and meetings to contribute information and advice.

There is a variety of mentoring program. Some of the programs are traditional one-on-one mentoring, peer mentoring, mentoring circles, and reciprocal or reverse mentoring (Catalyst, n.d.). A traditional one-on-one mentoring is commonly between a more senior employee and a more junior employee. For a long-term mentoring relationship, this program is most suitable. A mentoring, which involves an one-on-one mentoring between peers, is called peer mentoring. Peer mentors can give social support and exchange of knowledge to each other in a relaxed environment. Mentoring circle is a one mentor meeting with a group of mentees and is effective from resource point of view. Lastly, reciprocal or reverse mentoring involves a more junior person to serve the role as mentor to a more senior person to develop senior employees' knowledge of diverse outlook.

An example of a combination of traditional one-on-one mentoring and reciprocal or reverse mentoring is when Netsuite, an enterprise of software company, paired up with high-performing women at the company with mentors who work two levels above and in other departments (Big Think Edge, 2017). This is the one-on-one mentoring. It changes to reciprocal or reverse mentoring when the women help update more senior employees with the fast-changing technology, which improves the skills of existing high-level workers.

Organizations can detect top talent, support employees to set up their career paths, and produce an environment with equal opportunities to all employees, by producing a diversity mentoring program. Structured goal and the benefits of informal relationships can be combined by diversity mentoring. Organizations can record and measure the outcome of the program to the organization objectives to show the value of their programs, if the diversity mentoring program is arranged correctly.

Best practice tips for diversity mentoring programs are as follows: describe the purpose correctly, knowing your audience, create programs "Opt In," ensure that quality outdo quantity, invest in training and guide relationships, recognize stereotyping, and balance challenges with advantages (Tiao, 2015). By describing the purpose correctly, managers must provide straightforward objectives and goals to measure results easily. Managers must clearly understand the audience or groups that the manager support and be keen to their needs. Making a program that can stimulate the excitement and eagerness of the participant can benefit the mentoring programs. To dedicate the mentor's time to their mentees, it is recommended to set a maximum of two mentees. If the topic of stereotyping come up, mentors should be able to honestly and willingly discuss about the role of it.

## 8.3.4.4   EMPLOYEE ASSESSMENT AND BENEFITS

Employee assessment is also one of the tools in managing workforce diversity. By conducting employee reviews, assessment manager can examine the attitude of the employee, mainly how they interact with others. Issues such as only delegating tasks to coworkers of a certain race, and if an employee ignores ideas of people below or above a certain age, should be addressed when assessing employee performance.

Human resource management can also offer employee benefits to all employees to manage workforce diversity since diversity also includes gender, age, marital status, sexual orientation, etc. Employees benefit can also help in retaining qualified employees. The basic benefit program design is health and dental benefits. This can be a key to attract and retain employees if an organization can be creative, flexible, and generous in providing health and dental coverage.

Life and accident death or dismemberment is also one of the benefits an organization can provide. Basic coverage as factor of the employee's salary is often provided by the organization. Employee Assistance Plans, a counseling service for employees with personal difficulty, is another benefit that organization can provide. By providing this service, it can lessen the stress and problem felt by the employee, which can decrease absenteeism and turnover.

Another benefit is retirement benefits. An arrangement by an employer to supply their employees with an income when the employee is no longer receiving a regular income from working is called a retirement plan or a pension. Work–life balance is also one of the benefits given to employees. The benefits include flexible working and agile working policies, childcare vouchers, holiday entitlement, and flexible benefit scheme where employees can add up to 10 additional holidays per year.

Women's committees or network within an organization can be exceptionally important in influencing on the recruitment and retention of women. Some of the benefits given to women to retain them or recruit them in an organization are maternity leave and awards for women.

Examples of organizations or companies who give unusual benefit to their employees are Google, Twitter, and plenty of smaller companies (Jones, 2017). Lunches made by a professional chef, fortnight chair massages, yoga classes, and haircuts are some of the benefits that Google offer. Twitter provide three catered meals per day, onsite acupuncture, and improve classes for their employees. Some smaller companies give vacation expense reimbursement and free books to their employees.

## 8.4 CONCLUSION

As today's world is changing, it affects the marketplace directly, in which the economy is becoming increasingly universal and will make our workforce more and more diverse. Diversity in the workplace can be in terms of gender, age group, social class, personality, values, characteristics, religion, ethnicity, language, level of education, physical appearance, marital status, beliefs, lifestyle, ideologist as well as background characteristics like their geographic origin, economic status, and much more.

Workforce diversity has both challenges and opportunities for an organization. Despite with challenges that it might encounter, diverse work teams can actually brought high values to an organization. It is all subject to an organization itself on how they manage workforce diversity effectively by emphasizing the benefits they can gain from it, as well as their ability to respect individual differences that can brought out the best of their employees unique skills, creativity, knowledge, and experiences.

This chapter includes benefits and drawbacks of diversity management, the challenges, the possible solutions, and the required tools for managing a diverse workforce. It shows how managers, especially human resource manager, are handling and adapting workforce diversity in a workplace. Even though creating a successful diverse workforce may take time, especially in achieving the benefits, human resource management team should not lose their focus and interest in achieving their goals effectively and get rid of all the challenges occur.

Finally, every organization needs to take note and have a strong awareness of the importance of managing workforce diversity and to create their own diversity management plan that will fit their unique needs. Also, the most important thing that every organization should realize and keep in mind is that having a diverse workforce in a workplace is not at all a bad thing even though they will face more challenges at first, but the impact of having a diverse workforce is much more beneficial and it is very important for an organization to have it.

## KEYWORDS

- **workforce diversity**
- **equal employment opportunity**
- **human resource management information system**
- **IT emerging technology**
- **diversity management**

## REFERENCES

Ali, M. Impact of Gender-Focused Human Resource Management on Performance: The Mediating Effects of Gender Diversity. *Aust. J. Manage.* **2015,** *41* (2), 376–397. https://journals.sagepub.com/doi/abs/10.1177/0312896214565119.

Ashton. *The Dimension of Diversity*; 2010. Retrieved from http://ashtonfourie.com/blog1/2010/05/18/the-dimensions-of-diversity/.

Bedi, P.; Lakra, P.; Gupta, A. Workforce Diversity Management: Biggest Challenge or Opportunity for 21st Century Organizations. *IOSR J. Bus. Manage.* **2014,** *16* (4), 102–107. Retrieved from http://iosrjournals.org/iosr-jbm/papers/Vol16-issue4/Version-3/P01643102107.pdf.

Big Think Edge. *3 Successful Diversity Training in the Workplace Examples*; 2017. Retrieved from http://www.bigthinkedge.com/blog/3-successful-diversity-training-in-the-workplace-examples.

Boyaci, C. *Turistik Isletmelerde Haberlesme Teknikleri*; Antalya: Akdemiz Univeritesi Basimevi, 1996.

Business. *Equal Opportunity and Diversity*; 2017. Retrieved from https://www.business.gov.au/info/run/employ-people/equal-opportunity-and-diversity (accessed Aug 25, 2017).

Catalyst. *Optimizing Mentoring Programs for Women of Color*; n.d. Retrieved from http://www.catalyst.org/system/files/Optimizing_Mentoring_Programs_for_Women_of_Color.pdf.

Comaford, C. *75% of Workers Are Affected by Bullying—Here's What to Do About It*; 2016. Retrieved from https://www.google.com.bn/amp/s/www.forbes.com/sites/christinecomaford/2016/08/27/the-enormous-toll-workplace-bullying-takes-on-your-bottom-line-/amp/.

Dike, P. *The Impact of Workplace Diversity on Organisations*; 2013. Retrieved from http://theseus56kk.lib.helsinki.fi/bitstream/handle/10024/63581/Thesisxx.pdf?sequence=1&isAllowed=y.

Doyle, A. *Types of Harassment in the Workplace. Elements of Unlawful Harassment and Discrimination*; 2016. Retrieved from https://www.thebalance.com/types-of-harassment-in-the-workplace-2060886.

Foma, E. Impact of Workplace Diversity. *Rev. Integr. Bus. Econ. Res.* **2014,** *3* (1), 382. Retrieved from http://www.sibresearch.org/uploads/2/7/9/9/2799227/riber_sk14-026_382-390.pdf.

Heathfield, S. M. *How to Reduce Employee Resistance to Change*; 2017. Retrieved from https://www.thebalance.com/how-to-reduce-employee-resistance-to-change-1918992 (accessed Aug 27, 2017).

Heron, A. *More Women than Ever Are in the Workforce But Progress Has Been Glacial*; 2016. Retrieved from https://theconversation.com/more-women-than-ever-are-in-the-workforce-but-progress-has-been-glacial-54893.

Jones, K. *The Most Desirable Employee Benefits*; 2017. Retrieved from https://hbr.org/2017/02/the-most-desirable-employee-benefits.

Jonsen, K.; Tatli, A.; Ozbilgin, M.; Bell, M. P. The Tragedy of the Uncommons: Reframing Workforce Diversity. *Hum. Relat.* **2013,** *66* (2), 271–294. Retrieved from http://journals.sagepub.com.ezproxy.ubd.edu.bn/doi/full/10.1177/0018726712466575.

Kokemuller, N. *Negative Effects of Diversity in the Workplace*; n.d. Retrieved from http://smallbusiness.chron.com/negative-effects-diversity-workplace-18443.html.

Kowal, E.; Franklin, H.; Paradies, Y. Reflexive Antiracism: A Novel Approach to Diversity Training. *Ethnicities* **2013,** 13 (3), 316–337. Retrieved from http://journals.sagepub.com. ezproxy.ubd.edu.bn/doi/full/10.1177/1468796812472885.

Lim, G. S.; Werner, J. M.; DeSimone, R. L. *Human Resource Development for Effective Organizations: Principles and Practices across National Boundaries*; Singapore: Cengage Learning Asia Pte Ltd., 2013.

Madera, J. M. Best Practices in Diversity Management in Customer Service Organizations: An Investigation of Top Companies Cited in Diversity Inc. *Cornell Hospitality Q.* **2013,** *54* (2), 124–135. Retrieved from http://journals.sagepub.com.ezproxy.ubd.edu.bn/doi/full/10.1177/1938965513475526.

Martin, G. C. The Effects of Cultural Diversity in the Workplace. *J. Divers. Manage.* **2014,** *9* (2), 89–92. Retrieved from https://cluteinstitute.com/ojs/index.php/JDM/article/viewFile/8974/8934.

Robbins, S. P. *Management Forest*; Sydney, NSW: Pearson Education, 2003; pp 385–421.

Rospenda, K. M.; Richman, J. A.; Shannon, C. A. Prevalence and Mental Health Correlates of Harassment and Discrimination in the Workplace Results from a National Study. *J. Interpers. Violen.* **2009,** *24*, 819–843.

Rykers, K. The Impact of Diversity, Bias and Stereotype: Expanding the Medical Physics and Engineering STEM Workforce. *Australas. Phys. Eng. Sci. Med.* **2016,** *39* (3), 593–600. Retrieved from https://link.springer.com/article/10.1007/s13246-016-0473-7.

Sousa, K. J.; Oz, E. *Management Information Systems*; Singapore: Cengage Learning Asia Pte Ltd., 2015.

Tiao, S. *Mentoring Works: Diversity in Organizations*; 2015. Retrieved from http://chronus.com/blog/mentoring-helps-diversity-in-organizations.

# CHAPTER 9

# A MANAGEMENT INFORMATION SYSTEM FOR HUMAN RESOURCE MANAGEMENT WORKFORCE DIVERSITY

## ABSTRACT

Workforce diversity is the coexistence of people from various social, cultural, and ethnic backgrounds within the company. two levels of workforce diversity is apply; surface-level diversity and deep-level diversity. Surface-level diversity; the variation in one character that is easily distinguishable, such as gender, age, race, ethnicity that may induce stereotypes thinking. Deep-level diversity; the variation in one's preferences, personality, belief, and so on that need further involvement with one another to better understand. Here, workforce diversity can be an advantage to an organization's productivity and competitive advantage. In other hand, conflicts may also develop and unjust discrimination tends to happen. An organization's performance and competitiveness rely on its capability to adopt diversity and make benefits, and analyze workplace diversity issues, promote, and apply strategies, with various benefits; increased flexibility, increased opportunities for a wider service scope, and efficient and productive task activity. A diverse workforce will be able to use their skills and expertise, such as language and cultural intelligence, which allow organizations to do business internationally.

## 9.1 INTRODUCTION

### 9.1.1 DIVERSITY

Diversity is generally defined as variation or an assortment of differences. Meanwhile, workforce diversity, defined by Kundu (as cited in Kundu and Mor, 2017), is the "coexistence of people from various social, cultural, and ethnic backgrounds within the company." According to Robbins and Judge

(2013), there are two levels of workforce diversity, which are surface-level diversity and deep-level diversity. Surface-level diversity, as defined by them, is the variation in one character that is easily distinguishable, such as gender, age, race, ethnicity, and so on, which sometimes can induce stereotypes thinking toward them. Meanwhile, deep-level diversity is the variation in one's preferences, personality, belief, and so on that need further involvement with one another to be understood better (Robbins and Judge, 2013). Workforce diversity can be an advantage to an organization's productivity (Amaliyah, 2014), and also according to SHRM, competitive advantage (as cited in Green et al., 2015). However, at the same time, conflicts may also develop (Amaliyah, 2014) and unjust discrimination tends to happen (Robbins and Judge, 2013).

An organization's performance and competitiveness rely on its capability to adopt diversity and, therefore, make full use of its benefits. When organizations analyze their workplace diversity issues, promote, and apply diversity strategies, various benefits are seen, such as increased flexibility, increased opportunities for a wider service scope, and efficient and productive task activity. Organizations that are hiring a diverse workforce can generate a variety of solutions to solve problems in service and resources distribution, among others. Workers from a wide range of backgrounds have their own unique skills that they can tailor according to managers' and customers' needs. A diverse workforce will be able to use their skills and expertise, such as language and cultural intelligence, which allow organizations to do business internationally.

### 9.1.2   INFORMATION TECHNOLOGY

Both information technology (IT) and information system (IS) are distinctive, and it is necessary to demonstrate the understanding of the difference in both.

Organizations have employed IS even prior to the existence of IT. In fact, much of their systems back then did not utilize any IT.

IT consists of both hardware and software, which enable people to perform tasks like retrieval, process, store, deliver, and share of information. Information communication technology (ICT) and IT are usually used interchangeably to recognize the merging of both telecommunication and computer in ICT. As for IS, it was illustrated by UKAIS (1997) as the "means by which people and organizations, increasingly utilizing technology, gather, process, store, use and disseminate information."

The area of interest for IS researchers included the learning of theories and practices in connection with the social and technological phenomena which clarify the development, use, and effects of IS in organizations and society at the present time. IS is now as a part of the wider domain of human language, cognition, behavior, and communication. Subsequently, "IS will remain in a state of continual development and change in response both to technological innovation and to its mutual interaction with human society as a whole" (Ward, 2016).

With the development of modern technology at present, many organizations are now adopting cutting-edge technology. The use of computer and IT has increased, and now, it is used daily in day-to-day operations of almost all organizations. People in the business environment consider that there is no group or department in any kind of organization that has not been affected by technology at this age. Therefore, human resource (HR) managers should be able to adapt to changes in technology to be able to support organizational changes.

Normally, HR is viewed as a department where they conduct job interviews, hire new recruits, or dismiss workers. Currently, all aspects of an organization are run with the help of IT, and HR is not omitted. HR now uses IT where it helps HR managers plan and keep track of staff and workers in an organization. Nowadays, jobs in IT are continually multiplying as students in universities and technical institutes are seeking job opportunities in this industry.

According to Mayhew (2017), "the U.S. News World reported among the top 10 careers it was stated that computer software engineers are in demand because the work of designing, building, maintaining, and integrating those increasingly complex systems continues to be one the fastest-growing corners of the job market." He also mentioned that the U.S. Bureau of Labor Statistics reports indicate that computer science and IT will introduce more than 785,000 new jobs from 2008 to 2018. Statisticians foresee that the technology world will boom due to the demand for high-technology goods and services.

## 9.2   LITERATURE REVIEW

### 9.2.1   GENDER DIVERSITY

In 2015, 25% of computing-related occupations were made up of women (Ashcraft et al., 2016). On the other hand, men dominate the IT industry.

Social factors influencing career choice such as family, peer group, media, and gender stereotypes play a huge role in why females are not pursuing careers in the IT industry (Adya and Kaiser, 2005). Adya and Kaiser (2005) studied that parents, in particular fathers, are the major influencers of girls to choose IT careers. Print and electronic media influence and strengthen gender stereotypes that concentrate mostly on physical appearance rather than on motivating career options, especially for females (Adya and Kaiser, 2005).

According to Adya and Kaiser (2005), structural factors that influence career choice include teachers and technology access, both at school and at home. Women are most discouraged by teachers, guidance counselors, and male professors to pursue a career in the IT industry, according to Turner et al. (as cited in Adya and Kaiser, 2005).

In addition, Hewlett et al. (as cited in Ashcraft et al., 2016) discovered that the quit rate for women in the IT industry is 41%, compared to men 17%. A few of the reasons why women leave their IT jobs are the difficulty of working in male-dominated workplaces, and it is challenging to build and grow networks that allow them to advance and be recognized for their abilities, especially if they are in a highly professional field (Fisher, 2011). Furthermore, females on the whole value diversity are skilled more than males (Kundu, 2003).

## 9.2.2 AGE DIVERSITY

It was suggested in a statistics found in Bal, Reiss, Rudolph, and Balter's study (as cited in Railate and Cuitine, 2014) that young and old employees experience different treatments due to age issues. They also share the same thoughts with Rapoliene (as cited in Railate and Cuitine, 2014) that the latter tend to receive negative perception than younger employees. Bal et al. declared that older employees are not keen to changes and do not easily accommodate to new tasks. They also suggest that older age-group workforce is least involved in jobs concerning technology, innovation, and so on.

### 9.2.2.1 PRODUCTIVITY

Gellner and Veen (2009) believe that as age increases, productivity decreases. This is also agreed by Bal et al. who suggested that knowledge and skills deteriorate with age. Other authors, like Beck; Aaltio, Salminen, and Koponen (as cited in Railate and Cuitine, 2014), opposed. Beck believed that age is not the cause of the decrease of productivity, but it is caused when the skills become obsolete.

It was suggested by Mas, Morettin, Wiersema, and Bird (as cited in Gellner and Veen, 2009) that a workforce that consists of a diverse age group will offer a variation of opinions and solutions, as there will be an assortment of different experiences and preferences as opposed to a uniform workforce, whereby opinions might be similar and not as innovative. As Zenger and Lawrence concluded (as cited in Gellner and Veen, 2009), age-heterogeneous workforce is more productive than a workforce with the same age groups. A research conducted by Ely (2004) discovered that there is minimal evidence to support her hypothesis that age diversity would be negatively correlated to work performance.

### 9.2.2.2   INTERACTION

It is discovered that there are hassles in communication between age-heterogeneous workforce than age-homogenous ones as it is easier to communicate with individuals who are similar to each other as there are similarities between them, according to a social psychological research shared by several authors, like Harrison, Price, Gavin, Florey; Horwitz; Lazear; Page; Prat; Richard and Shelor (as found in Gellner and Veen, 2009). These may result in less interactions in a heterogeneous workforce, which is not favorable as suggested by Zenger and Lawrence (as cited in Gellner and Veen, 2009), because in business, frequent communication is needed to solve tasks. Therefore, age management and proper working conditions are needed to cater to the age-heterogeneous workforce to assure productivity (Railate and Cuitine, 2014).

### 9.2.3   CREATIVITY

A study by Florida and Gates (2001) revealed that a huge number of artists are an important indicator of a city's industry progress and growth. Ten of the top 15 "artistic" cities—that is, those with the highest number of residents that comprise artists, musicians, actors, writers, and so on—also rank among the nation's top 15 high-technology cities. The aforementioned areas include San Francisco, Seattle, Boston, Los Angeles, New York, and Washington DC.

### 9.2.4   SEXUAL ORIENTATION

Workforce at present has grown more diverse, including diversity sexual orientation (Ozeren, 2014). According to Ozeren (2014), gay, lesbian, bisexual, and transgender (GLBT) encounter obstacles like being imposed to a job dismissal, as they are considered as the "last acceptable and remaining prejudice" when juxtaposed with other visible diversity. However, this group of individuals seemed to have a strong function in a workforce. A study in the United States by Florida and Gates (2001) discovered that the significant indicator of a city's high-technology prosperity is due to a large gay community. The study discovered that the 5 cities with the highest number of gay populace are all among the nation's top 15 high-technology areas at the time, which are San Francisco, Washington DC, Austin, Atlanta, and San Diego. In addition to gays predicting the assemblage of high-tech businesses, they are also a forecaster of their growth (Florida and Gates, 2001).

### 9.2.5   RACIAL DIVERSITY

As stated by Proudford and Nkomo (2006), inequalities still exist between white and black people to both men and women in terms of wages earned and rates, work hours, career opportunities, and job promotions and incentives; sometimes they also encounter racism in their workplaces. Moreover, a research made by the Cabinet Office Strategy Unit (2003) stated that black people and minority ethnic workers tend to be more unemployed and have difficulties to be in a higher positions and getting promoted as compared to white people. A report on black and minority ethnic women made by the Fawcett Society (2005) informed that British Pakistani and Bangladeshi women obtained only about 56% of the average hourly wage as compared to white men (as cited in Gatrell and Swan, 2008). This shows that different ethnicities, especially men and women who are dark-skinned, are treated differently as compared to white-skinned people. This indicates that inequality still exists in the workplace and it affects businesses and workers alike.

### 9.2.6   CULTURAL DIVERSITY

Culture is complex as it is. Cultural diversity is described by Shachaf (2008) as an assortment of national cultures of team members, where one's native culture is considered to be from his or her home country. Researchers have identified differences in technology application and attitude of jobs in the IT industry between eastern and western cultures (Shachaf, 2008). On the other hand, a study by Richard (2000) reported that there is no correlation between cultural diversity and company performance; the effects are most likely to be decided by the plan of action a company aims for and by how business leaders, managers, and workers act in response to diversity and the management of diversity in that company. Alternatively, a culturally diverse workforce can be beneficial in terms of their pool of knowledge as they came from different cultures, which correspond to a variety of perspective, experience, and way of thinking (Martin, 2014). With that being said, there is also negative perspective to it, for instance, cultural barriers, miscommunication, and conflicts (Martin, 2014).

### 9.2.7   COUNTRY

The Indian populace differs in terms of religion, ethnicity, regional background, gender, color, education, language, disability, and socially disadvantaged—which include scheduled lineage and scheduled clans—among others (Kundu, 2003). Kundu (2003) also noted that shifting social patterns, transformation of population makeup, the need for socioeconomic improvement, and women's movement are several factors that create constraints on businesses to develop a diverse workforce.

### 9.2.8   DISABILITIES

The US Equal Employment Opportunity Commission defined a disabled person as a person that is physically or mentally impaired which significantly inhibits their daily activities (Robbins and Judge, 2013).

Robbins and Judge (2013) discovered that there is an increase in number of (the United States) workforce that consists of people with disabilities. Buciuniene and Kazlauskaite (2010) stated that disabled people are the most unfavorable people in a workforce. They found that, during the Soviet times, disabled people were separated in one environment in which they were provided with education and care, as well as a protected working environment, having minimal interactions with other people. Therefore, the public were oblivious as to how to interact with a disabled individual. Stereotypes surrounding disabled people as suggested by them include lower expectation in terms of job performance, unlikely to be hired due to assumed incompetencies, and people usually have strong unfavorable feeling toward people with disabilities and therefore are unwilling to work with them. Sin and Fong (2007) also agreed as they illustrated an example, whereby disabled staff in the "caring profession," that is, health sector, for instance, disabled nurses, are likely to be assumed to be incapable of working in the ward but are assumed suitable for receptionist tasks. However, it was also found that people with disabilities perform better than normal people; as quoted by Robbins and Judge (2013), they have "potency" and "dependability."

In addition, according to Stanley et al. (as cited in Sin and Fong, 2017), evidence shows that disabled people in fact can be a role model to others, especially to other handicapped people. Sin and Fong (2017) also illustrated that disabled people may also offer special skills, that is, being resourceful despite their disability.

Buciuniene and Kazlauskaite (2010) also identified one of the challenges that is associated with disabled people in the working environment; it is not easy to know the extent of their capabilities, that is, the duration of time they can sit down at a cash register, what weights they can bear, and so on. Further amendments to work schedule or task may be needed to cater to their capabilities. Buciuniene and Kazlauskaite (2010) also found that disabled people also tend to work in a certain organization that caters to the need of disabled people. They usually will be reluctant to work in an organization that may make them feel like they are competing with normal people. Moreover, according to them, if an employer wish to employ a disabled people, existing employees would have to adjust and learn how to communicate with them.

### 9.2.9  GEOGRAPHICAL ORIGIN

According to a research by Florida and Gates (2001), metropolitan areas with a high number of foreign-born or nonnative citizens rank high as technology hubs. Eight out of the top 10 urban areas with the highest percentage of foreigner residents were also among the nation's top 15 high-technology regions at that time, which were Los Angeles, New York, San Diego, San Francisco, Boston, Chicago, Houston, and Washington DC (Florida and Gates, 2001).

### 9.2.10  LANGUAGE

Language differences could result in miscommunication, loss of trust, and disagreements between employees (Shachaf, 2008). Forbes Insights jointly with Rosetta Stone (2001), through their research, learned that language barrier has a huge influence on business activities. Their research indicated that the majority (90%) of organizations struggled with language barriers in their daily work and the problem is getting even severe when 71% of business leaders were planning to run businesses in areas where English is not the common language and that 66% of multinational corporations rely on international virtual team and a total of 30% of businesses invest in foreign-language training.

## 9.3   RESULTS AND DISCUSSION

When human resource management (HRM) effectively manages its diverse workforce, the organization will benefit from a range of skills, abilities, and ideas (Green et al., 2015). Heathfield (2006) suggests HRM to be the provider of training, tools, consultation, knowledge, and so on to the organization to achieve success in their operations.

Etsy et al. stated that (as cited in Green et al., 2015) when diversity is perceived negatively, it can hinder productivity, damage relationships, and so on. They also emphasized that managers should not base their actions merely on stereotypes thinking and discrimination as it could also lead to legal conflicts. Devoe believes (as cited in Green et al.) that managing diversity is not only about accepting their differences but also to realize its value, encourage inclusiveness, and eliminate discrimination. The section below discussed how issues in a diverse workforce can be solved.

### 9.3.1   GENDER DIVERSITY

One way to make girls and women interested to work in the IT industry is by starting at an early age. Parents could expose girls at an early age to a computer and explain how it works, as well as doing interactive activities using computers or other technology to excite curiosity and instill interest in them. Involving the family members can also encourage career options indirectly but effectively; for example, a parent or sibling may not be an IT professional, but he or she may encourage girls to pursue or engage in careers recognized to be "masculine," that is, jobs in the IT industry (Adya and Kaiser, 2005).

Several ways, by which managers can retain women who work in the IT industry, are by making their accomplishments visible, encouraging them to take on roles and challenges, providing the opportunity for them to showcase their skills and abilities, providing ongoing and specific feedback, and treating employees as individuals and not as a representative of a group ("Top 10 Ways Managers Can Retain Technical Women", n.d.). They can also give rewards or recognition to those who perform well. Managers may also give their employees the opportunity to further polish their skills through training programs overseas.

Several ways, by which organizations can retain women in the IT workplace, are by conducting diversity and multicultural training for all IT employees, making educational changes to shift cultural perceptions of IT as "male domain" only, and designing HR policies for consistency, readiness, impartiality, and equality, for instance, job descriptions, compensation, benefits, public hiring, promotions, and reward programs (Tapia and Kvasny, 2004).

According to Barker et al. (2014), the benefits of gender diversity in the workplace are, but not limited to, improved financial performance, innovation, and productivity.

### 9.3.2   AGE DIVERSITY

In the literature review above, it was stated that elders have minimal involvement in tasks pertaining technology and so on. A friendly interface design for older employees may assist and encourage them to embrace technology. It is also necessary for older employees to be trained in using such technologies, to raise their confidence, and reduce anxiety, which may be the reason of their minimal involvement; as quoted by Buc˘i¯unien˙eanKazlauskait˙e (2010), if people are not sure what to do, they usually become anxious and therefore back out. In addition, as people become older, their well-beings are concerned. Watering (2005) suggested a decrease in cognitive skills, motor skills, hearing, vision, and sensory. This also must be taken into consideration, by managing the time they are involved with tasks pertaining computers and technology as to not tire them further. Xie (2003) also found that older people needed more assistance and time to learn about software as they are more prone to making errors than younger people.

### 9.3.3   RACIAL DIVERSITY

To promote diversity in the IT industry, IT managers can retain minorities by exposing ethnic minority IT workers to the same advancing responsibilities that have led their white colleagues to higher management positions;

encouraging open discussions to exchange views observed on discrimination that could handicap minority IT job promotions; giving prompt and precise assessment regarding job performance to make sure there is internal assistance to help advance minorities' IT career ambitions; supporting upward flexibility for minority IT workers through informal mentorship; offering scholarships and internships to minorities enthusiastic in an IT profession; giving minority IT workers authorization to take part in decision-making, problem-solving, developing strategies, and policy-making; and, last but not the least, reinforcing minority IT employee networks in the workplace (Tapia and Kvasny, 2004).

### 9.3.4   CULTURAL DIVERSITY

HR managers need a deeper comprehension of business strategies and other areas aside from the HR sphere in order to make their organization more culturally diverse, since it will lead to an improvement of the organization's social standing and financial performance (Richard, 2000).

On the other hand, a diverse workforce can implement a method aforementioned above, whereby employees may undergo training programs, or team-building activities that are organized by their company with other employees in their off-work hour to gain better understanding of each other and minimize cultural barrier.

### 9.3.5   SEXUAL ORIENTATION

Bower and Blackmon stated (as cited in Ozeren, 2014) that the management of invisible diversity like sexual orientation may be as imperative as managing other visible diversity. Managers' skills and role are vital to create a just environment for GLBT employees. One practice that manager may use to hear GLBT employees, as suggested by Dundon, Wilkinson, Marchington, Ackers, and Bell et al. (as cited in Ozeren, 2014), is by allowing free expression of discontentment and feedback (i.e., "anonymous complaint mechanisms"). Managers may create a forum catered to receiving feedbacks or any concerns from GLBT groups and may respond by making necessary actions to ensure that there are no unfair treatments toward them. All the data collected from the forum may be translated to meaningful information and may be used to review back on the organization's work policy; thus, appropriate amendments may be considered.

## 9.3.6 PREGNANCY

Women play an important role in the workforce. More than half of the women in the United Kingdom and United States work in the labor force. The increasing number of women who works indicated the importance of protecting them, especially during pregnancy (Salihu et al., 2012). Thus, by taking serious actions concerning pregnancy-related health is very important to ensure the health of both the women and the infants. Moreover, women can also be assisted by constructing family-friendly centers and breastfeeding rooms at workplaces (as cited by Low and Zohrah, 2013). Shepherd-Banigan and Bell (2013) argued that 41% of women in the United States received paid maternity leave for about 3.3 weeks with 31% of wage replacement. World Legal Rights Data Center (2012) stated that working women in the United Kingdom, France, and Australia are given between 14 and 52 weeks of paid maternity leave and job security assurance with wage replacement ranging from the national minimum wage to 75–100% of current earnings (as cited in Shepherd-Banigan and Bell, 2013).

## 9.3.7 DISABILITIES

As mentioned above, people with disabilities have potential and are dependable. To achieve their maximum performance and extract their hidden potential, their tasks should be aided with tools that are able to cater to their needs. One example would be by utilizing technology or equipment (i.e., computers, PDA) that are friendly to their condition, that is, voice recognition for visually impaired people.

Bučiūnienėan Kazlauskaitė (2010) suggested having one-to-one discussion with disabled employees about matters pertaining their work schedule as well as the nature of task that best suit them.

In addition, they also encourage seminars to create awareness on interacting and working with disabled employees and learn about their disabilities. Moreover, it is helpful to conduct sign languages class, to help other employees to understand better on how to communicate with people who have hearing impairment.

It is significantly important to assist people with disability as Bučiūnienėan Kazlauskaitė (2010) believed that they lack self-confidence. By providing them guidance, care, and assistance, they will have the courage to try. When they are confident, they are more willing share their experience, be open, and feel good about themselves, which is a positive development for them and can allow them to perform better at work.

## 9.3.8  CHILD CARE

It is inevitable that working women always have concerns regarding their children whenever they go to work. Women have their responsibilities both at home and work, and it is sometimes hard for them to juggle between both. It is vital for working women to have a balance between work and family as it will affect the quality of work that they will produce. If a woman suffers fatigueness from overwork, she will not be able to perform tasks efficiently and this may affect her job satisfaction and performance. More than two-thirds of businesses suggest courses to assist their employees achieve balanced work–life conditions—including the option to work part time and take leaves of absence—whereas only a small number of businesses offer programs devised particularly for parents, such as childcare subsidies and extended leave (Yee et al., 2016). It is best to provide a nursery or a day care in a workplace for female employees, as it will also generate revenue and job opportunities for the business. In addition, mothers will be in close proximity to their children and so they will not feel too worried about their children. They will be able to perform their jobs and accomplish tasks' productively if they know that their children are well taken care of.

In general, diversity is imperative to an organization's stakeholders, both internal and external. An organization that has a diverse workforce will enjoy greater benefits in the long run. A study by Kundu (2003) observed that the minority, disabled, and socially disadvantaged workers value diversity. These groups of workers responded by saying that companies should employ and retain more workers similar to them, as by doing so will highlight the importance of minority and socially disadvantaged representation in the workforce (Kundu, 2003). Workforce diverse businesses also enjoy greater employee satisfaction (Hunt et al., 2015).

However, managing workforce diversity increases costs because it requires training all levels of personnel in the organization. These costs might be from the workshops, seminars, classes, and programs conducted to encourage diversity in an organization, where these types of training are provided to all levels of personnel in the organization (Henry and Evans, 2007).

## 9.3.9   MANAGEMENT INFORMATION SYSTEMS

According to Rallaband (2009), MIS is a system that provides the management of a business with the right information, at the right time, which is essential to help assist in the decision-making process while also allowing the business planning, control, and working activities to be executed productively. MIS varies from the usual IS because of the main aims of these systems, in which it was designed to evaluate other systems concerning the business activities of the organization (Vuda and Srikanth, 2009). MIS help to facilitate coordination by integrating specialized activities and keeping each department or function of the organization aware of the problems of the other departments. In this way, they have concluded that MIS is a subset of the overall planning and control activities covering the application of humans, technologies, and procedures of the organization.

IS is one of the mechanisms to ensure that information is readily available to the managers in the form they want it and when they need it. "A system to convert data from internal and external sources into information and communicate that information in an appropriate form, to managers at all levels in all functions to enable them to make timely and effective decisions for planning, directing and controlling the activities for which they are responsible" (Bee and Bee, 1999).

Organizations found difficulties in organizing, gathering, storing, and distributing large amounts of data and information before the widespread use of computers. The developments in computer technology made possible for all managers to select and retrieve the information they require, in the form best suited for their needs and in time they want. This information must be current and in many cases is needed by many people at the same time. So, it have to be accurate, concise, timely, complete, well presented, and storable. Most firms nowadays depend on IT, but personal computers themselves will not improve organizational productivity: this only comes about if they are used efficiently and effectively. Putting in place the advanced IT emerging technology could support organization by provided data and information short and nicely, without extra budget to provide valid data from several of department within the organization.

"The need for MIS in decision making as it provides information that is needed for better decision making on the issues affecting the organization regarding human and material resources" (Adebayo, 2007). MIS may be viewed as a mean for transformation of data, which are used as information in decision-making processes. In result, MIS increases competitiveness of the firm by reducing cost and improving processing speed. Almost all business organizations normally have some kinds of IS for management. Accounting rules, stock control, and market monitoring systems are the most traditional and common examples. The power of technology has transformed the role of information in business firm. Now, information has become recognized as the lifeblood of an organization. Without information, the modern company is dead (Papows, 1999).

In contrast, "Leading management stated workplace diversity is not only depend on technology, but also related to the art of managing human resources to be continuing and facing with challenge in the 21st century" (Drucker et al., 1997). When technology makes it possible to telecommute, work from virtual offices, and communicate with businesses and individuals across the globe, the flexible work schedules are becoming more obvious due to duties and responsibilities that need to be accomplished from each of the employees.

Failure to achieve departmental responsibilities effectively could result in an unproductive and inefficient workplace. Poor HRM can have a negative effect on workplace productivity. In many cases, unproductive workers are not enthusiastic or engaged in their job duties and responsibilities. HRM can be defined as the effective selection and utilization of employees to best achieve goals and strategies of an organization, as well as the goals and needs of employees (Randy, 2013). HRM are directly involved with obtaining, maintaining, and developing employees, or they either provide support for general management activities or are involved in determining or changing the structure of the organization. Hence, it is crucial for every HR professionals to be able to fulfill their duties and hold a greater amount interest in employee concerns in terms of their needs and development.

Some employees, according to Owino (2015), felt intimidated with the introduction of IS as they believe that it has taken their job away, especially the senior officials, which causes discontentment.

However, from the study findings, it is clear that employees generally improve their performance whenever IS is used. An increase in access to resources can be attributed to the use of financial incentives by the company in a liberal way to motivate and recognize employee efforts to adapt to the use of new IS. It was proved that it allows the workers to access a better understand about the information leading them to respond to information quickly and effectively. At the same time, they found the information very accurate and up to date, which in fact are very important as these decisions affect the performance within the organization. In addition to this, IS also allows users to alternate and present information, as well as perform different tasks.

Moreover, employee motivation increased after introduction of the new IS, so did client rate of service while, on the other hand, employee complaints reduced; all are an indication that the company has been able to motivate the workforce to positively adapt to the use of IS. The new information management system also aids employees to become more creative and innovative, therefore improving their overall performance. The implementation of the new IS has, therefore, seen a general improvement in the employee performance at Kenindia Assurance Company Ltd.

A case study research made by Kenindia Assurance Ltd suggested that the performance of an employee should be further refined using MIS. Moreover, senior managers must have appropriate managerial skills in order to stipulate their core functions of management such as planning, directing, coordinating, organizing, and controlling. In order to surrender their fears, the rise in financial acquisition owing to improved employee performance must be a motivator to managers that IS is meant to usurp their official powers in place of enhance organizational performance. Power is a feel-good element permitted to require more than just financial motivation to address. According to Paul (2017), power refers to the possession of authority and influence over others. It categorizes as a tool that depends on how it is used on which it can lead to either positive or negative outcomes in an organization.

Paul also believed that continuous changes are necessary in an organization to be pertinent in current environment. This can prevent people being comfortable in a stagnant comfort zone and, in a way, derive new competitive advantage through these changes.

## 9.4  CONCLUSION

It is essential to have a diverse workforce in an organization (Saxena, 2014). A diverse workforce is valuable to an organization as they will gain competitive advantage and simultaneously be highly productive when the employees recognize and embrace each other's differences. Managing diversity allows every employee to have fair access to challenges as well as opportunities. It also helps to develop a just environment in the workplace (Green et al., 2002). Implementation of solutions to manage diversity however may cause an organization some costs. In today's world, one has to keep up with the latest trends since technology is very dynamic and fast paced. It is imperative that businesses keep up with the latest technological trends to achieve their goals. Since HR is the main component of any organizations, managers must learn how to develop, retain, and manage a diverse workforce.

## KEYWORDS

- **human resource management**
- **workforce diversity**
- **cutting-edge technology**
- **utilizing technology**
- **organization's performance**
- **competitiveness**

## REFERENCES

Adya, M.; Kaiser, K. *Early Determinants of Women in the IT Workforce: A Model of Girls' Career Choices*. Emeraldinsight.com, 2005. Retrieved from http://www.emeraldinsight.com/doi/abs/10.1108/09593840510615860 (accessed Aug 26, 2017).

Agarwala, T. Human Resource Management: The Emerging Trends. *Indian J. Ind. Relat.* **2002,** *37* (3), 315–331. Retrieved from http://www.jstor.org/stable/27767793.

Amaliyah. The Importance of Workplace Diversity Management. *Int. J. Sci.: Basic Appl. Res.* **2015,** *17* (2), 175–182. Retrieved from http://gssrr.org/index.php?journal=JournalOfBasicAndApplied&page=article&op=view&path%5B%5D=2582&path%5B%5D=1848.

Ashcraft, C.; McLain, B.; Eger, E. *Women in Tech: The Facts*. 2016. Retrieved from https://www.ncwit.org/sites/default/files/resources/ncwit_women-in-it_2016-full-report_final-web06012016.pdf.

Barker, L.; Mancha, C.; Ashcraft, C. *What is the Impact of Gender Diversity on Technology Business Performance?* 2014. Retrieved from https://www.ncwit.org/sites/default/files/resources/impactgenderdiversitytechbusinessperformance_print.pdf.

Buciuniene, I. Kazlauskaite, R. Integrating People with Disability into the Workforce: The Case of a Retail Chain, *Equality, Divers. Inclus.: Int. J.* **2010,** *29* (5), 534–538. DOI:10.1108/02610151011052816.

Checkland, P.; Holwell, S. *Information, Systems and Information Systems: Making Sense of the Field*; Chichester, UK: John Wiley and Sons, 1998.

Čiutienė, R.; Railaitė, R. Challenges of Managing an Ageing Workforce. *Proced.: Soc. Behav. Sci.* **2014,** *156* (26), 69–73. DOI:10.1016/j.sbspro.2014.11.121.

Ely, R. A Field Study of Group Diversity, Participation in Diversity Education Programs, and Performance. *J. Organ. Behav.* **2004,** *25* (6), 755–780. DOI:10.1002/job.268.

EP News Releases. *U.S. Bureau of Labor Statistics: Overview of the 2008–18 Projections*, n.d. Retrieved from http://www.bls.gov/emp (accessed Aug 28, 2017).

Fisher, J. *Information Technology: Where are the Women?*. Melbourne: Monash University, 2011. Retrieved from http://www.monash.edu/news/opinions/information-technology-where-are-the-women (accessed Aug 26, 2017).

Florida, R.; Gates, G. *Technology and Tolerance: The Importance of Diversity to High-Technology Growth.* 2001. Retrieved from http://webarchive.urban.org/UploadedPDF/1000492_tech_and_tolerance.pdf.

Gatrell, C.; Swan, E. Differences at Work: Race, Sexuality and Disability. *Gender and Diversity in Management: A Concise Introduction Gender and Diversity in Management: A Concise Introduction* [Online], 2008; pp 64–85. Retrieved from http://sk.sagepub.com/books/gender-and-diversity-in-management/n5.xml (accessed Aug 28, 2017).

Gellner, B.; Veen, S. *The Impact of Aging and Age Diversity on Company Performance,* 2009. Retrieved from http://www.ilera-directory.org/15thworldcongress/files/papers/Track_1/Track%201_W5_1415_Mohrenweiser.pdf.

Green, K.; López, M.; Wysocki, A.; Kepner, K.; Farnsworth, D.; Clark, J. L. *Diversity in the Workplace: Benefits, Challenges, and the Required Managerial Tools,* 2002. Retrieved from https://edis.ifas.ufl.edu/pdffiles/HR/HR02200.pdf.

Heathfield, S. M. *Beyond Hiring and Firing: What is HR Management?* 2016. Retrieved from https://www.thebalance.com/what-is-human-resource-management-1918143 (accessed Oct 2, 2016).

Henry, O.; Evans, A. Critical Review of Literature on Workforce Diversity. *Afr. J. Bus. Manage.* **2007,** *July,* 72–76. Retrieved from http://www.academicjournals.org/journal/AJBM/article-full-text-pdf/2D1C0DE16759.

Hunt, V.; Layton, D.; Prince, S. *Diversity Matters.* 2015. Retrieved from http://www.mckinsey.com/business-functions/organization/our-insights/why-diversity-matters.

Kenindia Company Ltd. *Kenindia Company Annual Accounts (1999–2008);* Nairobi: Kenindia Company Ltd., 2008.

Kundu, S. Workforce Diversity Status: A Study of Employees' Reactions. *Ind. Manage. Data Syst.* **2003,** *103* (4), 215–226. http://dx.doi.org/10.1108/02635570310470610.

Kundu, S. C.; Mor, A. Workforce Diversity and Organizational Performance: A Study of IT Industry in India. *Employ. Relat.* **2017,** *39* (2), 160–183. DOI:10.1108/ER-06-2015-0114.

Low, K. C. P.; Zohrah, H. S. Women and Human Capital—The Brunei Darussalam Perspective. *Educ. Res.* **2013,** *4* (2), 91–97. Available at https://ssrn.com/abstract=2232848.

Lim, G. S.; Werner, J. M.; De Simone, R. L. *Human Resource Development for Effective Organizations: Principles and Practices across National Boundaries.* Singapore: Cengage Learning, 2013.

Martin, G. C. The Effects of Cultural Diversity in the Workplace. *J. Divers. Manage.* **2014,** *9* (2), 89. Retrieved from https://search.proquest.com/openview/c6d62f72968b4b96316b135d604fcc96/1?pq-origsite=gscholar&cbl=2026889.

Mayhew, R. How Is Technology Impacting the Changes in the 21st Century Workplace? 2017. Retrieved from http://smallbusiness.chron.com/technology-impacting-changes-21st-century-workplace-3357 (accessed Aug 28, 2017).

Merchant, P. *5 Sources of Power in Organizations,* n.d. Retrieved from http://smallbusiness.chron.com/5-sources-power-organizations-14467 (accessed Aug 29, 2017).

National Center for Women and Information Technology. *Top 10 Ways Managers Can Retain Technical Women.* n.d. Retrieved 26 August 2017, from https://www.ncwit.org/resources/top-10-ways-managers-can-retain-technical-women/top-10-ways-managers-can-retain-technical.

Owino, P. Beatrice, J. Relationship between the Use of Management Information Systems and Employee Job Performance: Evidence from Kenindia Assurance Company Limited. *Br. J. Mark. Stud.* **2015,** *5* (2), 61–70.

Ozeren, E. Sexual Orientation Discrimination in the Workplace: A Systematic Review of Literature. *Proced.—Soc. Behav. Sci.* **2014**, *109*, 1203–1215. Retrieved from http://ac.els-cdn.com/S187704281305252X/1-s2.0-S187704281305252X-main.pdf?_tid=97ea956a-8c99-11e7-9f42-00000aab0f6c&acdnat=1503997876_68ff3dce96b358fb592ba871683 82be8.

Peppard, J.; Ward, J. *The Strategic Management of Information Systems: Building a Digital Strategy.* Chichester: Wiley, 2016.

Richard, O. Racial Diversity, Business Strategy, and Firm Performance: A Resource-Based View. *Acad. Manage. J.* **2000**, *43* (2), 164–177. http://dx.doi.org/10.2307/1556374.

Salihu, H.; Myers, J.; August, E. Pregnancy in the Workplace. *Occup. Med.* **2012**, *62* (2), 88–97.

Saxena, A. Workforce Diversity: A Key to Improve Productivity. *Proced. Econ. Finan.* **2014**, *11*, 76–85. DOI:10.1016/S2212-5671(14)00178-6.

Shachaf, P. Cultural Diversity and Information and Communication Technology Impacts on Global Virtual Teams: An Exploratory Study. *Inform. Manage.* **2008**, *45* (2), 131–142. Retrieved from http://eprints.rclis.org/15527/1/CulturalDiversity.pdf.

Shepherd-Banigan, M.; Bell, J. Paid Leave Benefits among a National Sample of Working Mothers with Infants in the United States. *Matern. Child Health J.* **2013**, *18* (1), 286–295.

Sin, C.; Fong, J. Are Caring Professions Restricting Employment of Disabled People?. *J. Integrat. Care* **2007**, *15* (6), 44–48. DOI:10.1108/14769018200700048.

Srinivasu, R.; Reddy, G. S.; Sreenivasarao, V.; Rikkula, S. R. Management Information Systems to Help Managers for Providing Decision Making in an Organization. *Int. J. Rev. Comput.* **2009**, *5* (1), 1–6.

Stone, R. *Reducing the Impact of Language Barriers*, 2011. Retrieved from http://businesshub.rosettastone.com/white-papers-business/forbes-insights-reducing-the-impact-of-language-barriers (accessed Aug 29, 2017).

Tapia, A.; Kvasny, L. *Recruitment Is Never Enough: Retention of Women and Minorities in the IT Workplace*, 2004. Retrieved from https://faculty.ist.psu.edu/lyarger/p84-tapia.pdf.

Watering, M. *The Impact of Computer Technology on the Elderly*, 2005. Retrieved from http://www.marekvandewatering.com/texts/HCI_Essay_Marek_van_de_Watering.pdf.

Wilkinson, A.; Bacon, N.; Redman, T.; Snell, S. *The SAGE Handbook of Human Resource Management*; London: SAGE, 2013.

Xie, B. Older Adults, Computers, and the Internet: Future Directions. *Gerontechnol. J.* **2003**, *2* (4), 289–305. http://citeseerx.ist.psu.edu/viewdoc/download?doi=10.1.1.461.9632&rep=rep1&type=pdf.

Yee, L.; Krivkovich, A.; Kutcher, E.; Epstein, B.; Thomas, R.; Finch, A.; Cooper, M. *Women in the Workplace*. 2016. Retrieved from https://womenintheworkplace.com/.

# CHAPTER 10

# THE INTEGRATION OF MIS–HRIS AND EMPLOYEE INVOLVEMENT

## ABSTRACT

HRD has main role to play in developing effective initiatives or policies in order to increase the employee involvement and participation with the help of the advancement of MIS for managing HRD, namely, HRIS, e-HRM, and HRMIS. These systems covered several HRM work processes such as recruitment, training, record keeping, benefits administration, performance management, employee self-service, and many more. The integration of information system and technology has become one of the emerging factors for the human resource professionals to become one of the top management's strategic partners by providing broader opportunities. The focus of HRM has started shifting from being the sole decision maker of the company into authorizing their employees to participate in producing the strategic decision-making as well as administrative works which relate to several human resource operations. The organization will be more dynamic, efficient, and productive especially in the process of engaging their employees in making strategic decision-making. A continuous advancement of HRIS and international rivalry among other organizations had been leading the organization to invest a huge budget to purchase the best HRIS application in order to gain and/or sustain their competitive advantages besides exploiting the competency and knowledge of their employees This chapter reveal and explain a comprehension of the impacts during the pre- and post-implementation of HRIS through IT-emerging technology as enabler, explore the best practices of HRIS in private and public sectors, analyze the advantages, the integration of HRIS, and its impacts toward employee involvement process.

## 10.1 INTRODUCTION

The importance of employees had been undeniable as the most valuable assets and investment of an organization; thus, in order to provide a positive work environment the leadership team must be more responsive and understand the needs of employees (Inanc et al., 2015). This can be done by giving them high trust, empowering them to initiate collaborative teamwork and problem-solving team, providing them with internal and external training, engaging them in decision-making and goal setting, etc. With the growth of advanced economy, it can be seen that there is an essential coexistence in the forming of a foundation in learning environment at work.

In the era of moving world, Bakotić and Rogošić (2015) said that employees have become one of the most important elements in creating and maintaining company's competitive advantage in today's competitive markets because they are the creators of innovations and support the organization to improve in order to stay competitive. Today's companies have become very meticulous in hiring people as employees who are proficient and enthusiastic in working and are able to improve the quality of the organization and increase its productivity (Bakotić and Rogošić, 2015). With that reason, organizations had been looking into the importance of the one of the popular trends in human resource management (HRM) called employee involvement to keep and sustain their business in this competitive market. Unlike in the past, employees were not just considered as constraint but as a problem-solving resource which to be activated through a variety of human resource techniques. This includes employee involvement processes which were said to assist an organization in creating better working conditions and high-trust relations between employees and management of the organization.

Nevertheless, human resource department (HRD) has main role to play in developing effective initiatives or policies in order to increase the employee involvement and participation with the help of the advancement of management information system (MIS). Some people may refer it as human resource information system (HRIS), electronic-human resource management (e-HRM) or human resource management information system (HRMIS), but the functions and features are most likely similar. These systems include several HRM activities or work processes such as recruitment, training, record keeping, benefits administration, performance management, employee self-service, and many more. In this chapter, the term HRIS will be widely used.. Aside from defining the terms such as employee involvement and HRIS in more details, this chapter will also explain a comprehension of the positive and negative impacts during the pre- and postimplementation of HRIS which was usually done by the HRM team.

Nowadays, the integration of information system and technology has become one of the emerging factors for the human resource professionals to become one of the top management's strategic partners by providing broader opportunities. The focus of HRM has started shifting from being the sole decision-maker of the company into authorizing their employees to participate in producing the strategic decision-making as well as administrative works which relate to several human resource operations. However, the organization usually needs to tackle several difficulties and issues during the transition processes and its implementation. HRIS has led the human resource's operations to be more centralized, convenient, as well as quicker which improves employee involvement, besides transforming the role of HRM into more likely as a decision support and negotiator. Additionally, the organization will be more dynamic, efficient, and productive especially in the process of engaging their employees in making strategic decision-making.

HRM is also responsible to exhibit a good leadership as well as becoming a supportive management agent. Information technology has, of course, changed the aspects of (HRM throughout the moving world. HRIS) has become the most central element in an organization's HRM. Communication and use of information have become vital to HRM and also to other activities in the company. Other important roles of HRM are providing suitable trainings for the employees; empowering employees in decision-making by giving proper guidelines; encouraging and training them to work in teams; and inspiring them to make own goals which are aligned to organizational objectives (Aluvala, 2017).

Currently, an increasing number of organizations has integrated their HRM with information system and information technology because of its great impact on the employee involvement process. Several advantages of HRIS are such as offering cloud computing where users can access the system anytime and anywhere; provide up-to-date information for immediate right decision-making; and provide automation on required or important processes which helps to eliminate tedious work processes and human errors.

A continuous advancement of HRIS and international rivalry among other organizations had been leading the organization to invest a huge budget to purchase the best HRIS application in order to gain and/or sustain their competitive advantages besides exploiting the competency and knowledge of their employees. Therefore, employees are expected to perform well besides increasing their creativity, motivation, productivity, and commitment in order to assist the organization on achieving greater profitability.

Therefore, the main objectives of this chapter are to study the benefits of employee involvement and its influences in HRM, analyze the advantages of HRIS, analyze the integration of HRIS and its impacts toward employee involvement process, and lastly, explore the best practices of HRIS in private and public sectors.

In the first part of this chapter will show the literature reviews that have been extracted from secondary research. In terms of employee involvement, HRIS, the integration of HRIS and employee involvement, the challenges of adopting and implementing HRIS, the impact of HRIS (e-HRM) and case studies to support the researches. Moving on, it will be further discussed and argued in the result and discussion part and finally the conclusion will wrap up on how employee involvement involves in HRM with the integration of today's advanced technology and system, with that it leads to the successfulness of an organization in this competitive world.

## 10.2   LITERATURE REVIEW

Under this part, the researchers will firstly explore the definition of employee involvement, its promising benefits on HRM, the challenges and issues that the management team needs to tackle, and point out few concepts of employee involvement which had been practiced by a typical organization. Secondly, the researchers will study the systems or applications under HRIS, its standard functions and benefits, as well as brief explanation on the categorization of HRIS software. Thirdly, the integration of HRIS and employee involvement will be analyzed with several studies made previously by other researchers to understand the limitation and challenges that the management faces during the implementation or adoption of HRIS into their HRM. In the final section of the literature review, the researcher will study three case studies on the implementation of HRIS into their HRM processes. The main objective is that in the end of this study, the researchers will be able to comprehend whether HRIS had assisted an organization to improve the employee involvement or the impact is still questionably at minimal and in need for more studies.

### 10.2.1   EMPLOYEE INVOLVEMENT

According to Bakotic and Rogosic (2015), employees have become the most important asset of today's organizations where an organization could not survive and achieve its goals without them. Bullock and Powell (n.d) defined employee involvement as an engagement, employee input, and direct participation of employees in order to assist an organization in achieving its main goals and meet the objectives by giving out their own opinions, expertise, as well as efforts toward making decisions and solving problems.

Meanwhile, Duran and Corral (2016) referred employee involvement as the opportunities that employees have to participate in making decisions which are either related to their own work (task discretion) or the issues of the company (organizational participation). Employee involvement only happens in the organization when employees have the power to perform and get involved in making decision with specific instructions and existing acquaintance that enable them to utilize their ability efficiently (Bakotic and Rogosic, 2015).

There are few concepts of employee involvement for organizations to be successful such as delegation, work teams, participative management, goal setting, empowering of employees, and lastly employee trainings (Aluvala, 2017; Bakotic and Rogosic, 2015; Wang et al., 2007).

By emphasizing the practice of employee involvement, better work cultures, processes, and systems can be offered by the organization to increase employees' motivation and utilize their valuable input. Innovative ideas and better employee commitment could be obtained if the employees are engaged in the organization's core activities and decision-making at all levels (Kokemuller, n.d). Similarly, Bullock and Powell (n.d) had agreed that in making an employee involvement process successful, three main activities must be involved: empowering the employees to participate in substantive decision-making; providing training or allowing them to experience decision-making, and lastly; providing either implicit or explicit incentives to the employees.

However, Kokemuller (n.d) brought up his concern on the boundary between management level and employee level and suggested that the managers need to formulate a disciplined structure and stabilize acceptable boundaries. He also stated that managers could be facing a communication complexity because employee involvement might have initiated potential inconsistency in decision-making (Kokemuller, n.d).

In another research, Duran and Corral (2016) had indicated four possible issues which are triggered by the strong implication of employee involvement: employee motivation, employee well-being, learning opportunities at work, and lastly, work and employment conditions. As recommended by Nabi et al. (2016), an organization needs to come up with new policies in order to enhance employee involvement process at top and middle management levels. Because of the remarkable advantages of employee involvement in

HRM, some companies in a number of countries (such as France or Luxembourg) had approved legal reforms of the social dialogue framework, a draft legislation had been made in Italy, and others had implemented a few initiatives to promote employee involvement (Duran, and Corral, 2016).

According to Weeks (2013), the emerging and improved HRIS will allow the decision-makers or planners (either the employees or management level) to access and generate required information immediately and conveniently without a hassle to approach the human resource department for the latest updates. Agyeman (2012) said that if the employees were given the highest realistic authority in decision-making, their involvement will also be increased.

Based on Agyeman's (2012) own research, he had pointed out at least five theories which explained employee involvement processes which had its own outcomes and issues. For the purpose of this chapter, only four theories will be elaborated. Firstly, he mentioned "Human Capital Theory" where the employees who are underperformed because of their lack of problem skills and poor working conditions will receive lower compensation. While those employees who are well-performed with better problem-solving skills and decision-making skills will be fortunate to receive higher compensation. As claimed by Williamson, this theory was also known as Compensating Differences Theory where employees will face the consequences of their own working conditions (Agyeman, 2012). This theory may be able to enforce or plant in the mindset of the employees to put more efforts to meet the main organizational goals, hence, increase the level of employee involvement.

The second theory that Agyeman (2012) had mentioned was called "Efficiency Wage Theories" which predicted that productivity of workers might be increased if they received higher wage. Also, the employees attempted to increase their efforts because of several reasons such as the increasing standard of living in one's country and the increase of unemployment rate. Therefore, Agyeman (2012) had concluded that a higher wage could minimize the rate of turnover and the costs of recruitment; also, judicious recruitment or increased number of effective trainings should be carefully revised by the management team.

Moving to the third theory which he referred as "Incentives and Complementarity" might be focusing on making better decision by setting decision-making rights with incentives (Agyeman, 2012). Just because of the employees felt motivated with incentives, a greater employee involvement could increase the productivity and gains of the company, in addition to be able to nurture the best practices within organization such as gain sharing, knowledge sharing, ownership plans, and others.

The fourth theory was referred as the "Conflict Theories." Some researchers described high-involvement system as "management by stress" (Agyeman, 2012). Some cases, the organizations increase their business efficiency and reduce number of employees through IT emerging technology, and also put employees in small work groups to generate the culture of knowledge sharing, loyalty, as well as discipline with punishment system (Agyeman, 2012). Such systems might encourage employees to decide or influence certain decisions which were seen as decentralizing decision-making by the employee involvement.

Lin (2006) elaborated employee involvement from other several aspects such as democratic conception, its impact and influence from the organizational levels, and internal marketing. On the first aspect, democratic conception, Lin (2016) had pointed out the concrete project of employee involvement in America, which eliminated the centralized authorization and HRM activities. By distributing and delegating the authority of making decisions among their employees, it could increase the employees' loyalty and improve their effort to achieve the organization's goals. This had created a win–win situation between two levels: management level and employee level. The second aspect was related to the organizational levels which Lin (2016) had explained that the employee involvement in America was focused on individuals or groups, whereas in Europe, the execution of involvement covered the entire organizational level. The influence of employee involvement relate to their participation on decision-making which was expected to have some improvement on the organization's productivity, employees working and employment conditions, and adjustment to their working tasks and job responsibilities. Lastly, the study was emphasized on internal marketing where the managers treated their employees as their partners. The employees were encouraged by the organization to imagine themselves as the company's customers when offering the products or services. Another researcher had agreed on these aspects because he believed that the employees would be giving their full commitment in giving good customer services if the managers treated them well and with full reasonable supports (Lin, 2006).

## 10.2.2  HUMAN RESOURCE INFORMATION SYSTEM

Never underestimate the power of people as it may lead to the collapse or success of one organization in order to accomplish their main objectives (Apostolou, 2000). This showed that the operation of an organization is related to the weaknesses or strengths of their employees, even though with the presence of fully automated business processes. As mentioned by Weeks (2013), since 1980s, the development of HRIS had been growing due to the needs of HRM automation especially in the large organization in order to manage their accumulating valuable human resource. Both the advancement of today's technology settings and higher employee skills had become the main drivers of employee involvement process in an organization (Inanc et al., 2015). As claimed by Weeks (2013), the HRIS is very useful to assist HRM in minimizing the repetitive administrative works processes and employees efforts; thus, the improved HRIS had promised a better HRM functions, specifically in managing employees' satisfactions, planning management (e.g., budgeting and facilities), payroll, effective training forecasting, or staffing management.

According to Lippert and Swiercz, human resource portrayed an important role in an organization to achieve an outstanding performance (Obeidat, 2012). Dessler and Al Ariss described HRM as a collection of strategies and implementations that surrounded the aspects of a human resource in an organization (Obeidat, 2012). Therefore, many researchers believed that the implementation of HRM in the organizations will leverage staffs abilities and competencies in order to maintain the competitive advantage of the organization. Furthermore, DeCenzo and Robbins identified some the functions of HRM such as the planning, recruiting, appraisal selection which offered some influences on decision-making; hence, it helps in improving the performance of the organization and employee involvement (Obeidat, 2012). In addition to that, Akhtar et al. agreed that the HRIS could include other HRM operations such as effective trainings, employee involvement, appraisals, and job opportunities (Obeidat, 2012).

Furthermore, Obeidat (2012) had further discussed the study made by Martinsons and Beulen which analyzed the main problems with HRM in the perspectives of an organization and employee. They had concluded that the impact of human capitals and value had its correlation with the success of an organization. Cathcart defined HRIS as a systematic line of action which involved collecting, storing, maintaining, retrieving, and validating data related to its human resources (Weeks, 2013). One example of HRIS application is called e-HRM, where a few studies of e-HRM had been carried out since 1995 which invoked the role of HRM and its transformation outcomes (Marler and Fisher, 2013). Unfortunately, as mentioned by Marler and Fisher (2013), based on the evidence from their studies, the benefits of e-HRM as an agent for creating a positive change for HR is extremely low because the implementation of e-HRM is not to achieve strategic goals but basically to enhance the welfare of employees such as better communication and HR services.

As elaborated by Khashman and Khashman (2016), HRM had several fundamental functions and roles which are usually integrated in HRIS software and applications such as "Performance Appraisal Application," "Recruitment Application," "Selection Application," "Job Analysis Application," and lastly "Communication Application." Due to the advanced technology, each application can be easily and conveniently accessed by the employees. The specification of HRIS of each organization is uniquely based on their own goals. In another research, HRIS consists of other important HR modules or operations such as "Employee Self-Service," "Collection and Monitoring," "Performance Management," "Trainings," "Record Keeping," "Payroll," and "Benefits Administration" (Dorel and Martinovic, 2011).

As added by Dorel and Martinovic (2011), individual HRIS software solutions could be divided into three categories. The first category characterizes HRM elements as part of ERP systems (such as "SAP HRMS"). Meanwhile, the second category referred as "Integral Software Solutions" is integrated with a few different elements for computerization of HRM (e.g., the widely used system called integral human resource information system). The last category consisted of partly software packages that only cover one task which is the cheapest and simplest of all three because it could only be able to automate and improve specific section of HRM (Dorel, and Martinovic, 2011).

However, the price of HRIS software package could reach or exceed $200,000 depending on its latest improvements on HRIS, mostly on its features and functions. Some organizations purchased the on-the-shelf HRIS software because the customized HRIS would cost the organization with a higher price tag. One of the most popular software that could handle HRM processes is called "PeopleSoft" because it provides interesting and important features, for example, training, career development, and motivators (which are also known as "Workforce Rewards"), allowing the employees in managing or self-enrolment for their own personal benefits information (Weeks, 2013). With these latest trends and features of HRIS, employees' involvement and participation could be improved.

The advancement of technology will continuously ensure the predominant use of several HRIS functions online, for instance, via the web, tablets, smartphones, or even with corporate email (Storey, 2014). He had also suggested several improvements on the functions of HRIS. Firstly, by improving the user dashboard type of view over data in order for the users to easily access the specific information, followed by refining the self-service modules which may include the advanced development of electronic-learning (e-learning) to attract and motivate younger generations (Storey, 2014). And lastly, he recommended the development and adoption of distance working where employees could do their work virtually from homes or other places where improved HRIS will be expected to well manage and track this category of employees which could increase their satisfaction for the working conditions (Storey, 2014).

Generally, HRD and middle managers had been bombarded with questions on HRM functions by those employees or top management who were opposing the ideology of new HRIS, in terms of its actual innovative capability, rigidity, flexibility, efficiency, as well as effectiveness (Bondarouk and Ruel, 2009; Duran and Corral, 2016). Therefore, the HRD and middle managers must prepare themselves with the benefits and issues of using HRIS in an organization. According to Altarawneh and Al-Shqairat, HRIS leads to the swift feedback and access to employee information as well as the upgrading of data or information control (Khan et al. 2017). Bamel et al. claimed that HRIS could assist in improving the employment services while reducing paperwork and inaccuracy of information. In addition, Krishna and Bhaskar's research indicated that HRIS could help to cut down manpower in the HR department because most processes are already automated. Beadles et al. concluded that HRIS had played an important role to enhance the effectiveness of channeling information and support the employees in decision-making (Khan et al., 2017).

Additional benefits had been highlighted by Mohapatra and Patnaik (2011), resulting from the implementation of HRIS in an organization, based on the international scale. The benefits are as the following:

1.  *Single data entry:* Employee data is logged into or updated in a database within the HRIS. These employee details will then be available to all departments such as payroll, production, or projects. This would minimize mistakes, getting rid of existing identical documentation, and thus preserved resources such as time and money. An example of an inaccuracy of a system would be the duplication of employee's record in both payroll department and HR department which will create errors especially during processing payroll. As a result, this type of mistakes would hinder the accuracy of processing or updating the attendance records because a similar information is

being used in other modules of HRIS such as employee compensation and benefits. Therefore data entry would be centralized into a single database where all departments can access the information at the same time in order to avoid unnecessary duplication of information.

2. *Improved and automated employee services*: HRIS can assist HR to generate immediate reports associated with the employee's rights of compensation and benefits. They can also be able to track the updated information on the employee's remaining benefits or balance of their total leave which they should deserve yearly. Conversely, in a traditional system, such information would be time consuming to be processed because the HRD had to retrieve the specific information from a various number of files and documents in order to generate such a report. In contrast to HRIS, the report can be produced by authorized users within a short period of time and location restrictions will not be a problem anymore because the HRM process is as easy as a click of a button.

3. *Preset reporting formats*: HRIS has convenient features integrated within the system to produce immediate reports. These reports provided details or statistical information about any HRM information, whether the reports are sorting out according to their position or other category-wise for examples wages given to staff members, staff turnover by position, years of tenure in the company, turnover report for every division, and each geographical detail and trend over a period of time. As a result, these type of reports aid the top management to understand the main reason for the employee attrition, and make the top management to consider on implementing several changes on the corporate policies which will hopefully improve employee well being, trust, relationship, and satisfaction. Also, in a global context, organizations can decentralize their HRM especially in decision-making with these types of reports. For example, in a large organization such as Food and Agriculture Organization of the United Nations (FAO), these types of preset reports helped the employee and management level to make better decisions and indicated whether the function of HR is required to be decentralized to handle local factors which are unfamiliar to particular geographic regions or otherwise. Different employees from different geographic areas would have a different response toward a certain issue or matter in terms of cultural differences besides country- or

region-specific law and order. In another example, several states in India have an existing regulation that the female staff members are not allowed to work in the office past 8 p.m. Therefore, HRD can utilize the HRIS attendance module to cross check if there has been a breach toward the mandate and regulation, which then initiate the HRD to direct preventive measures before the government or labor department would issue a warning memorandum to the company. As a result, this would prevent the brand value of the organization from being tarnished by reducing unnecessary hassle with the government authorities (Mohapatra and Patnaik, 2011).

### 10.2.3 THE INTEGRATION OF HRIS AND EMPLOYEE INVOLVEMENT

The integration of HRIS is mainly related to human resource functions which consist of input, maintenance of data, and output (Weeks, 2016). Employees' information such as appraisals, benefits, leaves, and trainings which is key in into HRIS by either by data entry, scanned documents, or is handwritten is called input function. Data management function is responsible to add, edit, or update the specific information to the database. Lastly, the generation of information into more beneficial information is referred to output function (Weeks, 2016). One of the priority concerns of an organization is the impact of employee involvement in HRIS pre- and postimplementation. Apostolou (2000) had predicted that with employee involvement in the HRIS, employees could enhance their responsibility, increase necessary authorization, and refine their jobs into more interesting and challenging tasks created by their competency and the desires of the organization.

However, Baroudi et al. (1986) had studied the impact of employee involvement on information satisfaction which was expected to show positive result in improving the quality of the system and the successfulness of IS implementation. Based on their finding on the measure of employee involvement, it showed no clear difference in the degree of actual employee influence in the design phase (Baroudi et. al., 1986).

Mazlan et al. (2016) in their current research investigated the participation of Malaysian public sector employees during the operation of the HRIS, and indicated that the majority of the employees felt that least involvement was initiated during HRIS pre- and postimplementation. Despite that, employees expressed their feeling that HRIS is well performed in terms of interactivity (Mazlan et al., 2016).

Technology is essential to attach typically spatially segregated actors and allow interactions among the employees regardless of their running in the same room or on other continents. In other words, technology acts as a medium with the purpose of connection and integration among the employers and the subordinates (Strohmeier, 2007). Therefore, with the help of the information technology, it can reduce the manual daily operations with automation systems.

According to Zuboff, the automation systems not only decrease the amount of routine work but it also offers more opportunities for each employee to utilize their thinking and cognitive capacities, that is, employees concentrate more on information interpretation in comparison to focusing more on administrative works (Gardner et al., 2003). These can help to increase the effectiveness and efficiency of the employees in using such systems. Hence, they will be more productive. In addition, HR professionals can access and determine personnel data statistics on the HRIS without any problem anytime and anywhere.

## 10.2.3.1   THE CHALLENGES OF ADOPTING AND IMPLEMENTING HRIS

However, as mentioned by Parasuraman and Colby (2001), some of the HR managers are having difficulties in encouraging their subordinates to adopt HRIS because quite a number of employees still refused and felt discomfort in adopting a new system because of their perception on the technology that will be taking control of their routine work. Others may feel overwhelmed with their feeling of insecurity because of their distrust on technology and were skeptical on its ability to work well (Parasuraman and Colby, 2001).

Parasuraman (2000) had classified the feeling of discomfort and insecurity into two categories: inhibitors or drivers of technology readiness (Fig. 10.1). The employees must possess two positive characteristics toward HRIS which are, namely, optimism and innovativeness, in order for the organization to embrace the usefulness and benefits of HRIS. This claim had been supported by Reitsema (n.d) in which he had recommended few solutions and precautions which the management team needs to enforce such as the clarification on security protocols, check security measures proposed by vendors, setting or changing of password, provide disaster recovery plan, and others. These recommendations will be discussed in more details under the discussion section.

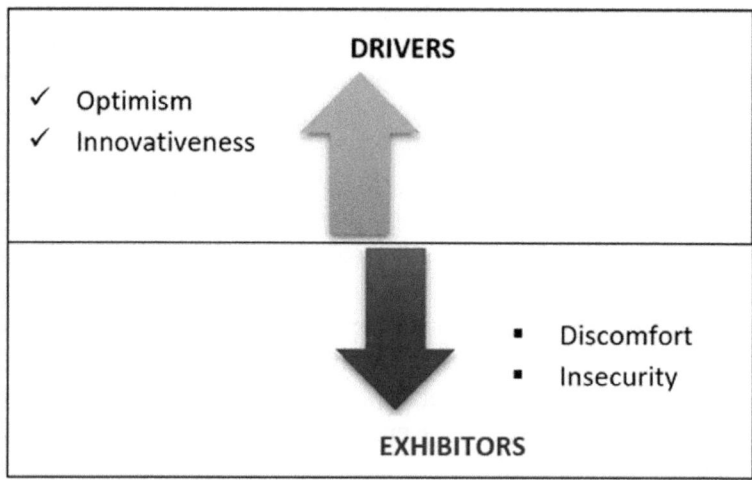

**FIGURE 10.1**    The positive and negative characteristics of employees toward HRIS.

Moreover, Dery et al. (2009) claimed that several barriers toward implementing or improving the HRIS were its ability to sustain the commitment of higher management level, financial resources, intricacy of HRIS, and acceptance to change amid the employees and relevant managers. Moreover, this implementation may lead to low levels of commitment from lower management due to their resistance to change or conservative corporate culture which does not allow any changes or transitions (Khan et al, 2017).

These challenges were also backed up by Bamel et al. and Al-Dmour and Al-Zubi. In addition, Ngai and Wat argued that the most notable challenge was the lack of financial support for such implementation but Al-Dmour and Al-Zu'bi and Bamel et al. suggested otherwise. In a Pakistani context, Ahmer contended that knowledge and awareness, if both were lacking, as well as the cultural opposition to any new change were the few problems faced by Pakistani companies during the implementation of HRIS (Khan et al, 2017).

### 10.2.3.2   IMPACT OF HRIS (E-HRM)

Stone et al. (2006) believed that the impacts of electronic-human resource (e-HR) systems are related to the flow of information, social interactions, sense of control, and acceptance of the system.

Ulrich had stated that Internet had changed the human resource activities by altering some strategies to attract and retain the employees. This is important because most organizations are dependent on the employee's abilities, skills, and knowledge in order to strive in the world of the challenging economy. Furthermore, Internet can provide the organizations to be more engaged, cooperative, and helps in converting the workforce requirement (Stone et al., 2006). As a matter of fact, the e-HR systems enable the individuals to do job application, change of job-roles, and improving their skills and abilities including their knowledge through web-based training systems. Besides that, the web-based systems also can help the employees of the organization to work geographically either in the office or even at home; at any time, any day, and any place. For instance, telecommuting can be done during the night.

However, despite with all the benefits of the systems, there are also other circumstances and impacts toward it such as the following:

1) *Information flows:* There is no negative impact toward the flowing of information in the organizations as it increases the rapidness in accessing the information to job applications online; making easier to access it conveniently. In addition, Nawaz (2012) had observed and analyzed that with the use of HRIS, it has shown a positive impact on the information flow in the organizations.

2) *Social interactions:* According to Nawaz (2012), the use of e-HR systems will replace the face-to-face communications among the employees and employers working in an organization. Therefore, the systems had shown a negative impact toward the attainment in achieving the organization goals or individual goals. Besides that, Cardy and Miller believed that by using such systems, reducing such face-to-face interactions may lead to negativity on trustworthiness between the superiors and the juniors (Nawaz, 2012).

3) *Systems acceptance*: Gueutal and Stone mentioned that the adaptation of using such systems involves the attitudes, feelings, and behaviors of an individual (the employee) who will be using the system (Nawaz, 2012). Incorporation to that, Dery et al. (2009) agreed that the user acceptance of using such systems was one of the barriers that was faced by an organization. They claimed that individuals restrain by their doubtfulness/distrust/suspicion, lack of expertise, inadequate engagement/dedication, and the distress of losing their jobs, losing entitlements, or rescheduling of shifts.

## 10.2.3.3   CASE STUDIES

In order to well understand the impact of HRIS in relation with employee involvement in an organization, the researchers had chosen three case studies which analyze two different perspectives and conditions. Case study 1 is based on the adoption of HRIS in private sector and measured its impacts on the employees who were involved in the implementation of HRIS, whereas case study 2 was taken from a recent research carried out by Mazlan et al. in 2016, which involved 385 employees from Malaysian public sectors. And the third case study, analyzed the benefits and challenges of utilizing HRIS where few responses were taken from three levels of management: higher management, middle management, and lower management level which consisted of HR professionals who worked in Pakistani organizations.

### 10.2.3.3.1   Case Study 1: The Adoption of HRIS in Private Sector

Since 2006, Malaysia Airlines (MAS) has shifted from a paper-based process of processing job applicants to implementing a HRIS as it improved MAS hiring process to ensure that their recruitment are of utmost quality and capability toward achieving their corporate goals.

This case study suggested that implementation of HRIS has to include involvement in a top-down approach in order for employees to be involved with the system from the beginning in terms of being in a task force or team together with the top management. This is to ensure that employees will have a more positive commitment toward the newly implemented HRIS. Top management involvement in the implementation process proved to a large influence toward the acceptance and involvement of the HRIS based on the study's multiple regression analysis (Razali and Vrontis, 2010).

### 10.2.3.3.1   Case Study 2: The Adoption of HRIS in the Public Sectors

A research had been carried out by Mazlan et al. (2016) on 385 employees from Malaysian public sector companies such as Universiti Utara Malaysia, Public Service of Malaysia who were routinely adopting HRIS in performing their daily operations. This research was carried out to study the perception of employees in Malaysian public sectors toward HRIS as well as employees' self-readiness to the advancement of information technology.

Daud (Mazlan et al., 2016) referred IS as the key factor which steers the success or sustainability of an organization; thus the operations must carefully design based on the users' requirement and specification. Based on their analysis, it showed that their adoption of HRIS has a direct positive significant relationship which had assured several benefits such as the quality of information system, user satisfaction, and service quality (Mazlan et al., 2016). The researchers had concluded that information quality, system quality, and information system user satisfaction have been the main contributors to the net benefit among the employees of Malaysian public sectors (Mazlan et al., 2016).

Also, Ridzuan and Amiruddin (2004) had added that the employees who have knowledge, competencies, and skills should be encouraged to use HRIS for knowledge sharing/transferring by initiating some of the following:

1. Empower employees by giving authorization on accessibility, creating, sharing, and use the information conveniently on any devices.
2. By integrating content management, tracking, and analyzing systems within the HRIS.
3. Encourage cultural change where people are more preferred to search information using IS because of easy accessibility, therefore motivating the employees to follow the organization's guideline and policy.

### 10.2.3.3.1   Case Study 3: Implementation of HRIS in a Pakistani Organization

A case study was focused on fast moving consumer goods (FMCG) sector of Pakistani organizations, consisting of HR professionals working in Pakistan. In total, 53 respondents were from the lower management, 40 from the middle-level management, and 6 respondents from the higher management (Khan et al, 2017).

The results from the case study claimed that the significant advantages of utilizing HRIS are "quick feedback and access to information" and "reduction of mistakes while using HRIS" as backed by several similar researches from Ngai and Wat, Bamel et al., and Altarawneh and Al-Shqairat. However, "improving employment services" is the benefits the least contradicting to Bamel et al.'s study on the usage, benefits, and barriers of HRIS in universities. It seems that "helping to make informal decisions" is not a benefit of using HRIS on contrary to Beadles et al.'s study, which claimed that decision-making is helped by the improvement to usefulness of information effect (Khan et al, 2017). Therefore, it can be concluded that Pakistani companies use HRIS mostly for these mentioned benefits as supported from this case study and previously mentioned researches.

Employee involvement and commitment as well as resistant to change of employees are some challenges during the implementation of HRIS. In addition, the "difficulty in changing the organization culture" may or may not pose as a challenge to the corporates. The results from the study had suggested that "inadequate knowledge and lack of expertise in IT" and "unavailability of suitable HRIS software" are insignificant reasons where employees are expected to be IT savvy in today's working environment (Khan et al, 2017).

Additionally, Ahmer indicated that top-management role has to be taken into account in order to be committed to the implementation of HRIS within the company. Occasionally, higher management may view the implementation of HRIS as a financial burden if it is proposed by middle or lower management. However, it would be the opposite if the higher management themselves endorsed the HRIS proposal to run their operations in Pakistan smoothly. Currently, companies can implement HRIS effortlessly according to their respective and specific requirements by outsourcing to the MIS developer (Khan et al, 2017).

According to this case study, database of employee details, monitoring employees' absences, appraisal of performance, and job evaluation and work design were the most significant HRIS applications being used. Meanwhile, Bamel et al. and Altarawneh and Al-Shqairat agreed that training and development application was least used by the employees (Khan et al, 2017). Thus, from their findings, it can be concluded that majority of the Pakistani companies used HRIS for the purpose of managing, monitoring, and tracking information of employees, employees' absence, appraisal of employee performance, job evaluation and work design, and lastly employment reward (Khan et al, 2017).

### 10.3 RESULT AND DISCUSSION

This chapter was intended to improve the researchers' knowledge and understanding of the relationship between HRM and IS/IT in an organization, specifically on one of the well-known trends of HRM, the employee involvement. Our study is based on several researches and case studies mainly available online. The results of our analysis had showed a number of HRIS benefits as well as its limitation. Although several limitation and challenges had to be confronted by the management team during the implementation of HRIS application, organization had decided to spend a large amount of their finances to reengineer their current work processes into better MIS. Firstly, the researchers will discuss the result of analysis of employee involvement in HRM during the implementation of HRIS which are presented in the Tables 10.1 and 10.2.

**TABLE 10.1**   Summary of Advantages of HRIS on Employee Involvement.

| Advantages | References |
| --- | --- |
| Engagement employee input | Bullock and Powell, n.d |
| Direct participation | Bullock and Powell, n.d; Inanc et al., 2015 |
| Opportunities to participate in decision-making | Duran and Corral, 2016; Bakotic and Rogosic, 2015 |
| Develop innovative ideas | Kokemuller, n.d |
| Better employee commitment | Kokemuller, n.d |

**TABLE 10.2**   The Limitation and Issues During the Implementation of HRIS in Relation to Employee Involvement.

| Limitations | References |
| --- | --- |
| The absence of lines of boundary between management level and employee level may lead to inconsistency in decision-making | Kokemuller, n.d; Nabi, Syduzzaman and Munir, 2016; Nabi, Syduzzaman and Munir, 2016; Duran and Corral, 2016 |
| Lack of employee motivation | Duran and Corral, 2016 |
| Neglected employee well-being | Duran and Corral, 2016 |
| Lack of learning opportunities (e.g., e-learning) | Duran and Corral, 2016 |
| Poor work and employment conditions | Duran and Corral, 2016 |
| Technological risks and inadequate security measures | Rietsema, n.d |

However, as explained by Duran and Corral (2016), their research had shown a strong evidence about the impact of employee involvement in an organization. He had concluded that there are several limitation and challenges that the organization needs to overcome in order to increase employee involvement. There are several ways that had been suggested to increase the employee motivation, for instance, by supporting employee well being, giving opportunities for them to attend trainings, and developing a good work and employment conditions (Kokemuller, n.d; Nabi et al. 2016; Nabi et al., 2016; Duran and Corral, 2016). But opposing Duran and Corral statement, Mazlan et al. (2016) had said that their research had shown a direct positive relationship of HRIS toward employees' involvement because it had increased their satisfaction, quality of information system, as well as service quality. Similar to Marler and Fisher's findings, e-HRM might be having least benefit in making positive change for HR, but its implementation within the organization had been proven that it can enhance the welfare of employees for the quality of information, services, and communication medium. Nevertheless, more studies should be carried out to investigate the success of this suggestion as employees have different preferences and needs depending on their working environment.

As mentioned earlier by Ridzuan and Amiruddin (2004), knowledgeable and skilled employees were encouraged to use HRIS for the purpose of knowledge management and content management. One of the reasons for this encouragement is to improve the quality of information of HRIS so that other employees can retrieve and track better information in order to make a better decision-making. The knowledgeable and skilled employees are authorized to access, create, share, and use the information in HRIS conveniently on any devices. Because it is easier to access, it can increase the employees' motivation but of course with proper organizational guidelines and policies.

An encouragement was also given to employees who had wider knowledge, competencies, and skills to share/transfer it into the HRIS system based on the following initiatives. First, Ridzuan and Amiruddin (2004) had recommended that employees should be empowered by providing a single access to use HRIS tools (such as analytical tools), data analysis (which can be accessed via smartphones), as well as other necessary tools such as calendar, what-to-do list, and email. Second, the content management as well as tracking and analysis system will provide an easier and more user-friendly interface to manage the specific information, and ensure that the information is well-secured and authenticated by the HRIS system (Ridzuan and Amiruddin, 2004). The last suggestion made by Ridzuan and Amiruddin (2004) was related to technology which was predicted to assist HRM in changing the traditional culture into more convenient system where data accessibility is simpler and the retrieval of information with reports or statistics format can be produced immediately. People would easy to accept the new system if they can see the remarkable benefits, and accept the new operations if they feel that the operations are easier to use/follow (Ridzuan and Amiruddin, 2004).

According to Obeidat (2012), in order to have the control on personal data, HRIS are created encompassing the knowledge management of the HRM. The organization will become more commercial and compelling in compared to others. Argryis et al. claimed that HRIS has made critical endowment to knowledge management by facilitating response actions among the employees which enable the organization to exchange ideas and discuss.

As mentioned by Bakovic and Rogosic (2015), in the concept of *delegation*, employee will have more chance in controlling and making choices freedomly. It improves the communication gap between the workers and the management. Apart from that, through the involvement in decision-making, problem-solving and participation in information processing lead to develop the psychological ownership among organization members.

Meanwhile, employee *empowerment* means employees are able to make decisions about their work and take responsibility for the results. It allows the employees to ruminate, perform, control work, take action, and make decision independently. By empowering employees, it helps to increase job satisfaction level, improves work quality, creates cooperative and productive teamwork, increases employees' loyalty and self-confidence, helps the organization in reducing cost by implementing employees' ideas, and enables organization to react more rapidly to the changes in the market (Bakotic and Rogosic, 2015).

As argued by Inanc et al. (2015), the research seeks to find out whether advanced technology settings or with higher employee skills will drive greater involvement in employee participation in the organization. Apparently, it has been stated that with today's advanced technology and system and skilled job scope, both key drivers had improved employees' learning opportunities to work in a team and participate in achieving organization goals. The effect of advanced technologies and skill level provide direct participation for learning processes that involved advanced technology which can easily monitor the process of learning (Inanc et al., 2015).

Since Bonito et al. and Holpp argued that skilled workers are more beneficial and have a greater level of involvement, it is wise to expect for the organization to increase number of *trainings* for these employees as it would initiate more employee involvement (Bakotic and Rogosic, 2015). These trainings are important to enhance their knowledge, skills, and performances as well as to engage them with their day-to-day work routines. However according to the case study of FMCG sector in Pakistan as mentioned in the literature review, training and development application of the HRIS was the least used application by the employees as stated by Bamel et al. and Alta-rawneh and Al-Shqairat (Khan et al, 2017). Thus, this contradicts with the previous point that employees are not undertaking trainings that are within the HRIS. The actual reason behind the lack of usage in the training application in the Pakistani context is unknown and the case study cannot be generalized to society at large since it is conducted in Pakistan only, specifically in the FMCG sector.

Bakotic and Rogosic (2015) stated that the organization should attentively focus more on learning and training in which it will let the employees to understand the whole process of the system and not restraining them from learning individually. This also can be part of their competitive advantage. Furthermore, training also can help the employees to obtain new expertise and knowledge to be utilized in the organization.

Aside from using old traditional methods of learning and training, e-learning system has been introduced by HRM that enables employees to enhance their skills and knowledge through the system according to individual preference. According to Wang et al. (2007), information systems have received much attention among which has been conducted to assess the achievement of e-learning systems in the perspective of an organization (Wang et al., 2007).

It is undeniable that resistance to change and company culture according to various scholars and researchers mentioned in the literature review play a significant role especially in barriers to the implementation of HRIS. Employees may perceive that the systemization of HR-related functions and duties will risk their employment positions since the HRIS is capable to take over their roles as compared to their initial traditional approach. The organization may favor to downsize their HR department as a result of the implementation of HRIS. Furthermore, if the corporate culture has been conservative in nature, it would pose a challenge to both higher and lower management as they are already accustomed to executing their duties in a traditional approach.

According to case study 1, a top-down approach is advisable toward implementing HRIS in the organization in order to involve the employees from the beginning. The top management is deemed to be the main influencer toward the acceptance of the new HRIS in order for employees to follow suit and provide positive support toward the implementation. The top management has to lead by example and not solely propose the implementation and leaving the lower management levels to figure out about the technicalities of HRIS (Razali and Vrontis, 2010). Therefore in order to ensure positive commitment from employees in general, they should be involved from the very beginning from pre- to postimplementation of HRIS.

Due to the complexity of the HRIS, employees may not have proper practical knowledge regarding its functions and the usage of it if they are not given training about or proper walkthrough with HRIS. Thus this barrier would result in lower levels of commitment with lower management and senior management (Parasuraman and Colby, 2001; Khan et al, 2017). However, this would depend on the organizational culture as mentioned in the previous paragraph as it would differ with an organization which practices openness to technology. It would be advisable for organizations to practice transparency when there is a proposed HRIS implementation by having a dialogue session with the whole organization in order to gather feedback on the opinions of employees and the specific needs that the HRIS has to fulfill in the company. It is crucial that higher and lower management at all levels of the organization are aware and knowledgeable about the implementation of HRIS and the technicalities involving it, either through workshops or seminars by the MIS developer as highlighted by Ahmer (Khan et al., 2017) and Mazlan et al. (2016). However, according to Khan et al. (2017), the employees have adequate knowledge and expertise in IT in a Pakistani context and thus, implementing HRIS was not a hassle for their case. On the other hand, one cannot neglect the fact that employees in today's world are not all IT literate and that specialized trainings as mentioned earlier are essential in order for employees to utilize the HRIS with ease.

In terms of finance-related matters, the insufficient financial support would have to be considered especially if the company is not operating on a large scale for, for example, small and medium enterprises (SMEs). According to case study 3, Ahmer argued there is a tendency of bias when allocating a financial budget to implement HRIS. It appears that higher management will have a negative perception if the similar HRIS is being proposed by middle or lower management, whereas they themselves are able to without restrictions (Khan et al., 2017). An organization should practice openness to ideas and appreciate feedback internally especially by employees in order to achieve a conducive and positive working environment. In addition, this will indirectly or directly contribute positively toward the organization's goals which the higher management should realize. Therefore, an organization should consider thoroughly the specific needs that HRIS can fulfill and provide for the overall benefit of the organization and its employees. In addition, the pros and cons of a certain HRIS have to be compared with other HRIS in order for the organization to invest strategically since the HRIS would be utilized in a long-term basis and not a one-off initiative.

As mentioned in the literature review, in most of the cases, HRIS can consist of one or more modules depending on the organization and employees' needs. The cost of HRIS is also depended on the modules and applications of the HRIS. After the implementation of HRIS, any modification or improvement may cost the organization a considerable amount of budget. Therefore, it is advisable if all parties, top management, middle management as well as employees work together to make the best decisions on the functions of their HRIS and its development. Apparently, Dorel and Martinovic (2011) had suggested that to make the HRIS cost effective in a long run, HRM teams must assess their needs carefully and choose the software or applications which can be fully utilized by the employees.

Based on the study investigated by Mohapatra and Patnaik (2011), the researchers also concluded that integration of HRIS had helped employees to minimize the time wasted. HRIS provide tracking system and filing system where employees can easily access important information quickly and conveniently, compared to the traditional methods where they need to retrieve the information from the physical filing system which was very time consuming (Mohapatra and Patnaik, 2011). HRIS can include a reminder, where employees are being reminded to attend a meeting, trainings, or even deadlines of their assignments. Employees are motivated to make decisions based on the quality information obtained from HRIS. Now, within HRIS applications, it may include training courses which sometimes referred to as e-learning, where employees can easily access and attend the assigned training from their homes or from their own office desks using their personal computer or smartphones which are connected via web. It had been proven that trainings can improve staff morale, motivation, and satisfaction.

In a global context with employees of different culture and backgrounds, HRIS can be used to decentralize HRM such as the example given in the literature review about the limited working time of Indian female employees as late as 8 p.m. Instead of staying back late at night to do her work, the female employees could access the HRIS system via web, her smartphone, or personal computer from home after office hours. Thus, this decentralization is beneficial to the employees because it support them by taking care of their needs, well-being, and working conditions, which can motivate and empower them to make their own decision independently, besides gaining their trust and respect. As for the organization, the decentralization of HRM into HRIS will ensure that company policies adhere to respective legislations related to employees' wellbeing (Bakovic and Rogosic, 2015; Mohapatra and Patnaik, 2011).

Other ways that the researchers can suggest to tackle the challenges and limitation of HRIS implementation toward employee involvement are by giving additional training courses online or offline where employees can access is using the self-service module at their own convenience. Pre- and posttraining courses should be provided during the implementation of HRIS to increase employees' acceptance on the system.

Not to forget, the management teams must frequently communicate with employees to show their support and concern either via HRIS dashboard, email, or communication application (e.g., chat space). The employees must not feel pressurized by the management because this may lead them to be resistant to change their mindset and way of doing works. Also, frequent communication between management level and employee level can enhance their relationship, respect, and mutual trust. This might be one of the best solutions to the issue which had been mentioned earlier by Duran and Corral (2016), to improve employee motivation, taking care of the employee well-being, and create favorable work and employment conditions.

HRIS should be prone to technological risks by keeping information secure throughout the implementation. A good HRIS application should allow its users to login from any devices. To tackle this issue, the management team must thoroughly and frequently revise their IT policies and infrastructure in order to maximize the HRIS security measures which will be elaborated more in the following discussion.

Based on the observation and analysis by Nawaz (2012), it has proven that the flow of information works well with the use of HRIS. With the use of the system, it automatically enhances the speed in accessing data and information, regardless the employee is at the office or not. In other words, employees access any information easily. However, limiting the access to information is also important especially when dealing with confidentiality. With the incline use of the cloud-computing, HRIS can authorize users to logon from any devices; hence, security issue is highly upsurge. Therefore, during the configuration stage of implementation, the superiors including the managers must be cautious about limiting the system and access control to information. Employees need to have permission if any modifications or adjustments needed to be done and such employees are only allowed to access their own personal information and not others. This is to prevent from breaches of information and identity theft from within the organization itself or external.

As recommended by Rietsema (n.d), one of the ways that HRM can do to maximize the security of HRIS is by increasing the security protocols. The management teams must first brief and train their employees on how important it is to protect their organization against external and internal security breaches. Rietsema (n.d) had mentioned that lax in internal security protocols has been one of the frequent causes of this issue that organization needs to tackle comprehensively. Employees usually lose their trust and engagement if they fail to understand the processes and how secure it is to share the information within the system. By providing intensive training on both employees and managers, it can help the HR to clarify the security protocols, help the management level to gain the employees' trust and engagement as well as increase their involvement on the HRIS adoption.

Rietsema (n.d) had also raised his concern choosing the best HRIS based on its vendors. The HRIS vendors must submit their adequate security measures during tendering process before the management level and IT experts verify and decide to purchase the HRIS system. Another recommendation that Rietsema (n.d) had proposed is to enforce their employees to change their password frequently because it is a healthy habit to maintain a secure information. Similarly, password is an example of supplementary security precaution that preserve confidential information from severe threats. Employees are discouraged to create an easy password for their personal identification to access the system which can be predicted easily by others. Some organization had even enforce their employees with IT policies to change their passwords as regularly as every 6 months, if not, they cannot have access to the HRIS system and further identification with IT experts will be needed to activate and retrieve their personal accounts.

The employees sometimes face issues while using the HRIS system such as system breakdown and being unable to use the system due to poor connection of Internet service. Therefore, it is important for an organization to enforce recovery plan policy by providing backup system or secondary recovery options to avoid sudden interruptions during their access to HRIS system (Rietsema, n.d). It is to ensure that sensitive information especially about employees are always secure and kept confidential solely to HR department or personnel handling the HRIS.

Employee appraisal has to be given a thorough consideration as reviewing a certain employee's performance could have a double-edge sword effect. Negative remarks about a certain employee would prompt management and the employees themselves to perceive negatively on how their performance is being reviewed. This would cause employees and even managers themselves to feel that their position is being threatened as a result. However, depending on the approach from both employee and management level on the employee appraisal, it can be looked from a positive perspective that it is an opportunity to improve on themselves continuously and not let negativism to be continued, rather be rectified together as an organization and working toward their mutual corporate objectives.

However, this study could only find out the similarities between public and private sectors under the benefits of HRIS, its challenges, and the solutions which had been done in both sectors. Further studies should be carried out in the future to determine whether there are any differences that had been faced by both sectors during the implementation of HRIS.

## 10.4   CONCLUSION

In this era, the view of employee has powerfully determined as one of the fundamental element in the point of competitive advantage in today's competitive markets. This is because they are the creators of innovations and support the organization to sustain its competitiveness. Each organization has different competitiveness as their employees have their own unique strengths. Thus, in order to improve the quality of the organization and increase in productivity, organization's HRM are strongly searching for employees who are highly proficient and enthusiastic in working. In addition, after hiring or with the existing employee in the organization, employee involvement has gradually also becomes part of an important element for them to survive in the competitive market.

Employee involvement creates an environment for employees to have the power in making decisions and actions toward their work. It is not just a goal, but has become an essential element to practice in most of the organizations. Rather, it is a guidance viewpoint about how people are facilitated on ongoing success and contribute to the continuous improvement of the organizations. Aside from that, it also increases the approachability and improvement of the employees in the organization to become more efficient and success, especially, with the concepts of goal setting, delegation, participative management, empowering of employees, work teams, and employee training—contribute directly to employee involvement success. Thus, involvement of HRM enables employees and managements to have good training, empowering them, and offering them sufficient information and rewards as an opportunity for continuous process improvement. It has been shown that HRM has an important role to play in employee involvement.

There is a raising demand of information systems in today's economy. A number of large organizations are going for such huge implementations in the support of employee involvement by HRM. This research through the case studies shows that implementing the system allows employees to involve with the system in working as a team together with the top management and has the positive impact in improving work task in the organizations. Aside from that, it also illustrates that there are some challenges during the implementation of HRIS in employee involvement.

To conclude, the researchers have agreed that benefits of HRIS have close relationships toward employee involvement processes. HRIS had improved employee's satisfaction and motivation by automating the HRM operations, improving its operational transactions, and involving employees in the development and implementation of HRIS. It can increase the feeling of ownership among the employees, thus increase their involvement in organizational processes and activities.

## KEYWORDS

- company's competitive advantage
- decision support
- appraisal application
- recruitment application
- selection application
- job analysis application
- communication application

## REFERENCES

Agyeman, R. D. An Investigation into the Effect of Employee Involvement Practice on Decision Making Process: A Case Study of Kumasi Anglican Senior High School. Master Thesis, School of Graduate Studies, Kwame Nkrumah University of Science and Technology, 2012. http://ir.knust.edu.gh/bitstream/123456789/7392/1/=--Richard%20Duah%20Agyeman.pdf (accessed June 2019).

Bakotic, D.; Rogosic, A. Employee Involvement as a Key Determinant of Core Quality Management Practices. *Total Qual. Manage. Bus. Excell.* **2015,** *28* (11–12), 1209–1226. http://dx.doi.org/10.1080/14783363.2015.1094369

Baroudi, J. J.; Olson, M. H.; Ives, B. An Empirical Study of the Impact of User Involvement on System Usage and Information Satisfaction. *Commun. ACM* **1986,** *29* (3), 232–238. https://www.researchgate.net/publication/220420410_An_Empirical_Study_of_the_Impact_of_User_Involvement_on_System_Usage_and_Information_Satisfaction

Bondarouk, T.; Ruel, H. Electronic Human Resource Management: Challenges in the Digital Era. *Int. J. Human Resour. Manage.* **2009,** *20* (3), 505–514. http://dx.doi.org/10.1080/09585190802707235

Dorel, D. Martinovic, A. B. The Role of Information Systems in Human Resource Management. *MPRA Paper* **2011,** *35286* (12). https://mpra.ub.uni-muenchen.de/35286/

Duran, J.; Corral, A. Employee Involvement and Participation at Work: Recent Research and Policy Developments Revisited. *Eurofound Articles.* 2016. https://www.eurofound.europa.eu/observatories/eurwork/articles/working-conditions-industrial-relations/employee-involvement-and-participation-at-work-recent-research-and-policy-developments-revisited (accessed Aug 23, 2017)

Gardner, S. D.; Lepak, D. P.; Bartol, K. M. Virtual HR: The Impact of Information Technology on the Human Resource Professional. *J. Voc. Behav.* **2003,** *63* (2), 159–179.

Inanc, H.; Zhou, Y.; Gallie, D.; Felstead, A.; Green, F. Direct Participation and Employee Learning at Work. *Work Occup.* **2015,** *42* (4), 447–475. http://dx.doi.org/10.1177/0730888415580650

Khan, H.; Hussainy, S. K.; Khan, K.; Khan, A. The Applications, Advantages and Challenges in the Implementation of HRIS in Pakistani Perspective. *VINE J. Inf. Knowl. Manage. Sys.* **2017,** *47* (1). http://doi.org/10.1108/VJIKMS-01-2016-0005

Kokemuller, N. (n.d.) Advantages and Disadvantages of Employee Involvement. http://smallbusiness.chron.com/advantages-disadvantages-employee- involvement-21399.html (accessed Aug 20, 2017)

Marker, J. H.; Fisher, S. L. An Evidence-based Review of e-HRM and Strategic Human Resource Management. *Human Resour. Manage. Rev. Elsevier* **2013,** *23,* 18–36. DOI: 10.1016/j.hrmr.2012.06.002.

Mazlan, M. S.; Ahmad Suhaimi, B.; Raman, S. The Moderating Effect of User Involvement and Self-readiness and Factors That Influence Information System Net Benefits Among Malaysian Public Sector Employees. *Int. J. Appl. Eng. Res.* **2016,** *11* (18), 9659–9673. www.ripublication.com

Nabi, M. N.; Syduzzaman, M.; Munir, M. S. The Impact of Human Resource Management Practices on Job Performances: A Case Study of Dhaka Bank Pvt. Ltd.; Bangladesh, *Human Resour. Manage. Res.* **2016,** *6* (2), 45–54. Doi: 10.5923/j.hrmr.20160602.03.

Mohapatra, S.; Patnaik, A. Sustainability in HRIS Implementation Through Effective Project Management. *Int. J. Project Organ. Manage.* **2011,** *3* (1), 78–90.

Nawaz, M. N. To Assess the Impact of HRIS in Facilitating Information Flow Among the Select Software Companies in Bangalore, India. *Res. J. Manage. Sci.* **2012.** ISSN, 2319, 1171. https://www.researchgate.net/publication/271132321_To_assess_the_Impact_of_HRIS_in_Facilitating_Information_Flow_among_the_select_Software_Companies_in_Bangalore_India

Obeidat, B. Y. The Relationship Between Human Resource Information System (HRIS) Functions and Human Resource Management (HRM) Functionalities. *J. Manage. Res.* **2012,** *4* (4), 192–211. Doi: 10.5296/jmr.v4i4.2262.

Parasuraman, A. P. Technology Readiness Index (TRI): A Multiple-item Scale to Measure Readiness to Embrace New Technologies. *J. Serv. Res.* **2000,** *2,* 307–320. Doi:10.1177/109467050024001.

Parasuraman, A.; Colby, C. L. Techno-ready Marketing: How and Why Your Customers Adopt Technology. Free Press: New York, 2001.

Razali, M. Z.; Vrontis, D. The Reactions of Employees Toward the Implementation of Human Resources Information Systems (HRIS) as a Planned Change Program: A Case Study in Malaysia. *J. Transnatl. Manage.* **2010,** *15* (3), 229–245. Doi: 10.1080/15475778. 2010.504497.

Rietsema, D. (n.d). Keeping Information Secure Through HRIS Implementation. https:// www.hrpayrollsystems.net/keeping-information-secure-throughout-hris-implementation/ (assessed Aug 28, 2017).

Ridzuan, D.; Amirruddin, K. In *The Impact of Information Systems on Organizations in Malaysia: Knowledge Worker Aspect.* WISICT'04 Proceeding of the Winter International Symposium on ICT, 2004, 1–6. (Retrieved from ResearchGate Journals Database).

Stone, D. L.; Stone-Romero, E. F.; Lukaszewski, K. Factors Affecting the Acceptance and Effectiveness of Electronic Human Resource Systems. *Human Resour. Manage. Rev.* **2006,** *16* (2), 229–244. https://doi.org/10.1016/j.hrmr.2006.03.010

Storey, J. New Perspectives on Human Resource Management. 2014. https://books.google. com.bn/ (accessed Aug 21, 2017)

Strohmeier, S. Research in e-HRM: Review and Implications. *Human Resour. Manage. Rev.* **2007,** *17* (1), 19–37. Doi: https://doi.org/10.1016/j.hrmr.2006.11.002.

Wang, Y.; Wang, H.; Shee, D. Measuring E-learning Systems Success in an Organizational Context: Scale Development and Validation. *Comput. Human Behav.* **2007,** *23* (4), 1792–1808. http://dx.doi.org/10.1016/j.chb.2005.10.006

Weeks, K. O. An Analysis of Human Resource Information Systems Impact on Employees. *J. Manage. Policy Pract.* **2013,** *14* (3). http://digitalcommons.www.na-businesspress.com/ JMPP/WeeksKO_Web14_3_.pdf

# CHAPTER 11

# HUMAN RESOURCE GLOBALIZATION AND MIS IMPLICATIONS

## ABSTRACT

Advancement of technology and knowledge over the years is so prodigious that the world has entered a new era of economic development. Globalization is defined as opening up of global economy with increase in integration of economies around the world through interaction with people, companies, and governments that are driven by international trade and investments aided by emerging information technology. The current state of globalization constitutes impacts on global markets and taken shape on the societies, governments, and organizational system which challenge will be faced by every country. While having numerous positive effects to an organization, technology alongside globalization also has changed the way HRM operates within an organization. HRM employees have become the primary source of information in most of the organizations. Information can quickly and easily be accessed and communicated via several channels of communication like websites or social media. Human Resource Information Systems (HRIS) now allow HRM employees to better utilize human resource (HR) plans, make decisions faster, clearly define jobs, evaluate performance, and provide cost-effective benefits that employees want. Technology helps to strengthen communication between the external community and employees.

## 11.1 INTRODUCTIONS

### 11.1.1 GLOBALIZATION

Globalization has started many years ago where the people and corporation made a trade to each other and invested in another country. Advancement of technology and knowledge over the years is so prodigious that the world has entered a new era of economic development. Economic growth enables countries to increase world trade and foreign investments at a faster rate and at farther places. As Scholte (1995) stated that "globalization stands out for quite a large public spread across the world as one of the defining terms of late twentieth-century social consciousness." Globalization is defined as opening up of global economy with increase in integration of economies around the world through interaction with people, companies, and governments that are driven by international trade and investments aided by information technology (International Monetary Fund, 2008).

## 11.1.2  HOW TECHNOLOGY AFFECTS HRM

While having numerous positive effects to an organization, technology alongside globalization also has changed the way HRM operates within an organization. HRM employees have become the primary source of information in most of the organizations. Information can quickly and easily be accessed and communicated via several channels of communication like websites or social media. Human Resource Information Systems (HRIS) now allow HRM employees to better utilize human resource (HR) plans, make decisions faster, clearly define jobs, evaluate performance, and provide cost-effective benefits that employees want. Technology helps to strengthen communication between the external community and employees.

## 11.1.3  MANAGEMENT INFORMATION SYSTEMS

Management information system (MIS) is a type of information system that incorporates hardware, software, and system in a joined arrangement and supplies information to the managers in a layout that is appropriate for analysis, observing, decision-making, and documenting (Asemi, 2011). These days, MIS is considered to be one of the very important functions of management. The system consists of information about individuals, documents, technology, venues, and other affairs within the organization as well as the surrounding environment (Nimani, 2010). The MIS acts as the role of information production, communication issues, and assists in the decision-making process. Therefore, MIS plays an important part in the management, control, and the day-to-day activities of an organization. For example, the MIS in human resource management (HRM) have numerous subsystems that take care of daily operations such as employees' registration and finances.

MIS works as a system that converts data from both internal and external sources into information and to communicate that information back in an appropriate form to the managers at all levels and in all functions. This will enable the managers to make timely and effective decisions for planning, directing, and controlling the activities for which they are responsible. However, despite the assumption that the MIS plays an important part in helping the business, the real value of MIS can be determined from how much it assists the managers in managing their work.

### 11.1.4   HRM AND ITS FUNCTION

HRM is one of the important departments that are used to identify a system in managing all employees' affairs in an organization. In order to manage the employment, HR department divided itself into many functional areas. However, according to Edward El Gubman (Impact, 1996), there are three major and basic functions in HR department, namely, recruitment, compensations and benefits, and job designing. These are the functions that will meet the basic needs of the employees to be hired.

Since revolutionizing of globalization, it brings a change to employability. The organization does no longer look on the basis of productivity, but the capacity of an individual to contribute knowledge and skills to the workforce (Chand, 2016). The HR will work together closely with management to identify and ensure whether employed staff are able to utilize their skills in leading management to achieve its goals. Therefore, HR's objective is to increase efficiency and prosperity of a company. In meeting with its objective, HR ensures retaining competency through the personal and operational development of the employees following globalization. It evades overseeing the employees as well as controls conflict in working environment conditions such as the development of technology, workforce demographic, and social issue.

### 11.1.5   CORRELATION BETWEEN GLOBALIZATION AND HRM

Globalization is an operation that gathers people from all around the world into a group and connects them through communication technology. Globalization allows HR managers to choose employees from a larger market in order to find suitable people for positions in the company. HRM plays an important role in making these decisions more effective and efficient.

Kapoor (n.d.) once mentioned that global organizations are making use of their data to produce solutions needed rather than depending on their own instincts. Therefore, the HR department of these organizations gathers data and information into data warehouses and data marts. These information includes the worker's details, staffing, and their welfares. By evaluating the past and current information, the business will be able to produce fact-based decisions.

## 11.2   LITERATURE REVIEW

### 11.2.1   FACTORS DRIVING GLOBALIZATION

Globalization provides the business with an opportunity to start a venture and expand businesses internationally. The empirical ideas of globalization are driven by many factors. Businesses' most fundamental purpose is to achieve economies of scale and higher marginal profits.

The primary factors which business will get benefits from are land regulations, taxations, and exchange rate. Business is required to meet specific products standards, health, and safety. A regulation and low taxations that sound promising to a business also influence on their profits, while an attractive exchange rate promises business gain a bargain in production (Harvey, et al., 2010).

However, trade liberalization is one of the influential factors businesses generally benefits. Trade liberalization creates openness in trade barriers that indicate reduction or removal of tariffs such as taxes for investors in embarking into a free-trade country. New Zealand is known to be a free-trade country which attracts foreign investors including investors from developing countries (Kitchen and Eagle, 2002). This agreement allows business to be able to produce a product at lower cost, and therefore, declining per unit cost means investors will find leverage in producing better output and performance.

With the world economy integrating together along the years of trading, new advancement has been constantly upgraded. The largest advancement took place in the 20th century over a wide range of activity (Encyclopedia Britannica, 2017). According to Harvey et al. (2010), technology has made a significant contribution to revolutionizing around business activities. The discovery of new systems, telecommunication, transportation, and equipment has changed the organizations' working technique. Businesses are aware of the change they have on their production and profits and therefore, it is inevitable for them to neglect the use of technology.

Development in technology has introduced the emerging of new services simultaneously. New jobs and companies are created and thus, there is a demand for specialist skills because of the change. As stated by Kitchen and Eagle (2002), the nature of business will place a business dependence on the access of resources other than profit aims. By having the right access to resources, business has a higher chance of success. Such as the health care business is looking for an access to better institution and facilities of a country so that its probability to conduct an experiment successful is high and this will reflect on the overall performance of the business (International Monetary Fund, 2008).

## 11.2.2   CHALLENGES OF GLOBALIZATION FACED BY HRM SYSTEM

In today's organizations, a global mindset is crucial, no matter where an employee is located. In the past, an organization's leaders, managers, and employees were needed to understand a culture only if they were going to another country to live. Now, employees work virtually across borders via technology, they work with a variety of ethnicities at home, and they interact with a globally dispersed customer base. So a global mindset and skills are necessary for all employees.

In spite of having multiple benefits, globalization still poses multiple challenges that need to be addressed. Several major challenges include:

### 11.2.2.1   HUMAN CAPITAL MANAGEMENT

#### 11.2.2.1.1   Employee Engagement (Employee Resists Change)

According to Mohan (2013), one of the Indian organizations has done a survey that indicates employee engagement (commitment) as one of the challenges that most organizations are facing.

Employee engagement is low because employee resists change and this is mainly due to the fact that different employees have different views or opinions. They have not adapted to change yet. Therefore, new systems will just make things even harder for some employee to cope. Technology illiterate might be an issue to some. As a result, employee engagement is weak.

#### 11.2.2.1.2   Employee Retainment and Recruitment

With the increase in number of companies around the world due to globalization, one of the challenges that an organization may face in HR is recruiting new employee to do specific task and this may be time consuming and may involve high cost.

Mohan (2013) also pointed the recruiting and retaining employee in an organization as another challenge. Retainment is difficult to HR managers as some skills that local employee have might not bring benefit to the organization and trying to find employee that can perform according to their standard adds more burden to the organization. Yet, company wants to retain talented pool of people. Changes in the workforce environment makes it hard to find labor and retain employee.

This somehow leads to bias or unfairness within the organization whereby an organization does not want to hire locals due to the lack of knowledge and skills; hence, local employees will encounter a crisis to find job rises due to local employees who are lacking in the skills,. It is due to the employers who are trying to find applicants who can perform the task accordingly. There are standards required by an organization.

In addition, more obstacles will be faced when entering a market where either the country is not well developed or emerging markets (Inonescu and Dumitru, 2011). This means that the HRM might send their employee from overseas in order to handle their workflow, this may cause a disagreement where the local employee might think that the HRM might be favoring their own staff from overseas than the host country.

### 11.2.2.1.3   Developing Leadership Capability

Workforce is the critical backbone—to face now and future challenges. Therefore, it is highly important to tackle the issues that involve the workforce itself. Another significant challenge to achieve great success in an organization is developing leadership capability. Changes in the process or regulations within an organization cause difficulty to demonstrate proper leadership skills as HR managers need to keep up every time.

### 11.2.2.2   TECHNOLOGY CHALLENGE

Technological change and advancement is said to be one of the most salient factors impacting organizations and employees today. With the increasingly developing use of technology, HRM practices are changing. HR managers must find some competitive advantage through technological advancement that will give them competitive edge over their competitors.

### 11.2.2.2.1   *Communication Breakdown*

Communication is important when it comes to workflow of the organization as language plays an important aspect of communication. However, due to globalization this might be jeopardized. Kapoor (2011) once mentioned that international organization needs to create an environment where workers can communicate easily in order to achieve a common goal.

However, a message can be misinterpreted despite using advanced technology. Mistranslation could occur. As people become more receptive toward the usage of social media as a platform to communicate globally, organizations will try to use the same as their way to communicate with the staff. This could be as a result of language differences as the world becomes more globalized.

### 11.2.2.2.2   *System Knowledge*

As the world becomes more globalized, organizations also become global-ized in order to be ahead of their competitors hence, when this occurs changes will be made without a doubt in the HRM practices and its system. Due to these changes, problems arises within the organization.

Since globalization improved technology, new systems emerge thus to learn and be able to use certain system is highly important. In order to use new technologies more efficiently, it usually requires literacy and a more specialized trainings or workshops. Dahlman (n.d) even added that for any organizations that are behind the technological frontier, acquisition of existing knowledge is expected to yield higher productivity.

There are several channels of technology and knowledge transfer. Direct foreign investment, licensing, technical assistance, importation of tech-nology as embodied in capital goods, components or products, copying and reverse engineering, and foreign study are the key channels to a successful knowledge acquisition.

On the other hand, the knowledge is also an issue behind communica-tion breakdown. It is because there is not enough factual data or information present and this leads an organization to capture right information that it can use to solve the problems.

### 11.2.2.3   *CONFLICT OF CULTURAL DIVERSIFICATION*

As the technology has made advancement in communication and transpor-tation, the network has made it easier for people to connect one another across other regions through a medium. Business can now transfer cultural ideas and knowledge with the access of wireless and satellite technology (Brown, 1999). Its transformation has shaped globalization into another level whereby cross-culture is influenced by not only one but both physical and virtual activities.

The influence of cross-culture in modern globalization is bringing in different values, norms, and languages from different cultures. Diversification in the culture will be a potential issue when business is unable to deal with understanding the differences. This is due to the people from different countries has a distinct meaning of taste, gesture, preferred color, treatment, age, attitude, and opinion. In some cases, it may develop into a sensitive issue in the context of ethnic, racial, economic, politics, language, and religious view when it is interpreted differently by a different country.

Study of other cultural values will be a challenge when business resists recognizing the difference in the people culture. It will lead to a clash in host country's cultural belief and practice. A business practice such as the marketing advertisement of its media concept of image, signage, colors, and gesture is well received in some countries but may be conflicting to the core value of host countries (Brown, 1999). For example, a cheek kiss in western business is considered to be a friendly gesture, but to conservative countries, it is an illegal intimacy.

When hosting a business in a foreign country, one of the challenges business will face is language. It is because the nuances of a language differ in style and content which it may mean something else in another country. In business, there could be a problem with translating one language into another when it is done by the person whose mother tongue is not their first language and therefore, it can alter or shorten the meaning in the translated version. This will confuse managements whether the company has presented them enough information. Language in status will also be a problem to a business where business email is done by addressing their first name or last name. The short ideas of addressing name in a formal style will be offensive and disrespectful to some countries such as in Germany, Italy, and Spain and hence, it brings sensitivity issue to local management (Holden, 2001). As a result, this will set up a barrier in delivering a proper message and communication.

Hurn (2013), stated that having a linguistic fluency of more than one or two foreign languages is an asset to a business. It gives business a competitive advantage in the international market as the business will not have a problem in socializing and engaging with the local people.

## 11.2.2.4   LAW AND POLICY

Understanding legal procedures, standards, and principles of a country is substantially important prior to entering the foreign market. It is the international, criminal, and civil laws and regulations such as code of conduct which a business requires abiding. The challenges of having regulatory differences between countries will determine whether the nature of the business is suitable to fit into their legal system even if it is an e-commerce based.

Living in the modern global market, businesses emphasize on using e-commerce such as website, social media, and other to share, transfer, and do online commercial trading. Internet provides a remote control of business for conveniently handling the task and trade internationally; however, when a business overlooks into the legal rights, it will face uncertainty risks of unknown legal issues such as copyrights and can be confined to punishment and court case. Because of the issue of the illegitimate use of contents and freedom of speech, the government is regulating the laws on nonenforcement of intellectual property rights, the legality of electronic contract, and legal liability of internet service providers (ISP) for third-party use (Frynes, 2002).

A policy is a governed method of doing right things to protect social rights. It is to ensure social compliance to the rules made by government in a place of a host country. The policy in the organization constitutes the documented rules that prevent from any mishap behaviors and activities which can be caused to an unethical issue. If business disobeys the rules hosted by the foreign government, it will face challenges of policy risk. (Frynas, 2002). Therefore, business needs to know whether selling and hiring workforce in the foreign country has offending laws such as racism, sexism, and indecent materials of civil and human rights.

## 11.2.3   IMPACTS OF GLOBALIZATION ON HRM SYSTEM

### 11.2.3.1   HR PROFESSIONAL COMPETENCE

#### 11.2.3.1.1   Employee Engagement (Resentment Toward New System)

Engagement is weak among the employees mainly due to several reasons rooting from the main obvious cause, that is, diversity. As Mohan (2013) once mentioned that in order for an organization to be able to manage engagement, the managements need to manage diversity. Diversity refers to the different group of employees within an organization that have different motivation, aspiration, and expectations moving toward achieving the same goals of the organization. Without employee understanding, an organization will encounter labor resentment and reluctance to change and therefore, conflict arises which can stop other production in the department that are interrelated to the organizational operation.

Everyone is motivated differently depending on the desire, needs, and wants within an individual that incline them to work in a specific behavior. Moberg and Leasher (2011) stated that motivation is an "internal psychological state that stimulates a person to engage in a particular behavior," it is logically argued that employee motivation will be directly related to their job performance. To have a motivated labor workforce, HR managers should provide a meaningful work place, allocate task and responsibilities, align intrinsic and extrinsic motivation values, and access to right resources as when employee becomes motivated, they will be more productive and engaged in their work. The right motivation will make employee to become more receptive toward new systems and therefore, less resentment toward new practices despite of globalization. Therefore, they will become more open to change.

Furthermore, diversity within an organization causes a mismatched opinion and this can create workplace ambiguity, confusion, and conflict. So in order to avoid confusions, misinterpretation, or mistranslation, HR managers should consult a specialist on the subject of corporate culture who would add value by providing a roadmap for a change (Moberg and Leasher, 2011).

### 11.2.3.1.2   *Implication in Response to Employee Retainment and Recruitment*

"Organizations becoming more and more globalize, the number of variables that must align for new organizational initiatives increases exponentially, and that makes mastery of change management more challenging" (Friedman, 2007). Therefore, it is crucial for HR managers to add value to the cultural sensitivity and understanding.

High performance employees are highly important for the growth of any organization. Employees with excellent job performances commonly demonstrate the capability to be the future generation leaders and be responsible for company's future business growth. Recruiting and retaining the key talent is very important to guide future business success.

As mentioned by Mohan (2013), HRM should provide genuine opportunities for career advancement because shortage of real opportunities for advancement and professional growth within organizations will only results in the feeling of stagnation.

Trade liberalization of bringing down the tariff has essentially given a chance for an organization to explore cheap labor and it also encourages bringing in the people of different skills into a country. It allows the market to choose employees among the pooling talent. This will increase competition among domestic countries as market size increases. People can be laid off upon freeing of country regulations. Hence, job security is becoming transparent. Incompetent workers will face struggle in finding and landing for a job. Therefore, HR needs to recruit manager specialist to undergo training with existing employee so that employee can have equal opportunity both in career advancement and knowledge acquisition to be secured in their position.

### 11.2.3.2   TECHNICAL KNOWLEDGE

#### 11.2.3.2.1   Implication in Response to Communication Breakdown

Information technology can bring numerous improvements to the organizations. In order to empower local management without losing necessary control, HR managers have to develop reliable channels for rapid communication between headquarters and overseas units so that business relays on the important information flow from overseas operations to the headquarters faster. Its gives the company a competitive advantage over its rival.

Communication facilitates HR to implement new strategic system in order to create a healthy organizational culture (Kipkoech, 2015). Therefore, communication is one of the very important strategic factors to be addressed.

#### 11.2.3.2.2   Acquisition of Knowledge

Change in the way a business works and introduce a new system use could raise some issues to the employees. Employees may come across unfamiliarity with the use of the new system and therefore, they will encounter a culture shock in the change that business has brought in. A drastic change in the system will take time for employees to learn and adapt to the change. The system may alter the culture of the business which it reflects in the process and the procedures. Thus, some employees may find it reluctant to learn because of the huge gaps. It will affect to the productivity of the output to slow down. Acquisition of knowledge will help both managers and employees to cope up with the changes occurring within the organization and prevent misuse of the system.

Hence, managers are required to identify the impact it will bring and its influences on the business activities. HR needs to create new capabilities to support global business on its process and procedure of the system adapted and come with strategies to tackle with the issues. While implementing these strategies HR has to ensure that the employees are flexible, managers have the learning capabilities, and the network is suitable for the business.

### 11.2.3.3   INCONSISTENT VALUES AND BEHAVIOR

Globalization creates an inconsistency in values and behavior because it combines cultures, languages, and perspectives together. And this majorly influences on strategic decision-making within an organization. A distinct different in cross-culture has impacted on the norms and belief of the people. Businesses need to be aware of the sensitivity content used in every aspect of the business operations (Frynes, 2002). Misinterpretation of message will affect the effective flow of communication, business function will be interrupted and thus, it decreases the productivity of an organization.

Local business practices, employment laws and HR practices, effective communication across nationalities, and gaining the trust of individuals with a diverse backgrounds and even languages must be comprehended by HRM professionals. The need to comprehend several content expertise, influence skills, and personal credibility must be taken into consideration.

### 11.2.3.4   LEGALITY AND FRAUD

#### 11.2.3.4.1   Security Exposure

HR managers need to become more knowledgeable about the legal issues occurring in all parts of the world that business is interested in. It is not only the law in the human rights that management is exposed to but also legal rules in the business. HR managers has to be familiarized with the legal issue and challenge in hiring employability because some of the laws are protecting the civil rights in work that managers must be aware of (Messmer, 2013).

Moreover, business has to look into the rules of using the system such as the terms and conditions of adopting a foreign system. Some of the systems used are vulnerable to the copyright issues. Managers who share sensitive information around the companies may have a privacy concern of information from leaking out to public. Server that is not highly safeguarded will be prominent to hacking by virus and this will let competitors to get away with the confidential information (Cardenas, n.d). So, managers have to be aware of using an open source due to its features being vulnerable of getting confidential information and selling it to their competitors (Microsoft, 2013).

### 11.2.3.4.2   Restrict Policy

A restriction in the business system put an obstacle to the business from negotiating and selling to other regions hosted in foreign country. The system they use in homeland may not be supported by the company.

## 11.3   RESULTS AND DISCUSSION

### 11.3.1   EFFECTS OF GLOBALIZATION/GLOBAL INFORMATION SYSTEM ON HR PRACTICE

The increase in trend of globalization has an impact to all the workplaces thus a lot of businesses are undergoing certain changes. However, due to number of changes some complication may occur for the good and the bad as mentioned in the literature review. In addition, the practices that the HR uses affect the major factors in the department that is shown below:

#### 11.3.1.1   INCREASE IN DIVERSITY WHEN IT COMES TO HIRING

As there is an increase in diversity of employees within the company, the HR department has to adopt new policies and guidelines that are suitable for all workers. According to McFalin (n.d.) when it comes to comparing different culture and mindset it is a very sensitive issue globally. This means that when creating an environment that is suitable for workers with different cultural background and ethics it needs to be equal and not contradicting to each other.

Therefore, in terms of recruitment using information system the HR department can use an E-recruitment system that can help assist in hiring new recruitments especially choosing candidate with better varieties. A case study of using the recruitment system is mentioned by Greiner (2004) using a recruitment system called "ORSEE" which the public can access and enter their personal data and the system automatically filters the candidates and then email the participants. This study shows that using this system automatically finalizes the proper candidate so there will not be any nonsense registration thus it is efficient instead of using the CV that needs to be checked manually before getting the candidate.

For HRM to assist in this system, the manager needs to have training in terms of using this system in order to understand the function thus they need to be more aware of the changes in recruitment and develop them systematically. Moreover, even though the system automatically recruits and selects the member automatically, however, it is the manager's final decision to finalize the recruitment.

### 11.3.1.2   VARIETY CULTURE WITHIN THE DEPARTMENT

As mentioned above where there is an increase in diversity there will be complications occurring, this is where training implicated by the HR managers plays a major part in order to prevent any discrimination and protects the employee welfare. In order to create the environment for all workers to work together, the higher management has to appreciate and understand the differences in culture thus providing training and understanding so that employees would understand one another (McFalin, n.d). Therefore, the companies will have a consistent work in a stress-free environment thus the employee can reach foreign standards.

In order to inform all the employees about this awareness and reach a certain standard, an online training can be applied such as their individual knowledge and teamwork. In addition, this system also can grade the employee performance.

For HRM to apply this system to their department, the managers will need training and conduct the evaluation annually in order to make sure everyone meet the standards that the managers have set. In addition, the managers should also be considering and understanding the difference in culture that should not be biased. As mentioned by Leider (2006), the management needs to reduce the conflict by contributing and reshaping IT values that can be recognized by the department. This statement shows that in terms of information system, the managers need to be proactive and more engaged toward the employees.

### 11.3.1.3    INCREASE TECHNOLOGY AND STANDARDS IN THE WORKFORCE IN DEVELOPING COUNTRIES

In order companies in developing countries to follow the international standards they have to keep up in terms of technology and the work mindset of their foreign competitors. Spero (1996) mentioned that undeveloped nations will fall behind by 10 years if they remain vigilant to global economy. This means that without any knowledge and training, local employees in foreign companies will face difficulties keeping up with international standards.

The mechanism that can assist the employee to use the new technology, the HR department can conduct their training using a seminar in terms of introducing technology. On the other hand when there is an update in technology that is not massive, the department can conduct an online course for them in order to update them.

For HRM to play a part in the training it is important to have the knowledge of the whole system so that managers can conduct sessions to assist the employee about using the new system, especially in the older generation and staffs that is IT iliterate.

### 11.3.1.4    CHANGE IN POLICY AND EMPLOYEE WELFARE

When mentioning about welfare, this includes the employee working condition, health and safety, and finally perks and benefits.

When it comes to the system used for the HR department, they can use the system such as "Kronos Workforce Ready" or "ClearCompany HRM." Both of these systems are available to the public to use in order to keep track on the employee payroll and even their welfare system. However, majority of the largest companies will use an in-house system that is not mentioned to the public.

As HR managers, they need to hire more locals in order to accommodate the host country policies and also this may benefit them as they might have an improvement in working conditions in a company which is globalized. In addition, it is recommended for the manager to research regarding the country policies before making the decision. Also, the organization will gain a competitive advantage in the host country in terms of hiring more specialists in their department.

### 11.3.1.5   INCREASE IN PROFESSIONAL DEVELOPMENT AND SPECIALIST

When talking about professional development it refers to improve current employees' skills and knowledge through education or training. As mentioned by Bradley (n.d.) some international companies provide scholarship in terms of education and training in order to further their current employee capabilities. This can increase the variety of skills that the HRM can employ and benefit for the company. As explained, in terms of globalization it is to provide local employees an opportunity to improve their capabilities that is not widely available in their country.

### 11.3.2   UNDERSTANDING GLOBAL MARKET

When business tries to set up business operations and process, business is required to consider global culture, law and policies, and other risks that are present that could ultimately cause problems for their own company.

Companies that are focusing on operating only nationally will find themselves with challenges such as limitation of resources. Whereas companies that are operating in the global market, they will have opportunities to increase their market share and customer base. Organizations that understands the impacts of the global market will be able to develop a good opportunity to successfully market their products and services to a larger market. Global marketing if done properly will have the capacity to propel the company's reputation. Global marketing thus pays a critical role especially for companies that cater to general goods and services for instance foods and books.

Nowadays global organizations have turned into a passageway for cultural understanding leading to the inflow of fresh new ideas, development of communication between governments as well as the enhancement of product and services. However, there are several factors that organizations must be able to understand prior to introducing themselves in the global market.

### 11.3.2.1   CULTURE

One of the main factors for companies to understand is the importance of language. When entering the global market, the company faces situations where the employees may need to conduct their business in a language that could be unfamiliar for them. Examples could include using that country's language when advertising and also meeting with clients from such countries. The language barrier could miscommunication thus making the message that needed to be delivered lost in translation. This causes customers from that country to be uninterested in the products or services the business has to offer.

Another way could be the difference in how the businesses in that country work. The pace and conditions for conducting business could be different. For example, big companies in the United States may have the tendency to push forward negotiation in order to not waste time, however, different countries may find creating relationship and networking are more important than just reaching an agreement for the business. Other countries may also consider body movement and hospitality rather than just figure of speech. Companies can cater to this problem with training. From there, companies are able to refrain from any etiquette mistakes and the use of local customs might build a better relationship when performing day-to-day activities.

### 11.3.2.2  LAWS AND POLICIES

Different countries operate with different set of laws and regulations. If the companies are not aware of the specific laws and regulation that a certain country needs to operate a business, this may lead to an abundant of paper works and problems which may causes difficulties it poses in daily operations of the business. It is advisable to be familiar about the laws and regulations at the countries that the company has been keeping tabs on beforehand. An example would be that the European countries have very strict laws regarding how we make use of our customer's information. With that, it is recommended that companies hire lawyers or people that are knowledgeable about that country's laws and regulations ahead of time to avert any problems that makes the companies vulnerable as well as taking a toll on the business time and finance.

Another problem may arise when companies are finding and working with suppliers or distributors located in that country's market the companies wishes to enter. Without prior research, there may be insufficient information to determine whether the companies are making a right choice based on ethical factors. Companies may also need help with marketing and advertising from the locals which require the companies to be working with them. These cases may also lead to run ins with the laws. Therefore, companies must be able to choose individuals who are trustworthy and qualified to work with.

Bribery can also be used as an example in this case. Different countries have different standpoints of what they consider bribery is. In some cultures, accepting gifts or favors which are related to doing business are considered normal. People from high-ranking positions may also accept these things which mostly lead to the influencing factor when making decision as well as conducting the business at an unethical way. Some countries have an accurate and fixed policies regarding bribery and protocols avoiding these circumstances to happen.

### 11.3.2.3   RISKS

An important fact that all organizations must be aware of is that the market is always unstable and unpredictable. There are numerous risks that could lead to consequences of making one's company vulnerable. Examples of such risk includes financial issues, unpredictable fluctuations, the country's political setback, changes in the country's policy, new or increasing competitors, the country's red tape, and so much more.

Take, for example, the financial factors. A crucial factor could include the desired country's currency or fluctuation rate. These factors can cause the companies to lose their earning. The problems that they may bring will impact on how the company will run its business in the future. For instance, companies may have the need to manage negotiations better with specifying some terms in the contracts made and also carefully planning future projects.

Social and political changes can also lead to instances affecting the way the company operates. Social factors play an important role when doing business. These days when choosing a product or a service to buy some customers have the tendency to choose companies that they feel should speak out to them and can cater their needs. Word by mouth plays an important role in securing customers for your business. Therefore, companies use social media to connect with their customers more and maintain that relationship. Another example concerning social views could be when the company conducts day-to-day activities in immoral ways such as human slavery. These things can lead to bad press and response from the public which may result in the downfall of the business. Political backlashes can also produce grave consequences. Take, for example, with what happened between Mexico and Trump. Trump gave out orders to build a border between Mexico and the United States. In return, the Mexicans are planning to boycott all businesses that originated from the United States which could lead to such business losses.

Companies can try to prevent these things from happening by doing a detailed assessment of the factors under consideration that could lead to the company being vulnerable. However, companies must also be more aware of the people that they are engaged in doing business with. The people that work or are associated with the company is likely to play a big role in the operations of the company. Therefore, companies must take extra precautions when thinking about expanding their business to the global market as there are many factors that they need to take into consideration.

### 11.3.3   MANAGEMENT INFLUENCE ACROSS BUSINESS ACTIVITY

The main purpose of the management is to help certain organization achieve its objectives and prevent any activities that will obstruct the efforts reaching those goals. The management therefore is given the task to assign people their tasks that will help the organization achieve its goals. This will affect the productivity and the organization's growth.

These days, the management must be able to accommodate different strategies to numerous challenges and opportunities as well the development in technology in order for the organization to continue operating. Due to the constant changes affecting the organization, the management must be able to have fast thinking and responses in order to allow the company to grow and actively compete with different organizations.

### 11.3.4   ADVANTAGES AND DISADVANTAGES OF ADAPTING GLOBALIZATION ON HRM SYSTEM

The HRM centralizes around managing the workers and how their choices will affect the worker's progress. Companies heavily depend on HRM system to make this possible; however, the HRM system is capable of producing both pros and cons that heavily will affect the company.

#### 11.3.4.1   ADVANTAGES OF ADAPTING TO GLOBALIZATION ON HRM SYSTEM

##### 11.3.4.1.1   Recruiting Employees

Due to the growth in globalization, the organizations of all sizes are communicating with their customers and stockholders from a wide range of different cultures, languages, and upbringings. As a result, numerous HR managers are selecting workers with various foundations. Organizations following this strategy will be able to benefit having individuals in the workforce that their customers can identify with, and realize that having a group of diverse people assist in providing opinions and impacts in an organization. Hiring can now be made online in which applicants apply through the business website. This hiring process can now be done both online and offline.

### 11.3.4.1.2  The Push of Professional Development

Another advantage on HR management caused by globalization is the advancement of professional development. Professional development involves providing employees with chances to accomplish career-related objectives. A few organizations give assistance to their workers to acquire a university degree, others send their employees to meetings or networking occasions and trainings. Professional development is essential to globalization since it produces a win–win condition. The employees feel just as the organization is concerned with giving a scope of abilities and capabilities for their staff. In the same manner, the organizations are able to gain from the additional abilities and network the employees who benefits from what the professional development programs are able to offer.

### 11.3.4.1.3  Increasing the Significance on Training

Much alike professional development, the increasing significance of training is due to globalization in HRM. Training, nonetheless, tends to concentrate around the necessities and skills of the workers in the organization. For example, the organization can provide language lessons to give their call center staff an advantage in telephone sales. It may also educate their workers on how to utilize a new global software program. The importance of training aims to give the organization a competitive advantage in the worldwide marketplace by sharpening their employees' diversification.

### 11.3.4.1.4  The Administration of Laws Across Jurisdiction

One more impact of globalization on HRM is the requirement that organizations must know and implement the laws of various jurisdiction to the appropriate business. The government sets out various tax and work laws which companies in the United States must follow, yet there may also be local and regional laws that apply to the organizations that work in various states or nations. For instance, selling items in Europe may imply that an organization needs to place a value-added tax on its goods. Employing workers at branch areas in various locations might change the condition on the lowest pay permitted by the law, tax allowance or working hours. Understanding these laws is important for the organization due to the fact that any violation of these laws will have a severe influence on the company's financial welfare as well as on its reputation.

### 11.3.4.2    DISADVANTAGES OF ADAPTING GLOBALIZATION ON HR SYSTEM

#### 11.3.4.2.1    Invasion of Privacy

Over usage of HRIS leads to violation of privacy of employees, management, and business authorities. Regardless of the government and state laws giving legal security regarding privacy in the workplace, if the HRM system is invaded by unauthorized people, your team's personal information can be easily obtained by other people. Despite the fact that the system is secured using password protection, this obstacle can be easily overcome by people who are good with technology from either inside or outside the organization. Strengthening the system and increasing the security of your information are needed.

#### 11.3.4.2.2    Security Impeachment

Comparable with the breach in privacy protection are the gaps in the security system that is guarding the HRM software. Also, electronic media is weak against attacks by viruses from any place on the internet. Getting a virus can immobilize the HRM system badly enough to render it unusable for an uncertain time. Additionally, for purchasing a strong backup application, a simultaneous decision ought to be made to devote into a virtually attack-proof program that defends the HRM data from infringement.

#### 11.3.4.2.3    User Error

Inseparable with the utilization of HRM systems is the means for making mistakes. Since computers and their related programs are only as productive as their human users, inaccuracy in data entry can and do happen. In HRM systems, such blunders can have severe outcomes. Best case scenario, minimal mistakes, for example, a couple of incorrectly spelled worker's name may happen. At worst, errors in data entry could obstruct the progress of the business. For instance, if the HRM system shows various incorrect dates assigning when to perform crucial employee's performance reviews with coming promotions, the unfavorable outcomes may involve a prevalent decline in employee morale, which could influence the levels of productivity.

### 11.3.4.2.4   Employee Assessment

HRM systems can be useful in choosing workers for specific positions or for promotions, based on performance scores and other information. Nonetheless, the human factor is taken out from these types of tasks by the system. For instance, while evaluating employee's talent, the system may report high scores in sales, new records, or marketing attempt; however, it may not be able to discover how the employee deals with the customers.

### 11.3.4.2.5   Down Time

Combined with all computer and programs are the unavoidable issues that results in technical difficulties. HRM system are also vulnerable to down time, with potentially critical results. These can comprise prompting an important business activity to a standstill if employee's data are not accessible. For instance, the HR experts have assigned a day committed to open enrollment for another worker's health plan, the failure to get the HRM system data could not only postpone the enrollment in the health program, however, could possibly force the open enrollment out for weeks or months, leaving qualified staff members with no coverage. These type of matters can be limited by procedures calling for ongoing focused technical system maintenance.

## 11.4   POTENTIAL SOLUTIONS (IN RESPONSE TO THE CHALLENGES OF GLOBALIZATION)

### 11.4.1   THINK GLOBAL, ACT LOCAL

Company has to understand that global market is a very competitive industry. Although it can be an opportunity, a business that develops a product aimlessly in foreign country might face future uncertainty. Business marketing its product accustomed to the locals' customs and language will have better chance of sustaining. Business can still introduce their own products but in different design and packaging. By doing so, it can restraints itself from any sensitive issue that may leads to cultural conflicts related to the products.

Additionally, it will make the product more appealing to the host market. The products living in the culture of others impose a meaning which the people can be more related to it. Businesses that just begin to operate in the country may find it easy to enter and so, it will be able to adjust and allocate its finance properly.

On contrary, act global and think local goes beyond operations and production into workforce affair. Business needs to find suitable employee who has a well-rounded knowledge of the market. Human resource will take the in charge to search suitable candidate to take the role. HR is best to initiate selecting local workforce so that it could minimize the climax issue on the bias, which can impediment any social conflicts. Finding a local workforce can be a challenge because a foreign company is perceived as having different working system. HR has to find a way to tailor the benefits to their culture such as holidays and rewards so that the people working in the company will not feel remorse.

### 11.4.2   INFORMATION SYSTEM UPDATE THROUGH TRAINING

A lot of changes have occurred due to globalization. To state some of them which are changes in diversification, technology, culture, and policy/ employee welfare. These changes can either cause advantages or problems to the company. Therefore, HRM will play an important role in overseeing other employees regarding the problems mentioned in order for the company to not face any dire consequences.

### 11.4.2.1   EMPLOYEES DIVERSIFICATION

As globalization is becoming more widespread, companies have to deal with the expanded diversification within the workforce. This factor requires the companies to learn how to deal with diversity as well as implement new rules for workers. The increase in diversification that has produced may benefit, as companies can acquire new understanding about the various individuals that comes from different background from the management and marketing standpoint. Although it is recommended to have a variety of people working for the companies, it may not always be considered a good thing. People who are working under HRM are responsible in selecting and hiring qualified individuals to work with the company. The HRM can make use of IS of the company to see which are the skills needed by the different department in the companies in order to hire the right people. Managers also have the job to retain employees in order for the company to not acquire any loses in their productivity.

### 11.4.2.2 CULTURE

When dealing with companies involved in globalization, culture will play a big part in the identity of the workers in the company. Different culture has different values and sometimes the difference of these culture is very noticeable. Another reason to be cautious about is that different things symbolizes different meanings that may cause misunderstandings. Marketing styles and ethical behavior may vary with different countries. Therefore, in order to avoid mishaps, employees are required to attend training in order to be prepared. Strengthening teamwork can also lessen misunderstandings as workers are more at ease with each other. When choosing representatives to perform business in different countries, HRM can make use of the company's IS to search for their employee's performance management to assist in electing suitable workers that will do well.

Due to globalization, cultural, religious, and ethnic diversity are present. Therefore, in order to protect the companies and their employees from discrimination, managers have to implement new policies as well as offering training to other employees in order that there is acceptance between one another and that the company can maintain its working environment. This can lead to the employees' appreciation to other cultures and viewpoint in the companies.

### 11.4.2.3 TECHNOLOGY

With the emerging new markets and the increased competitions, technology plays a big role in business. Technology can be seen as a competitive advantage while doing business. As time pass by, technology is becoming more prevalent and important in the daily operations of a business. With globalization in place, there are chances that not every country operates the same types of technology. With the advancement of technology and the varieties it provides, IS plays an even greater role in the business. A concern that could occur is that managers could have difficulties in accessing important information if the IS is not standardized. Employees must have the knowledge to use its company's IS. Managers will also play their role by assisting and consulting the employees, more specifically the newer ones on how to use the IS in order for their work to become more effective.

Technologies are also used as a communication device. Companies can use technology to ease business transactions and communicate with customers to get feedback. Nowadays, a lot of companies are adapting to use social media or the internet as the means of marketing and performing parts of their operations. Workers must be attentive to these things for the company to gain a slight advantage over others.

### 11.4.2.4   POLICY/EMPLOYEE WELFARE

Companies need to understand the importance of policy issues and employee's welfare due to equity and political reasons. It is said that with the rise of globalization, it has increased the standard of working, however, that is not always the case as some companies promote inequalities. A way to tackle this is to educate the employees using e-education and providing training. Managers must also be aware of any changes or difficulties that occurred either within or outside factors that concerns the companies. Example would be that some countries have strict regulations about the types of business practices allowed in their countries that often include HR and pension restriction and policies if the company chose to work with a foreign workforce.

### 11.5   CONCLUSION

To draw together all the different aspects of globalization that affect HRM practices is demanding; however, this paper will cover most of the aspects ranging from factors, challenges, impacts, and effects. Some are more tentative than others. Obviously there is much more that can be done on the aspects discussed in this chapter. The topics are summarized below to provoke further discussion and build more depth understanding on the underlying cause of some of the issues that arises due to globalization.

- *The HR managers rely on efficient communications and information system, plus excellent logistics* to get information in and out of countries without misinterpretations or fraud and delivered to employees in a matter of hours or days, rather than weeks or months. This has led to a speedup in production and distribution systems.

- Due to globalization, *new trends have started to evolve more* giving the HR managers more and newer roles to fulfill in their company. One such role is to become a change manager where the HR managers lend a helping hand to companies for them to adapt to change. Employee as well as HR managers ability to be receptive toward change and new systems must be enhanced.
- By *understanding globalization and the effects* it brings can help managers to prepare the companies to operate in the global market.
- *Globalization* is the *root cause* for changes in the employment environment. Companies now have employees working from a diverse background. Hence, it is the job of HR managers to provide necessary training with the primary focus of bringing all the talents together to increase the value of the organization.
- HR managers now have the challenge to think globally while acting locally in the world of business. Therefore, the HR managers must be able to adjust its methods, procedures, policies as well as providing training to assist the expansion of the organization to the global market.
- The occurrence of globalization have brought in the changes that has affected companies and their employees. Training that are provided by the HR managers plays an important role which allows the employees to be more adaptable to the changes occurred by globalization.

In conclusion, globalization is inevitable trend businesses follow. There are challenges every HR managers will meet from the diversity of cross-culture, technology change, human capital management, and legal policy. Therefore, management needs to adapt strategies with HR playing as the important roles. The impact of the issues will give HR an insight of where the problem lies and will help seeking a solution in tackling the problem it affect to the HRM system. However, managing human capital will be the most challenging factor to be tackled but HR manager still need to adapt to globalization change despite how challenging the issues they faced. Even change in practices, it may not be well accepted in the mindset of the employees but with times, employees can be educated and be more receptive toward accepting change.

HRM and its system is the critical backbone to the success of an organization.

## KEYWORDS

- **development of technology**
- **workforce demographic**
- **social issue**
- **marginal profits**
- **leadership capability**
- **IT emerging technology**

## REFERENCES

Asemi, A. The Role of Management Information System and Decision Support System for Manager's Decision-Making Process. *Int. J. Bus. Manag.* **2011,** *6* (7), 164-173.

Baldwin, R. E.; Winter, L. A. Challenges to Globalization: Analyzing the Economics. University of Chicago Press: Chicago, US; 2004.

Bradley, J. Effects of Globalization on Human Resources Management, 2017. Smallbusiness. chron. http://smallbusiness.chron.com/effects-globalization-human-resources-management-61611.html (accessed Aug 26, 2017).

Brown, K. D. Globalization and Cultural Conflict in Developing Countries: The South African Example. *Ind. J. Global Legal Stud.* **1999,** *7* (1), 225–254.

Cardenas, H. (n.d.). Laws Affecting Human Resource Management.<complete details>

Carl, D. *Technology, Globalization, and International Competitiveness: Challenges for Developing Countries*, 2017.

http://www.un.org/esa/sustdev/publications/industrial_development/1_2.pdf (accessed Aug 27, 2017).

Chand, S. Importance and Functions of Human Resource Management (with Diagram): Your Article Library, 2016.

http://www.yourarticlelibrary.com/human-resources/importance-and-functions-of-human-resource-management-with-diagram/32363/ (accessed Aug 23, 2017).

Friedman, B. A. Globalization Implications for Human Resource Management Roles. *Employee Resp. Rights J.* **2007,** 157–171.

Hurn, B. J. Response of Managers to the Challenges of Globalization. *Ind. Commerc. Train.* **2013,** *45* (6), 336–342.

Ionescu, A.; Dumitru, N. R. Multinational Companies Under Globalization Context. *Romanian Economic and Business Review*; Bucharest, Romania, 2011; Vol. 7 (1), pp 86–95.

Leasher, M.; Moberg, C. R. Examining the Differences in Salesperson Motivation Among Different Cultures. *Am. J. Bus.* **2011,** *26* (2), 145–160

Messmer, M. Key Federal Laws Affecting HR. *Dummies;* 2013. http://www.dummies.com/education/law/key-federal-laws-affecting-hr/ (accessed Aug 27, 2017).

Scholte, J. A. Globalisation and Modernity. *International Studies Association Convention* 1995.

George Frynas, J. G. The Limits of Globalization—Legal and Political Issues in E-commerce. *Manag. Decis.* **2002,** *40* (9), 871–880.

Greiner, B. An Online Recruitment System for Economic Experiments. *Munich Personal RePEc Archive* **2004,** *63,* 79–93

Gubman, E. L. The Gauntlet is Down. *J. Bus. Strat.* **1996.**

Harvey, M. et al. Globalisation and Corporate Real Estate Strategies. *J. Corp. Real Estate* **2010,** *12* (4), 234–248.

Holden, R. Managing People's Values and Perceptions in Multi-cultural Organisations: The Experience of an HR Director. *Emp. Relat.* **2001,** *23* (6), 614–626.

International Monetary Fund. Globalization: A Brief Overview, 2008. https://www.imf.org/external/np/exr/ib/2008/053008.htm (accessed Aug 21, 2017)

Kapoor, B. Impact of Globalization on Human Resource Management. *Inf. Syst. Decis. Sci.* **2011.**

Kitchen, P. J.; Eagle, L. Towards a Globalized Communications Strategy: Perceptions from New Zealand. *Market. Intell. Plann.* **2002,** *20* (3), 174–184.

Kipkoech, A. K. Influence of Information Communication Technology on Monitoring of Strategic Plans in Top 100 Mid Size Companies in Kenya. *Int. J. Sci. Res.* **2017,** *6* (6), 344–356.

Mcfalin, K. The Effects of Globalization in the Workplace, 2017. *Smallbusiness.chron.* http://smallbusiness.chron.com/effects-globalization-workplace-10738.html (accessed Aug 26, 2017).

Mohan, B. J. C. A Study on HRM Practices and Its Impact on Globalization of Indian Business. *IOSR J. Bus. Manag.* **2013,** *9* (4), 34–36.

Nimani, M. B. The Role of Information Systems in Management Decision Making—A Therotical Approach. *Information Management;* Frankfurt, Germany, 2010.

Leidner, D. E. A Review of Culture in Information System Research: Toward a Theory of Information Technology Cultural Conflict. *MIS Quart.* **2006,** *30* (2), 357–399.

Microsoft. What Are Software Restriction Policies, 2013. https://technet.microsoft.com/en-us/library/cc782792(v=ws.10).aspx (accessed Aug 27, 2017).

Spero, J. E. (1996, September 26). The Challenges of Globalization. *Mtholyoke.* Retrieve on 23th August 2017 from https://www.mtholyoke.edu/acad/intrel/spero.htm

The 20th Century. In Encyclopaedia Britannica Online, 2017. https://www.britannica.com/technology/history-of-technology/The-20th-century (accessed Aug 26, 2017).

World Economic Outlook. *A Shifting Global Economic Landscape,* 2017 [online]. https://www.imf.org/external/pubs/ft/weo/2017/update/01/pdf/0117.pdf (accessed Aug 21, 2017).

UK Essays. *Globalization and HRM,* 2017. https://www.ukessays.com/essays/management/globalisation-hrm-management.php (accessed Aug 24, 2017).

INC. (n.d.). Human Resource Management. https://www.inc.com/encyclopedia/human-resource-management.html (accessed Aug 24, 2017).

# CHAPTER 12

# DIGITAL TECHNOLOGY IMPACT ON DECENTRALIZED WORK SITES

## ABSTRACT

Advancement of digital technology has enabled businesses to run tasks and store information more conveniently, especially for those workers that are allocated at decentralized work sites. A decentralized work site is an alternative work location which may be in a different geographical region than the organization's main office. Hence, advancements in information technology (IT) facilities and affordable prices to acquire organizations allow more and more workers to be able to work remotely. On the other hand, most businesses are leveraging on this technology as a competitive advantage. The advancement of IT also enables job offerings for human skills in using and maintaining the facilities, which allow business functions especially human resource management (HRM) to be more streamlined in the sharing and transferring of information. The challenges in HRM today is exist. This chapter discus and reveals on telecommuting, the perspective from management information systems (MIS) which includes six areas of focus: (1) data, (2) software, (3) hardware, (4) people, (5) telecommunication, and (6) policies; then highlight the benefits and challenges of telecommuting from both managers' and employees' perspectives.

## 12.1  INTRODUCTION

In this modern technological era of the 21st century, advancement of digital technology has enabled businesses to run tasks and store information more conveniently, especially for those workers that are allocated at decentralized work sites. A decentralized work site is an alternative work location which may be in a different geographical region than the organization's main office (MBASkool, 2017). The term "telecommuting" describes the work arrangement whereby decentralized workers continue to work from home or other alternative sites through technologies which help them to stay connected with the headquarter. After initial review of existing literature works, it was discovered that there are no generally agreed definitions or terms for the concept of telecommuting. However, the different terms or conceptualizations have three similar characteristics such as having the flexibility of work schedule, the act of working away from the office, and still being connected electronically to the organization while doing so (Allen et al., 2015). Hence, the term telecommuting will be used interchangeably with other similar terms such as telework and remote work in this chapter.

Telecommuting was initially introduced in the United States during the 1970s by Jack Nilles who came up with the idea of allocating works in different states to telecommuters instead of having to relocate local employees to other states to prevent unnecessary waste of resources (Gan, 2015). In present days, technological advancements have facilitated the widespread usage of telecommuting (Allen et al., 2015). In contrast, during the 1980s, the early invention of the internet, personal computers and mobile phones were expensive infrastructures to acquire. However, due to significant economic growth, these infrastructures and equipments have notably dropped in prices as more demand to telecommute throughout the years caused them to become affordable (Avery and Zabel, 2001). Hence, advancements in information technology (IT) facilities and affordable prices to acquire them allow more and more workers to be able to work remotely. According to U.S. Census Bureau (2012), the number of workers working from home rose by around 3.9 million people between the year 1999 and 2010. Nowadays, it is common to see people holding sophisticated devices especially smartphones to communicate with not only colleagues, but also with friends and family members. Therefore, these devices highly complement telecommuting resulting in the transformation of traditional work arrangements to one that allows working from anywhere around the globe.

On the other hand, most businesses are leveraging on this technology as a competitive advantage. Many use telecommuting almost exclusively for its flexibility to achieve full business potential (Evan et al., 2010). The advancement of IT also enables job offerings for human skills in using and maintaining the facilities, which allow business functions especially human resource management (HRM) to be more streamlined in the sharing and transferring of information (Kizza, 2013).

As convenient as it seems, telecommuting has caused new challenges to arise for HRM. For instance, it is difficult to track the productivity and quality of a teleworker as well as his or her performance evaluation. Furthermore, should compensation policy be treated as the same between nontelecommuters and telecommuters? Other challenges include ensuring the health and safety of the workers, especially when they were assigned to an isolated area with higher risks of danger. Some of these challenges in HRM today still exist, as managers continuously seek to address them. In 2013, a news surfaced concerning Yahoo's Chief Executive Officer (CEO), Marissa Mayer, who changed telecommuting policies in order for Yahoo workers to stop working from home in order to enhance productivity and work closely to each other (Mauldin, 2013). However, this news caused a public uproar, forming bad impression on Yahoo, and stating that Yahoo is moving backwards to the Stone Age, and not having an open mindset (Goudreau, 2013). Moreover, Weise and Swartz (2013) suggested not making the same mistakes as Yahoo! did. This chapter will first review studies on telecommuting, the perspective from management information systems (MIS) which includes six areas of focus: (1) data, (2) software, (3) hardware, (4) people, (5) telecommunication, and (6) policies. Last but not least, the benefits and challenges of telecommuting from both managers' and employees' perspectives are also discussed.

## 12.2   LITERATURE REVIEW

### 12.2.1   DIFFERENT TERMS AND DEFINITIONS FOR TELECOMMUTING

Due to the insufficiency of a standard definition and conceptualization of this particular work arrangement, it is difficult to have a solid comprehension about it as different researches' results cannot be compared (Allen et al., 2015). Telecommuting has been referred to as remote work, telework, and virtual work among other terms. However, Offstein et al. (2010) have differentiated telework from telecommuting, in that telework is a much broader concept. They argued that telework is more than just eliminating the commuting aspect of work but also involves the foundational reengineering in the design of work and processes (Offstein et al., 2010). For simplicity's sake, this chapter will utilize the different terms of telecommuting interchangeably to mean the same concept of working away from the office with the benefit of flexible job hours that is enabled by information systems.

### 12.2.2   DEFINITION OF MIS

MIS consists of all the parts which collaborate in the processing of data and output them in order to contribute to an organization's core objective

(Sousa and Oz, 2015). In an organization, MIS consists of data, software, hardware, people, telecommunication, and policies. Data refers to the input that systems process to output information. Softwares are programs that tell the hardware how to perform the fundamental functions of taking in data, processing it, output the information, and storing data and information. Hardware includes a computer, the peripheral devices, as well as information transmission devices. People includes MIS experts and users who build information systems, analyze information requirements of an organization, maintain software, and utilize hardware. Telecommunication refers to the software and hardware that enables quick transferring and receiving of multimedia in digital form. Policies are regulations to achieve reliable and optimal operations in the processing of data. These can include safety measures and priorities in the distribution of software programs (Sousa and Oz, 2015). For managers that have decided to implement telecommuting in their organizations, they must address three core technology issues (Pearce, 2009). These are communication, connectivity, and the sharing of information.

### 12.2.2.1   DATA

Teleworkers need materials that are downloadable from the internet so that they can access corporate resources even when they are away from office (Pearce, 2009).

### 12.2.2.2   *SOFTWARE*

Softwares that facilitate employees to work together over the internet in real time are called groupware programs (Sousa and Oz, 2015). These applications enable the articulation of staffs' ideas by expressing them through multimedia, and remove the need for them to commute and be located together. Collaboration applications are tools that help to enhance the productivity of employees working together on projects that require human interaction (Sousa and Oz, 2015). These softwares can be used for the sharing of documents, calendars, email, audio and video conferencing, and project management.

Applications and websites that facilitate the sharing of content are critical in telecommuting (Pearce, 2009). Examples of such softwares are e-storage systems, portals, and social bookmarking system. Premium softwares that facilitate collaboration on the internet are products such as Microsoft Windows SharePoint Service which is a portal that enables telecommuters to share data, communicate through e-mail, set up meetings, and gain access to other internal assets (Pearce, 2009). Another related instrument that is helpful for teleworkers is file sharing. Remote file sharing and storage is becoming widespread.

In order to fortify connectivity in telecommuting, thin client technology is instrumental as it allows organizations to maintain a high level of control over where operations are executed (Pearce, 2009). This technology is a mixture of software and hardware. Telecommuters' computers or laptops which contain only a limited number of programs, as well as powerful servers, that facilitate most of the processing in thin client makes up the hardware part. The software part consists of the client program that allows users to access corporate softwares and documents. Interestingly, these resources which may potentially be confidential are not stored on the user's laptop as the thin client only acts like a window that enables telecommuters to access the internal network. Once the telecommuter finishes his or her work and signs out, all documents and programs are retained in the internal network, and their window is terminated (Pearce, 2009). Thin client softwares can be expensive or cheap. An example of a free albeit a limited one is RealVNC. A more costly option is Citrix Systems' business-grade thin client software which allows teleworkers to access a simulated desktop from anywhere as long as they have a working internet connection (Pearce, 2009).

### 12.2.2.3  HARDWARE

A set of wireless hardwares are imperative for enabling internet connectivity for remote working. These can include but are not limited to wireless modem, broadband, credit card transactions, phone calls, and portable headsets that are Bluetooth enabled (Pearce, 2009).

Perhaps the most important requirement of telecommuting is having access to high-speed internet at home as even the simplest program needs such an infrastructure. Otherwise, the productivity of teleworkers will be significantly lowered (Pearce, 2009). There are two major kinds of high internet bandwidth networks with varying speed between the two (Pearce, 2009).

Cloud storage is a mode of storage whereby data stored is accessible from a remote location through the use of internet (Sousa and Oz, 2015). With this technology, employees can access and consolidate their data storage while teleworking (Sousa and Oz, 2015). Additionally, it is more reliable and safe to store confidential documents in the cloud than on Universal Serial Bus (USB) drives which are prone to theft.

Internal corporate networks provide access to corporate assets such as intranet portals, printers, and network shares. The core benefit of utilizing safe internal networks is that workers do not have to worry about data breach when working with confidential data (Pearce, 2009). Virtual private network (VPN) enables telecommuters to access the corporation's network using their laptop or computer at home. This would have been impossible for telecommuters as the traditional methods of accessing internal corporate networks require the staff to be on-site (Pearce, 2009).

## 12.2.2.4  PEOPLE

As teleworking is vastly different from traditional work mode, the recruitment of suitable leaders and employees is critical to the success of telecommuting (Offstein et al., 2010). Traditional leaders tend to overstress the logic of "management by walking around" which focuses on leading through physical proximity. This approach is counterintuitive in telecommuting as employees are spread over a geographical region or even across different time zones (Offstein et al., 2010). Instead, one of the key best practices of managing telecommuting is emphasizing on outcomes instead of the techniques since it can be strenuous to monitor the process in telework (Offstein et al., 2010).

Other qualities of a great leader in telework are being talented at communicating virtually and physically, excellent at scheduling, and often calling their telecommuters on the phone to complement virtual communication in order to help build a social context for their job assignments (Offstein et al., 2010). Additionally, successful leaders spend more assets and efforts to build rapport within their group as they know social support is more difficult to achieve when face-to-face interaction is infrequent (Offstein et al., 2010).

Staff who are extrovert and socially agreeable to a greater extent tend to be the ones that are offered telecommuting vacancies because organizations understand that these are the people that are vocal, and can attain virtual presence through their personality (Offstein et al., 2010). In comparison to hiring tech-astute and introverted staffs who are prone to become insignificant more frequently in a telecommuting environment, hiring individuals who are extroverted, independent, and agreeable may better forecast the success rate of virtual tasks (Offstein et al., 2010).

Furthermore, teleworkers require technical and managerial support that is available all the time (Pearce, 2009). This is because remote computers can break down anytime and there will not be any on-site technicians present to fix them.

## 12.2.2.5 TELECOMMUNICATION

Communication technology is the core part to allow remote workers in any part of the globe to communicate (Stephanie et al., 2015). It is noted by research papers that interactions and feedback between subordinates and their supervisors are key to job performance of the employee (Tsai and Chuang, 2009). Utilizing new communication technology is also essential for teleworkers due to the minimal face-to-face communication in order to achieve favorable performance and productivity (Watson and Belanger, 2007). With increasingly better technologies available, organizations have integrated the technology to allow workers have access to more variety of communication channels that vary in their richness of transmitting information (Stephanie et al., 2015). Email is a text-based communication that is still widely used in every organization (Stephanie et al., 2015).

According to Digital Company Stats (2017), there were 4.3 billion email users worldwide. Users are allowed to have continuous conversations for the ease of reference (Marvick, 2001). Furthermore, there are also audio and video conferencing (Stephanie et al., 2015). Audio conferencing especially voice over internet protocol (VoIP) is increasingly used by workers as it is cheap and convenient for users to contact colleagues at remote locations with internet access (Shaw and Sharma, 2016). It has similar features as a typical telephone call except the fact that the cost is cheaper and it can only be accessed via internet (Shaw and Sharma, 2016). However, the quality of call for VoIP is directly proportional to the internet bandwidth, where high bandwidth means better quality (Vaishnav, 2006). As for video conferencing, it is one of the most favored communication tools used by remote workers as it is the only mean to have visual interactions with coworkers through camera and the internet (Stephanie et al., 2015). Video conferencing is highly preferred by remote workers who spent longer hours or time teleworking but it also depends on the preference of the individuals (Veerraju and Rao, 2011). Some workers who prefer to keep their identity and looks unknown are more susceptible to use emails or any other text-based communication (Stephanie et al., 2015). Therefore, telecommunication channels are known to enhance organizational performance and communication (Stephanie et al., 2015).

## 12.2.2.6 POLICIES

There is a sheer obligation for organizations to install proper policies to complement the demands of the organization with the demands of the teleworker that is working at a different geographical location (Bayrak, 2012). Furthermore, organizations may install the appropriate standard guidelines and procedures to ensure that the work rate of teleworkers is sustained (Wright and Oldford, 1993). In reality, there are plenty of organizations that face the issues of implementing the right policies required to control and to productively meet the requirements of teleworkers (Hemphill, 2004).

### 12.2.2.6.1  *Information Security*

According to Information Security Standard Document (2008), information security is an essential element when establishing telecommuting. Leakage of organizational information might pose a serious issue to the organization depending on sensitivity of the information (Leung, 2009). Based on a study by one American think tank from a pool of 56 American companies interviewed, it showed that average leak of each organization amounted to more than USD 4.7 million (Leung, 2009). Information leakage could occur in many ways such as due to an unauthorized party, such as a hacker, and technical software issues. Therefore, both the manager and the teleworker should be aware of the risks and ensure the safety and confidentiality of organization information (Information Security Standard Document, 2008). As such, information security must be carried out by both parties to ensure its effectiveness.

### 12.2.2.6.2  *Rules and Regulations*

Telecommuters are obligated to comply to the terms and conditions for telecommuting which involves the security, safety, contract termination, equipment, and other expenses covered by the company (Ye, 2012). From the perspective of Federal Government of United States of America (2011) regarding teleworking policies, it is mentioned that the criteria on employee's eligibility of teleworking must be well written to ensure that it complies with the public sector's requirements. On the other hand, the eligibility to telework depends on management decision as teleworking is not mandatory for employees but rather, it is voluntary (Long, 2010). Ye (2012) discussed that the employees could be dismissed from teleworking at anytime with rational explanation by informing their manager, or when the employees do not meet the expected performance.

In some studies done on telework policies by the Federal Government of United States of America (2011), teleworkers are required to provide isolated telework space to prevent any interruption while also maintaining a safe working environment with locked file cabinet when the teleworkers are accountable for equipment and data security measures. Ye (2012) also stated that when telecommuters are in their telework space, companies are not liable for expenses such as utilities, home insurance, and maintenance except for office supplies. Additionally, The Department of Defense's Computer Accommodations Program (CAP) in the United States also gave support to teleworkers with disabilities by providing them with assistive devices to aid them on completing their job tasks (Federal Government of United States of America, 2011).

### 12.2.2.6.3   Compensation Policy

In accordance to compensation policy relating to pay system for teleworkers, the Federal Government of United States of America (2011) discussed that the pay is to be based on the location of the employee's work site with three different work-site conditions as follows. Firstly, the teleworkers who managed to report themselves physically at least twice every 2 weeks to the original work site. Secondly, the teleworker whose work site is based at home and does not require to report at least twice every 2 weeks. Lastly, teleworkers whose work site varies from time to time and who have to perform work within the same geographical area do not require reporting to the original work site.

Furthermore, there is a slight difference in applying premiums such as night pay and Sunday premium to teleworkers and nonteleworkers. Night pay for a nonteleworker is generally computed by 10% of that employee's basic rate of pay. In contrast, a teleworker will not be compensated when they work overtime during the night. Subsequently, a teleworker is also not entitled for Sunday premium unless he or she are arranged to work on Sundays regularly. Meanwhile, a nonteleworker will be compensated with 25% of the basic pay for tasks performed on a Sunday.

Additionally, Ye (2012) indicated that telecommuters and nontelecommuters are compensated with worker's compensation insurance in the event of injuries occurring during work. This statement was supported by the guideline for telecommuting by the public sector of the United States, which states that local employees are entitled to Federal Employees' Compensation Act when any individual suffers from injuries on the incident during working hours.

### 12.2.2.6.4    Evaluation

Since monitoring telecommuters' performances could be critical to the organization, the public administration of the United States suggested that companies should consistently evaluate teleworkers and nonteleworkers equally with specific guidelines such as the performance management system. Furthermore, evaluating in telecommuting programs could be accelerated when company's teleworking policies are clearly stated and easy to understand without difficult terminologies (Allen et al., 2015). Moreover, supervisors or managers being constantly in contact with telecommuters may be beneficial for assessing an employee's current progress and could also lessen that teleworker's feeling of being excluded in the workplace (Allen et al., 2015). In order to evaluate performances through frequent communication, it is suggested to have a common IT system such as a company intranet or a shared email folder (Lautsch and Kossek, 2011).

In the research of Stone et al. (2015), advanced technology is used to evaluate employees' performance management process such as performance measurement and performance feedback. This is also termed as electronic performance management (e-PM) system. Instead of evaluating through traditional methods such as in written handwritten form, e-PM systems could track employee performance in progress and the system would prompt the results to the managers to discuss about further improvements needed or any expectations on future performances (Stone et al., 2015). On the other hand, there are certain limitations with e-PM systems such as inaccuracy of ratings regarding an employee's performance (Stone and Lukaszeweski, 2009). In other words, it is concerned with the telecommuter's response to electronic feedback, which may not be fully sincere and trustworthy as behavior changes in electronic responses (Stone et al., 2015). Additionally, telecommuters might dishonestly report the number of hours they were working when managers overlooked the evaluation system (Allen et al., 2015).

### 12.2.2.6.5    Training

Teleworkers who wish to level up their skills can undergo training programs to hit certain criteria set by the policies (Elnaga and Imran, 2013). With the advancement of technology, organizations can now easily deliver trainings to remote workers via online mediums such as video conferencing, also known as e-training (Johnson et al., 2008). e-training is convenient and flexible in nature. Some organizations provide virtual reality training for the workers to experience real life scenarios and live learning (Larsen, 2014). In addition, communication management for remote workers is also integrated into the e-learning so that trainees are able to ask questions and receive feedbacks almost instantly (Horton, 2004). However, studies found that telecommuters often feel social isolation as they could not interact with colleagues like they did during face-to-face training (Johnson et al. 2008). A research study by Benbunan and Arbaugh (2006) pointed out that trainings with social interaction instill higher performance and satisfaction level than those who do not.

### 12.2.2.6.6  Health Benefits and Risks

The government of the United States stated that both the manager and telecommuter are responsible on ensuring the health and safety while telecommuting. It was also stated that managers should provide and discuss the safety checklist with teleworkers, and to follow up with any reports of accidents during work. Tavares and Isabel (2015) stated that telework does have an impact on health. However, due to the scarce amount of researches regarding the trade-off between telework and its benefits and costs, it is rather difficult to conclude that the benefits of telework were greater than the disadvantages in terms of the worker's health. One study found that "depriving people of control over their lives is indeed damaging to their health" (Marmot, 2013, p 1), Thus, telework which grants people the ability to better control their life suggested that telework can be more beneficial than harmful when compared to the costs. This statement was also supported by in the work of other researchers such as Csey and Gryzwacz (2008). Konradt (2003) found that there is less stress when the concern on daily work–home transportation is dismissed. Bloom (2015) suggested that teleworking enables greater work life balance due to higher flexibility. While Sardeshmukh et al. (2012) stated there is higher life control and satisfaction with telework. On the other hand, Tavares and Isabel (2015) further argued that there are still several health issues regarding telework such as musculoskeletal problems, stress, overwork, isolation, and depression regardless of the benefits.

### 12.2.3  BENEFITS OF TELECOMMUTING

Generally, teleworking provides benefits at the individual, organizational, as well as societal level. Through extensive review of past literature works, advantages of telecommuting from several sources are consolidated below.

## 12.2.3.1   INDIVIDUAL

The successful implementation of telecommuting has become favorable to the employees due to its flexibility (Coenen and Kok, 2014). Furthermore, telecommuting also enables an individual to save both time and transportation cost as a result of less commuting (Mahler, 2012). Additionally, teleworking also benefit employees in the form of greater job satisfaction, flexibility to schedule their work around their personal peak efficiency period or around family needs, and better control of work–family boundaries (Lautsch and Kossek, 2011).

Allen et al. (2015) further argued that flexibility of work scheduling positively affect employees' satisfaction and motivation in accomplishing their task. Hence, productivity level is more likely to increase when employees are motivated in performing their job (Perez et al., 2002). Perez et al. (2002) further indicated that employees have higher productivity level when the organization appreciates highly motivated teleworkers and receive higher compensation. However, the impact of telecommuting on productivity depends on whether the task at hand is of creative or dull nature (Dutcher, 2012).

## 12.2.3.2   ORGANIZATIONAL

From an organization's perspective, telecommuting enhances employee productivity, job satisfaction, and reduces turnover (Pearce, 2009; Lautsch and Kossek, 2011). Furthermore, organizations can benefit from telework in the form of cost savings as less office space is needed (Pearce, 2009; Mahler, 2012). Further, these works can be distributed to geographic locations where the costs of real estate are cheaper. Due to the rise in globalization, pursuing this work mode is also beneficial as businesses are able to work internationally and possess more working hours in a work system that is dispersed globally (Lautsch and Kossek, 2011).

Telecommuting also increases the pool of potential employees in multiple ways (Allen et al., 2015). One way is by providing a window of opportunity for individuals with disabilities to be employed. These people are allowed to work in settings already well-suited to their requirements (Mahler, 2012). Secondly, teleworking can also increase opportunities for those who have decided to live on the outskirts of bustling cities and on the countryside (Mahler, 2012; Allen et al., 2015). In this manner, teleworking makes an employer more attractive to potential employees who may not be able to work in a traditional work arrangement (Pearce, 2009).

Another way in which teleworking complements organizations is through providing business continuity during unfavorable weather conditions, epidemics, or other catastrophic events such as natural disaster or man-made crises such as terrorism which can interrupt the flow of organization operations (Pearce, 2009; Mahler, 2012; Allen et al., 2015). This is due to the semiautonomous nature of teleworkers (Pearce, 2009).

### 12.2.3.3 SOCIETAL

Mahler (2012) also argued that telecommuting contributes to an organization's green initiatives as it reduces air pollution by eliminating or severely reducing the need to telecommute. A government research from the United States found that when 20,000 civil servants telecommuted only once a week, they saved more than 380,000 L of petrol, 3 million commuting kilometers, and 37,000 kg of carbon dioxide emissions weekly (Pearce, 2009).

### 12.2.4 CHALLENGES OF TELECOMMUTING

Despite the positive aspects of telecommuting, there are some negative aspects to it as well. As telecommuting changes the work of those who have chosen to adopt this particular work mode, it also requires more efforts on the managerial side. This is because managers of teleworkers must now accommodate and implement budding organizational policies with regards to this flexible work form that is growing (Lautsch and Kossek, 2011). One of the most prominent challenges for organizations implementing teleworking is information and security protection (Pearce, 2009). Examples of potential breaches are loss or theft of computers, e-mail spams, and hackers.

### 12.2.4.1 INDIVIDUAL

On the employee perspective, Mokhtarian and Salomon (1996) found that teleworkers were concerned about being in isolation. Bailyn (1988) also agreed to the statement above, and observed that teleworkers are more concerned on isolation than nonteleworkers. Kurland and Cooper (2002) highlighted that there are two types of isolation which are professional and social isolation. Professional isolation occurs when teleworker felt they were forgotten from promotions and the organization's regards due to their workplace location. Social isolation occurs due to having different geographical workplace and thus, teleworkers are unable to have informal interactions with other colleagues. This could alienate teleworkers from other colleagues (Kurland and Cooper, 2002).

## 12.2.4.2   ORGANIZATIONAL

From an organizational perspective, Shamir and Salomon (1985) found that telework results in the difficulties in monitoring teleworkers. In another research conducted by Tomaskovic-Devey and Risman (1993), it was mentioned that managers worry they might lose control over employee's action when they have the authority to work at home. Managerial control over employees can be diminished as employees are working away from the office without constant supervision. Mokhtarian and Salomon (1996b) pointed out that from their researches, certain managers disapprove telework as they are unable to monitor teleworkers behavior.

Furthermore, when workers get distributed into decentralized work sites, computers, office furniture, and telecommunication equipment must be relocated (Pearce, 2009). Moreover, new and different kinds of resources are needed as teleworking demands newer technologies to be installed such as a central customer database that can be accessed from home. Other possible issues are the hostility between telecommuters and nontelecommuting colleagues, deterioration of team chemistry, and the breaking of social network in the office (Pearce, 2009). Additionally, when telecommuters are spread over a geographical region, there can be a depletion of some efficiency for certain organizations as they are not able to share costly assets anymore (Pearce, 2009). Examples of such assets are high-costing photocopying, location-based licenses for real-time information in the financial industry, as well as special computers.

## 12.3   RESULT AND DISCUSSION

In this section, the subtopics under the literature review section will be discussed with analyses, arguments, and elaborations where they are necessary.

## 12.3.1   TELECOMMUTING

Allen et al. (2015) noted that due to the abundance of the different terms and definitions used for the concept of telecommuting, it hindered their understanding as the results of papers done by different researchers could not be compared directly. Similar obstacles were faced by the authors of this chapter which caused initial confusion. Additionally, due to the existence of the various terms for telecommuting, the literature review may not have been extremely exhaustive due to the potential of missing out on certain keywords when searching for the relevant journal articles. Nevertheless, adequate research was performed to form a solid review of the relevant literature in this chapter. As was stated earlier in the literature review section, this chapter defines telecommuting as working away from the office with flexible working hours, and uses the various terms interchangeably.

## 12.3.2   MANAGEMENT INFORMATION SYSTEM

According to Sousa and Oz (2015), MIS consists of six major synergetic components that process data and produce information in order to achieve an organization's vision. Hence, this chapter examined the nature of telecommuting with an MIS perspective by investigating the data, software, hardware, people, telecommunication, and policies which specifically enable telecommuting.

### 12.3.2.1   DATA

In general, the scope of data required by telecommuters is technically same as the data required by traditional workers. For instance, a teleworker may require consumer preferences in order to create a better marketing plan. The same data is also required by a traditional worker if he or she were to perform the same job. However, Pearce (2009) stated that telecommuters require data that is downloadable from the internet so that they are able to gain access to organizational resources at home. If the consumer preferences were recorded on paper filed at the office, the teleworker will not be able to gain access to it from his or her home. As such, while the scope of data between teleworkers and nonteleworkers are alike, the medium in which the data is stored must be accessible to the former when they are not at the office.

### 12.3.2.2   SOFTWARE

Aside from the standard productivity applications being utilized, there are more specialized softwares that enable and enhance the experience of telecommuting. For instance, as teleworkers will be working away from their colleagues and managers, collaboration tools can facilitate employees working on a group project (Sousa and Oz, 2015). This is due to functions such as the sharing of a document whereby group members can work on it at the same time from remote locations. Google Drive, a combination of cloud storage and cloud computing, is one such tool. It allows several users to work together on a presentation slide or a word document which is saved online automatically (Sousa and Oz, 2015). Furthermore, the participants can check what changes are made by each member. All these factors allow organizations to cut down on commuting expenses and save time. In fact, this chapter was completed mainly by the authors through Google Drive, proving the effectiveness of such a tool for teleworking.

Other softwares allow the organizations to monitor internet usage, keystrokes, and mouse activities in order to help managers better monitor telecommuters (Pearce, 2009). There are other ways to track telecommuters' work activities depending on the nature of their jobs. In the example of a call center, a system known as Symposium uses VoIP to monitor how long a phone has been ringing before it is answered. Additionally, it is able to track the amount of outgoing calls, the duration for each call, and the amount of calls responded to determine the productivity of the staffs (Pearce, 2009).

### 12.3.2.3   HARDWARE

As was aforementioned in the literature review, one of the most essential elements required to enable teleworking is access to high-bandwidth internet at home (Pearce, 2009). Nowadays, this requirement is extremely easy to satisfy as more than 70% of internet connections at home are high-speed connections (Pearce, 2009). Therefore, organizations often do not have to worry about this part. Nevertheless, if a potential teleworker lives in the rural area where connectivity infrastructure may not be as advanced as those in the city centers, the organization should consider whether to allow that employee to telecommute or not.

On the other hand, even though cloud storage has many benefits for an organization considering telecommuting, it has its disadvantages. For instance, the confidentiality and safety of business data saved in the cloud remains a glaring problem (Pearce, 2009). If there is any defect with the cloud storage service provider, confidential business data may risk being compromised. Furthermore, organizations would be more reliant on networks, and incur higher related costs because data storage is not on-site (Pearce, 2009). Hence, organizations should conduct a cost–benefit analysis when choosing to use in-house storage or outsourcing it to cloud storage service providers.

As for VPN, organizations must know that staffs have the capacity to duplicate confidential files to their computers while they are connected to the VPN. Once a staff disconnects from the VPN, control over the safety of confidential files and the staff's computer is no longer under the control of the organization (Pearce, 2009). Similarly with cloud storage, organizations should consider the risks and benefits of VPN before utilizing it.

### 12.3.2.4   PEOPLE

Due to the different work nature of telecommuting, certain characteristics are more preferable in potential people who will telecommute for the organization. For instance, a leader who focuses on the process or technique of a job instead of the outcome would be unsuitable to lead teleworkers as it can be difficult to monitor them from a distance (Offstein et al., 2010). Additionally, organizations would also prefer telecommuters to be more outgoing, agreeable, and independent as this increases their chance of succeeding when working from a remote location (Offstein et al., 2010). This is important because introverted employees tend to volunteer themselves into telecommuting positions.

As teleworkers' computer or laptops can be malfunctioned while they are working from home, organizations need support technicians who know how to troubleshoot and repair them from a distance. Administrators who can utilize remote management services are able to perform upgrades to off-site computers, and track their conditions and activities (Pearce, 2009).

### 12.3.2.5   TELECOMMUNICATION

As mentioned by Stephanie et al. (2015), communication is the cornerstone for telecommuters or decentralized workers where they can interact with the headquarter through communication technology. There are many aspects where communication plays a part in the organization. Teleworkers with higher sense of social interaction preferred to use communication media to make their involvement and physical presence felt in work interactions (Fonner and Roloff, 2010). Teleworking increases the chance that workers feel isolated due to less interaction with other coworkers (Cooper and Kurland, 2002).

Moreover, as mentioned in the literature review, some types of communication media for telecommuting are emails, audio and video conferencing. First of all, email is widely used by teleworker as it is inexpensive, convenient, and the messages are able to reach recipients instantly. However, communicating through email has its risk of conveying the wrong message or information, or to the wrong person by mistakes (Ferrazzi, 2013). Every individual has different styles in delivering his or her message, and the recipients may misinterpret the intention of the information received. Furthermore, the recipients may also interpret the tone of the sentences differently (Ferrazzi, 2013). Overall, email is still a preferred way of communication for teleworkers.

Other communication mediums such as audio conferencing like VoIP and video conferencing contain similar features. Both mediums rely on the internet to function and allow users to interact in a similar manner as they would in person. Furthermore, they are relatively inexpensive and are able to connect with multiple users at the same time. These functions allowed remote workers to easily communicate with their colleagues and managers without being physically at the same location. Hence, this helps the organization to save time and traveling expenses. However, many organizations still prefer to meet in person for important or high-level meetings due to confidentiality and complex decision-making (Arvey, 2009).

Furthermore, it is also important to observe the body language during a meeting to determine a person's skills and to build trust between coworkers (Morgan, 2012), which neither audio nor video conferencing could portray. In addition, audio conferencing only allows the voice of users to be heard. As for video conferencing, people cannot look directly into the camera and at the display screen at the same time causing the absence of eye-to-eye interaction as they otherwise would have in a face-to-face meeting (Giger et al., 2014). In the case of Google, a large IT company that sells innovative products and services, they prefer face-to-face interactions between workers and avoid too much of telecommuting to enhance innovation and creativity of new products (Schmidt and Rosenberg, 2014).

Last but not least, it is also important for teleworkers to transfer knowledge. Without telecommunication, it is impossible to transfer knowledge and exchange information between teleworkers and the organization (Golden and Raghuram, 2010). Therefore, telecommunication enables teleworker to work efficiently even from a distance, save time in transferring information, and helps to foster a collaborative spirit between teleworkers and organizations.

### 12.3.2.6 POLICIES

#### (a) Rules and regulations

Teleworking policies in the United States indicated that the organization should ensure that guidelines and procedures are properly applied in order to comply with the requirements. Such matters include obtaining the equipment, compensation agreements, teleworking hour agreements, suspension agreement, and related matters. The eligibility and selection process to be a teleworker may be critical as one of the telework policy stated that an employee is not eligible to telework if the performances does not fulfill the desire requirements on the written agreement. Furthermore, the selection of teleworker does not apply to employee whose duty is to work at the original workplace every day. As an illustration, an on-site activity that involves face-to-face communication with client cannot be conducted remotely (Federal Government of United States of America, 2011).

According to the studies in the literature review conducted by Long (2010), it is mentioned that the condition to be a teleworker is up to company management decision in the United States. Moreover, a case study conducted by Bernardino et al. (2012) pointed out some of the criteria on employing telecommuters. They include the knowledge and ability to deal with technology for effective communication, a mindset to be able to deal with multiethnic group of people, and a characteristic of being able to work independently (Bernardino et al., 2012). Subsequently, the rules and regulation of Federal Government of United States of America (2011) also stated that if the place of teleworking were at the residence of the employee, it is essential to have a separate space to avoid distractions. Furthermore, the agreements between telecommuters and the company should list down in detail on what equipments are provided for teleworkers such as laptops, office supplies and other related equipment. It is clearly stated that most of the organization are not responsible for any expenses at the residence of teleworkers such as for utilities, home insurance, and maintenance fees.

## (b) Compensation policies

Meanwhile, compensation policy on telecommuting is also necessary to state clearly in written agreement. As identified in the literature review, the teleworker's pay in the United States is based on the location of his or her work site. It is categorized into three different types of conditions such as teleworker who report themselves back to the regular work site at least twice a week, employee who work at their residence, and employees whose workplace are not fixed but perform the job task within the same geographical area. The different pay rates for these three conditions were implemented to ensure that the employees get compensated by an equal amount of effort they work and to be able to compensate employees accordingly to local prevailing rates (Federal Government of United States of America, 2011).

In comparison to the compensation policies between Brazil and the United States, it appeared that policies in different countries can be different. In the case study conducted by Bernardino et al. (2012), they stated that are no differences in Brazil's compensation system between telecommuters and nontelecommuters. For instance, telecommuters are subjected to over-time compensation law in Brazil which means that telecommuters who work overtime are compensated with amount of pay agreed in teleworking agreement (Bernardino et al., 2012). This is unlike the teleworking policies in the United States whereby telecommuters will not be entitled for any premium if they work overtime or on a Sunday, while nontelecommuters are paid 10% and 25%, respectively, on top of their basic rate of pay (Federal Government of United States of America, 2011). Hence, teleworking policies apply in varied ways across different countries and organizations. Therefore, implementation of teleworking should be conducted carefully by following the appropriate teleworking policies to manage both telecommuters and nontele-commuters in the best regards (Lautsch and Kossek, 2011).

## (c) Information security

From the literature review, it was concluded that information security is a critical factor for telecommuting. If an organization did not apply safety measures on information security, information leakage might occur and result in significant damages to that organization. From a research done by an American think tank, many organizations faced the issue of information leakage. Based on their interview result, it was shown that the involuntary information disclosure of each company costed them more than USD 4.7 million on average (Leung, 2009). Thus, it was recommended that organizations should either set up or follow certain policies before implementing remote work arrangements. An example of existing ones is a set of information security policies set specifically for telecommuting by Emory University and Emory Healthcare. Under this set of policies, both the manager and telecommuter are responsible and should work together to achieve maximum effectiveness of the safety measures. The set of policies was divided into two sections: managerial and telecommuter aspect. For the managers, they are required to (1) ensure that all devices used for telecommuting have antivirus software installed and to be up to date, (2) ensure all devices are protected with firewall, (3) provide management of computer system to telecommuters, and (4) provide the latest security patch and version of softwares and programs used for telecommuters. On the other hand, telecommuters should follow these instructions: they must (1) not attempt to override security measures, (2) connect to a secure network and utilize VPN, and (3) prevent accessing email and organization softwares using untrusted network (Information Security Standard Document, 2008).

## *(d) Health benefits and risks*

After reviewing the current literature works, it is imperative for both manager and telecommuter to be responsible for the health and safety conditions during work assignments. Therefore, both parties must be aware of the risks resulting from telecommuting and take several actions to minimize them. Nevertheless, telecommuting provides both health benefits along with risks as well. Konradt et al. (2003) stated that telecommuting enables employees to be less stressed as the concern on daily work–home transportation is dismissed. The frustration and stress of being in the middle of a traffic congestion is resolved. Other than that, telecommuters can get more rest as more time is gained from not needing to commute daily. These factors can contribute to a positive impact for the health of telecommuters. Bloom (2015) further stated that a telecommuter can achieve a better work–life balance due to better flexibility in working period. Flexibility in working hours enables a telecommuter to schedule personal working hours based on his or her family and self-needs.

On the other hand, Tavares and Isabel (2015) pointed out that telecommuting can result in several health risks such as musculoskeletal problems, stress, overwork, isolation, and depression. Musculoskeletal pain can arise under several circumstances such as long hours of the usage of a computer, long hours of continuous work, extreme position of forearm and wrist, and repetitive movements. Crawford et al. (2011) stated that these factors might cause musculoskeletal pain in the neck, wrist, shoulders, hand, and lumbar area. Generally, most workers working with computers face the same risk of musculoskeletal issues. However, the possibilities of getting musculoskeletal issues is higher for telecommuters as there is no interaction between any colleagues so they might accidentally ignore breaks and work for a longer period of time (Sang et al., 2010). Additionally, it was concluded that telecommuting can lead to isolation and depression in the aforementioned statement. Therefore, it was recommended for managers to make sure that telecommuters spend a minimum of 20% of the work period located in the organization's office to avoid the feeling of isolation (Fairweather, 1999).

## (e) Evaluation

Aforementioned in the literature review by the public sector of the United States in 2011, evaluating teleworkers and nonteleworkers must be treated with equal conduct. They suggested evaluating teleworkers' performance by implementing a performance management system. This system assesses employees' performance on their work developments and progresses to ensure that they achieve the company's strategic goals (Suhardi, 2015). In other words, it is also described as the action of measuring work progresses of employees where the performance is known as the outcome they deliver (Brudan, 2010). Performance management system often involves identifying individual's goals and objectives, measuring employees' accomplishment on a given task, the recognition of employees' performance, and providing feedback to employees with corrective measures (Suhardi, 2015).

A study conducted by Allen et al. (2015) also indicated that teleworking policies are necessary to be written clearly in the contract to facilitate teleworking programs. The terms and conditions can specify the requirements for evaluation in a telework program and for the organization in general. Some examples of these criteria are number of hours, days per week, and time required to work for telecommuters (Allen et al., 2015). For instance, KPMG, an auditing company, includes various metrics to evaluate the teleworkers identified in their contract such as the satisfaction level received from clients, the team, or coworkers; quality of work that employee delivered; individual performance assessment by senior management; and punctuality on handling tasks (Allen et al., 2015).

Eventually, a more frequent level of communication between the managers and the teleworkers could eliminate some of the issues on evaluating teleworkers. In the case study conducted by Lautsch and Kossek (2011) for instance, one of the largest corporations in the United States reported to have connected with their teleworkers at least 32 times per week and mentioned that it is important to make time available to the telecommuters at any condition. One of the telecommuting employees of that company responded on the relevant matter stating that he was well informed in the loop of workplace as he was frequently in contact with the coworkers by email (Lautsch and Kossek, 2011).

Moreover, with the advancement of technology in the 21st century, evaluating employee's performance became simpler through electronic performance management system, which is also known as e-PM (Stone et al., 2015). It was believed that this system could simplify evaluation process, save time and cost, and decrease efforts needed to evaluate teleworker performance (Stone et al., 2015). By conducting e-PM in the organization, it enables managers to easily evaluate teleworkers' ongoing task with the help of multirater feedback and the results would prompt the managers for further evaluation and improvement in teleworkers' performance (Stone et al., 2015). Furthermore, the research done by Stone et al. (2015) indicated that the teleworkers are more likely have more confidence in the feedback process provided by the computer compared to the direct feedbacks by managers. The computerized feedback was more reliable as it will be directly focusing on teleworkers' ongoing task, instead of managers own intent (Stone et al., 2015). With the application of e-PM, it could also enable employees to stay motivated and achieve greater job satisfaction while telecommuting. Nevertheless, researchers also mentioned that e-PM might induce less accurate responses toward the rating process compared to traditional systems due to teleworkers' misunderstanding in the computerized feedback process. Furthermore, they may have a change of behavior when responding (Stone et al., 2015). However, this could be overcome by advancing evaluation with video conferencing with teleworkers to enable them to further clarify the nature of their responses (Stone et al., 2015).

### 12.3.3   BENEFITS AND CHALLENGES OF TELECOMMUTING

During the 1970s, telecommuting was already practiced among the private companies due to its ability to address certain work issues outside the main organization (Allen et al., 2015). Even now, telecommuting has been increasingly carried out by more organizations as a result of its many benefits as well as the advancement in technology. From the literature review above, it was concluded that telecommuting will be able to generate advantages in three different aspects: on the individual, organizational, and societal levels. These benefits will be discussed further in the following section while some of the negative aspects of telecommuting will also be explored.

#### 12.3.3.1   INDIVIDUAL

From the aspect of employees, telecommuting has diminished the transportation expenses and travel time as a result of less commuting time to the main office (Mahler, 2012). This also prevents individuals from wasting valuable time while being stuck on the road when commuting. Further, the reduction on travel time also enables individuals to have more resting time, which can prevent fatigue and stress, and allows them to further have a better quality of life. Other than that, expenses such as petrol and cost of the office attire will be minimized, which would allow for more financial savings for telecommuters.

Furthermore, a telecommuter with flexible work schedule can boost productivity, work–life balance, commitment, job satisfaction, and employee retention (O'Leary, 2013). It will also grant employees the flexibility and authority to control the schedule of their working time according to their preferences (Harpaz, 2002). This enables telecommuter to schedule their personal peak productive time in order to have more family time, and attain a healthier balance between work and personal time. Additionally, this has a positive relationship in reducing stress and family conflict as well as giving telecommuters more control on their work (PGi, 2014). Aforementioned in the literature review by Perez et al. (2002), it was stated that level of productivity will increase and staff will be more motivated when telecommuting. Further, Tavares and Isabel (2015) mentioned that telecommuters can achieve a higher productivity as there are less interruptions and distractions from colleagues. Another finding by Grant et al. (2013) proved the reliability of this statement and through his interview it was shown that remote workers should have the capability to avoid distractions in order to work productively. Other than that, office politics can also be avoided which can provide telecommuters with a more enjoyable working environment that allows for a higher concentration level during working hours.

Regardless of the benefits, telecommuting is not totally flawless and faces several challenges. As mentioned in the literature review, it was highlighted that telecommuting might cause a telecommuter to feel professional and social isolation. Kurland and Cooper (2002) also found that the issue of professional isolation was more prevalent in telecommuters than in traditional employees. This is due to the nature of remote workplace as there is a minimal interaction between telecommuter, manager, and other colleagues. Thus, it raises the concern of a telecommuter being neglected for promotion opportunities as he or she is not physically around the organization's workplace. Perin (1991) validated telecommuters' concern and found that the visibility of an employee does have a correlation between organizational politics and promotion opportunity. It was further explained that employees who work physically closer with their manager might encounter favoritism and obtain a higher chance of getting a promotion. Another concern of telecommuting is social isolation. The nature of telecommuting does not provide a social–work relationship between colleagues especially one with face-to-face interaction (Crawford et al., 2011). The continuous long hours of working alone can cause the telecommuter to feel isolated and be in solitude. In the worst case scenario, it can cause depression in a long run.

However with the help of current technology, social isolation can be minimized. Nowadays, internet which runs with optical cables provides a fast and stable network (Tavares and Isabel, 2015). Therefore, video conferencing software which relies on network speed has become a more widespread communication tool. This tool can lead to more quality interactions between telecommuters, colleagues, and managers (Brumm, 2016). With video conferencing, telecommuter can seek information and advice from colleagues anytime provided that both parties are connected to the internet. In this way, it can enable telecommuters to feel more as a part of the team and therefore reduce the feeling of social isolation.

Regarding the point that telecommuting can enhance individual productivity, this might not be universal. Coenen and Kok (2014) found that when telecommuters are situated at different work sites and potentially at different working hours, this could negatively affect the level of productivity when there is no interaction and cooperation between employees in the company (Coenen and Kok, 2014). Kelliher and Anderson (2010) also concluded that when teleworkers are unable to communicate with their coworkers during business hours, they are less likely to be productive when undertaking tasks. A research also found that telecommuters with a task that has higher interdependence with others may result in negative productivity when remote work arrangement is utilized (Turetken et al., 2011). Furthermore, in the study conducted by Dutcher (2012), it was stated that productivity level in telework is highly affected by the nature of the task at hand. For instance, performing dull tasks will lead to low productivity whereas productivity level will be higher when solving creative tasks. Nevertheless, it was argued that telecommuters tend to produce work task more efficiently, and are likely to have a lower turnover rate compared to nontelecommuters (Allen et al., 2015).

## 12.3.3.2   ORGANIZATIONAL

From the organization's perspective, telecommuting can provide a solution to managerial challenges such as maintaining employee job satisfaction and retention (Pearce, 2009; Lautsch and Kossek, 2011). When employees have the authority to control the schedule of their working time based on personal needs, they achieve better work–life balance and better management of family relationship. These are the factors that can contribute to higher job satisfactions and reduction in turnover rate. Another benefit of telecommuting that Pearce (2009) mentioned was that office costs can be reduced through telework as less office space is required. An example of this is when the American telecommunications company, AT&T, saved roughly USD 60 million a year from reducing office space by up to 30% through implementing telecommuting strategies (Pearce, 2009).

Another benefit of telecommuting is the ability to expand the pool of potential employees that can be hired by an organization (Allen et al., 2015). People with disabilities are able to participate in the workforce through telecommuting as it offers accessibility to work at home, flexibility on scheduling, a barrier-free workplace, and the elimination of discrimination (Alkan and Ciftci, 2013). The rapid advancement of IT and communication infrastructures are also factors that contribute to the employability of people with disabilities (Alkan and Ciftci, 2013). An example of that is e-training which enables employees to be skilled and receive the relevant training via electronic medium. This can be advantageous to assist the disabled employees in gaining adequate job skills.

However, even though e-training is more convenient for telecommuters, this system is still new and many people felt that they have no control over the learning process and its flexibility (Stone et al., 2015). Hence, it is hoped that future improvement on technologies can enhance the e-training experience for remote workers, especially for the disabled individuals. West and Anderson (2005) concluded that many disabled individuals have the ability and desire to work at home. Moreover, some of them may have good working skills and a strong work ethic. Although it has been argued that telecommuting might cause the feeling of isolation for disabled people, another study conducted by Virginia Commonwealth University and MITE (2001) stated that 90% of the disabled telecommuters did not experience isolation with this work arrangement. Further, they even proposed that they have achieved a healthier work–life balance (Alkan and Ciftci, 2013). Therefore, telecommuting may serve as a great opportunity for many organizations to recruit more talented and skilled employees who would otherwise have not been considered in the first place.

Other than that, Shamir and Salomon (1985) found that telecommuting has given rise to the challenge of monitoring and controlling employees' performances. Due to the characteristics of remote workplace, it is likely to be impossible for managers to physically supervise and provide guidance to telecommuters. Besides that, measuring productivity can also be challenging according to past studies. One of the limited ways managers can measure telework productivity is by giving them a deadline for their tasks. Then, it is up to the teleworkers to finish the tasks on time and produce quality works. After all, it still depends on the individual remote workers whether they are going to undertake their work seriously. Mokhtarian and Salomon (1996a) stated that some managers disapprove of telework as they are unable to monitor the behavior of telecommuters.

Additionally, managers might also face some difficulties when assessing a telecommuter's performance during promotion decision. Another issue is that the relocation of computers, office furniture, and telecommunication equipments have to be performed when workers are spread out (Pearce, 2009). Furthermore, in order to ensure the effectiveness of telecommuting, high-end technology such as a central customer database is required to allow telecommuters to access important customer information from any geographical area (Pearce, 2009). Therefore, organizations are required to bear the cost to acquire and maintain such databases. As workers are decentralized, there is also a diseconomy of scale in terms of sharing costly office equipments and organizational assets between employees (Pearce, 2009). A solid illustration is in the securities industry which requires on-site permits to access instantaneous financial information. Such an industry would not be able to afford to acquire these costly infrastructures for each of its telecommuting employees (Pearce, 2009).

### 12.3.3.3   SOCIETAL

The benefits of telecommuting can be observed from the society's point of view as well. Mahler (2012) claimed that telecommuting can reduce air pollution by eliminating the need to telecommute daily. As the need to drive has been eliminated or significantly reduced, carbon dioxide emitted from vehicles will also be reduced. A study on Xerox Corporation, an American multinational company, showed that when more than 8000 Xerox employees practice full-time telecommuting, it resulted in a reduction of more than 40,000 metric tons of gas emissions. In addition, more than 17 million liters of gas was saved due to not needing to commute to work (Fell, 2015). This suggests that telecommuting does indeed have a positive impact on the environment.

However, Arif et al. (2013) found that despite the reduction in fuel consumption, telecommuting can generate more energy usage at home. In traditional workplace, every employee shares the consumption of electricity in the building such as for air conditioning and lights. In the case of telecommuting, however, each individual consumes electricity separately. The total amount of energy used by each telecommuter during working time will be greater as compared to traditional workers who share the same office. Therefore, it was argued that telecommuting can increase the overall energy usage which is linked to additional carbon emissions resulting from the production of this increased consumption in energy. As such, it is quite complex to determine to what extent does telework contribute to the improvement of the environment.

## 12.4  CONCLUSION

To conclude this chapter, telecommuting has been adapted in many organizations due to technological advancements, and its advantages not only to the organizations but also for employees and the society as a whole. As a result of the different nature of telework when compared to traditional work arrangement, an organization must be vigilant in making sure that it possesses and is using the adequate resources to telework. As a framework used in this chapter, organizations may benefit from looking at the perspective of MIS by breaking the concept down to its six elements. This allows them to better inspect their needs in each element, and fill in the gap where necessary to facilitate and enhance the experience of telecommuting. In other words, businesses thinking about adopting telework should ensure that the appropriate and adequate data, software, hardware, people, telecommunication, and policies are in place to maximize the effectiveness of remote work arrangement.

## KEYWORDS

- **digital technology**
- **decentralized work sites**
- **telework**
- **voice over internet protocol**
- **Information Security Standard**
- **computer accommodations program**

# REFERENCES

Alkan, M.; Ciftci, C. Attitudes of People with Disabilities Towards Teleworking as an Employment Opportunity: It's Modelling in Terms of the Turkish Case. *Afr. J. Bus. Manag.* **2013,** *7* (4), 227–243. DOI: 10.5897/AJBM11.1472.

Allen, T. D.; Golden, T. D.; Shockley, K. M. How Effective is Telecommuting? Assessing the Status of Our Scientific Findings. *Psychol. Sci. Public Interest* **2015,** *16* (2), 40–68. journals.sagepub.com/doi/pdf/10.1177/1529100615593273.

Arif, M.; Darr, M.; Orgren, M.; Sun, C. *Assessing the Potential for Telecommuting at UC Berkeley*; 2013. http://sustainability.berkeley.edu/sites/default/files/GHGReductionPotentialUCB_Telecommuting_GraduateCourse268E%20Final%20Report_2013.pdf.

Arvey, D. R. Why Face-to-face Business Meeting Matters. *The Hilton Family,* 2009. http://www.colestraining.com/wp-content/uploads/2009/10/Meetings-Why-Face-to-Face-Business-Meetings-Matter-a-white-paper-by-Dr-Richard-Arvey-1.pdf.

Avery, C.; Zabel, D. *The Flexible Workplace: A Sourcebook of Information and Research*; Quorum Books: Westport, CT, 2001. https://books.google.com.bn/books?hl=en&lr=&id=9Kz7Hke4PAC&oi=fnd&pg=PT1&dq=The+flexible+workplace:+A+source-+book+of+information+and+research&ots=bjEZS4R0wH&sig=feX2R1dq0ONkcMvljXoP_V396m4&redir_esc=y#v=onepage&q=The%20flexible%20workplace%3A%20A%20source-%20book%20of%20information%20and%20research&f=false.

Bailyn, L. Freeing Work from the Constraints of Location and Time. *New Technol. Work Employ.* **1988,** *3* (2), 143–152. DOI: 10.1111/j.1468-005X.1988.tb00097.

Bayrak, T. IT Support Services for Telecommuting Workforce. *Telemat. Inform.* **2012,** *29* (3), 286–293. http://dx.doi.org/10.1016/j.tele.2011.10.002.

Bernardino, A. F.; Roglio, K. D. D.; Corso, J. M. D. Telecommuting and HRM: A Case Study of an Information Technology Service Provider. *J. Inf. Syst. Technol. Manag.* **2012,** *9* (2), 285–306. http://www.jistem.fea.usp.br/index.php/jistem/article/view/10.4301%252FS1807-17752012000200005.

Bileviciene, T.; Bileviciene, E. Telework Organisation Model as Method of Development of Disabled Persons' Employment Quality. *Perspect. Innov. Econ. Bus.* **2010,** *5* (2) (1231-2016-100695), 71–74. http://ageconsearch.umn.edu/record/92385.

Bloom, N.; Liang, J.; Robert, J.; Ying Z. Does Working from Home Work? Evidence from a Chinese Experiment. *Q. J. Econ.* **2015,** *130* (1), 165–218. http://www.nber.org/papers/w18871.pdf.

Bricout, J. C. Using Telework to Enhance Return to Work Outcomes for Individuals with Spinal Cord Injuries. *NeuroRehabilitation* **2004,** *19* (2), 147–159.

Brudan, A. Rediscovering Performance Management: Systems, Learning and Integration. *Meas. Bus. Excell.* **2010,** *14* (1). http://www.emeraldinsight.com/doi/abs/10.1108/13683041011027490.

Brumm, F. Telework is Work: Navigating the New Normal. *Cornell HR Review,* 2016. http://www.cornellhrreview.org/wp-content/uploads/2016/05/CHRR-2016-Brumm-Telework-1.pdf.

Coenen, M.; Kok, R. A. W. Workplace Flexibility and New Product Development Performance: The Role of Telework and Flexible Work Schedules. *Eur. Manag. J.* **2014,** *32* (4), 564–576. http://www.sciencedirect.com/science/article/pii/S026323731300159X.

Cooper, C. D.; Kurland, N. B. Telecommuting, Professional Isolation, and Employee Development in Public and Private Organisations. *J. Organ. Behav.* **2002,** *23* (4), 511–532. DOI: 10.1002/job.145.

Crawford J. O.; MacCalman, L.; Jackson, C. A. The Health and Well-being of Remote and Mobile Workers. *Occup. Med.* **2011,** *61* (6), 385–394. DOI: 10.1093/occmed/kqr071.

Dutcher, E. G. The Effects of Telecommuting on Productivity: An Experimental Examination. The Role of Dull and Creative Tasks. *J. Econ. Behav. Organ.* **2012,** *84* (1), 355–363. http://www.sciencedirect.com/science/article/pii/S0167268112000893.

Elnaga, A.; Imran, A. The Effect of Training on Employee Performance. *Eur. J. Bus. Manag.* **2013,** *5* (4). http://pakacademicsearch.com/pdf-files/ech/517/137-147%20Vol%205,%20 No%204%20(2013).pdf.

Fairweather, N. B. Surveillance in Employment: The Case of Teleworking. *J. Bus. Eth.* **1999,** *22* (1), 39–49. https://link.springer.com/article/10.1023/A:1006104017646.

Federal Government of United States of America. *Guide to Telework in the Federal Government*; 2011. https://www.telework.gov/guidance-legislation/telework.

Fell, S. *How Telecommuting Reduced Carbon Footprints at Dell, Aetna and Xerox*; 2015. https://www.entrepreneur.com/article/24529.

Ferrazzi, K. How to Avoid Virtual Miscommunication. *Harvard Business Review*, 2013. https://hbr.org/2013/04/how-to-avoid-virtual-miscommun.

Fonner, K. L.; Roloff, M. E. Why Teleworkers are more Satisfied with Their Jobs than are Office-based Workers: When Less Contact is Beneficial. *J. Appl. Commun. Res.* **2010,** *38* (4). DOI: http://dx.doi.org/10.1080/00909882.2010.513998.

Gan, V. *What Telecommuting Looked Like in 1973*; 2015. https://www.theatlantic.com/technology/archive/2015/12/what-telecommuting-looked-like-in-1973/418473/.

Giger, D.; Bazin, J. C.; Kuster, C.; Popa, T.; Gross, M. *Gaze Correction with a Single Webcam*; 2014. https://graphics.ethz.ch/Downloads/Publications/Papers/2014/Gig14a/Gig14a.pdf.

Golden, T. D.; Raghuram, S. Teleworker Knowledge Sharing and the Role of Altered Relational and Technological Interactions. *J. Organ. Behav.* **2010,** *31* (8). DOI: 10.1002/job.652.

Goudreau, J. Back to the Stone Age? New Yahoo CEO Marissa Mayer Bans Working From Home. *Forbes,* 2013. https://www.forbes.com/sites/jennagoudreau/2013/02/25/back-to-the-stone-age-new-yahoo-ceo-marissa-mayer-bans-working-from-home/#38dc49151667.

Grant, C. A.; Wallace, L. M.; Spurgeon, P. C. An Exploration of the Psychological Factors Affecting Remote E-Worker's Job Effectiveness, Well-being and Work-life Balance. *Employ. Relat.* **2013,** *35* (5), 527–546. DOI: 10.1108/ER-08-2012-0059.

Harpaz, I. Advantages and Disadvantages of Telecommuting for the Individual, Organisation and Society. *Work Study* **2002,** *51* (2), 74–80. https://doi.org/10.1108/00438020210418791.

Hemphill, B. Telecommuting Productively. *Occup. Health Saf.* **2004,** *73*, 16–18. https://ohsonline.com/Articles/2004/03/Telecommuting-Productively.aspx.

Information Security Standard Document. *Information Security Requirements for Telecommuting Arrangements*; 2008. https://it.emory.edu/MEDIA/Teleworking.pdf.

Kelliher, C.; Anderson, D. Doing More with Less? Flexible Working Practices and the Intensification of Work. *Hum. Relat.* **2010,** *63* (1), 83–106. https://doi.org/10.1177/0018726709349199.

Kizza, J. M. *Ethical and Social Issues in the Information Age*; Springer-Verlag: London, England, 2013.

Konradt, U.; Hertel, G.; Schmook, R. Quality of Management by Objectives, Task Related Stressors and Non-task-related Stressors as Predictors of Stress and Job

Satisfaction Among Teleworkers. *Eur. J. Work Organ. Psychol.* **2003,** *12* (1), 61–79. DOI: 10.1080/13594320344000020.

Kurland, N.; Cooper, C. Manager Control and Employee Isolation in Telecommuting Environments. *J. High Technol. Manag. Res.* **2002,** *13* (1), 107–126. https://doi.org/10.1016/S1047-8310(01)00051-7.

Lautsch, B. A.; Kossek, E. E. Managing a Blended Workforce: Telecommuters and Non-telecommuters. *Organ. Dyn.* **2011,** *40*, 10–17. https://www.researchgate.net/publication/256923983_Managing_a_blended_workforce_Telecommuters_and_non-telecommuters.

Leung, K. Information Leakage & Data Loss Prevention. University of Waterloo, 2017. http://uwcisa.uwaterloo.ca/Biblio2/Topic/KarenKarYanLeung.pdf.

Long, S. D. *Communication, Relationships and Practices in Virtual Work*; Information Science Reference: New York, USA, 2010.

Mahler, J. The Telework Divide: Managerial and Personnel Challenges of Telework. *Rev. Public Pers. Adm.* **2012,** *32* (4), 407–418. http://dx.doi.org/10.1177/0734371x12458127.

Marmot, M. The Art of Medicine. Europe: Good, Bad, and Beautiful. *Lancet* **2013,** *381* (9872), 1090–1091. http://www.thelancet.com/pdfs/journals/lancet/PIIS0140-6736(13)60749-7.pdf.

Marwick, A. Knowledge Management Technology. *IBM Syst. J.* **2001,** *40* (4), 814–830. DOI: 10.1147/sj.404.0814.

MBASkool. *Decentralized Work Sites*; 2017. http://www.mbaskool.com/business-concepts/human-resources-hr-terms/15320-decentralized-work-site.html.

Mokhtarian, P. L.; Salomon, I. Modeling the Choice of Telecommuting: 2. A Case of the Preferred Impossible Alternative. *Environ. Plan.* **1996a,** *28*, 1859–1876. https://doi.org/10.1068/a281859.

Mokhtarian, P. L.; Salomon, I. Modeling the Choice of Telecommuting: 3. Identifying the Choice Set and Estimating Binary Choice Models for Technology-based Alternatives. *Environ. Plan.* **1996b,** *28*, 1877–1894. DOI: https://doi.org/10.1068/a281877.

Morgan, N. 5 Fatal Flaws with Virtual Meetings. *Forbes*, 2012. https://www.forbes.com/sites/nickmorgan/2012/10/02/5-fatal-flaws-with-virtual-meetings/#26e25a49704f.

Offstein, E. H.; Morwick , J. M.; Koskinen, L. Making Telework Work: Leading People and Leveraging Technology for Competitive Advantage. *Strateg. HR Rev.* 2010; Vol. 9 (2), pp 32–37. DOI: https://doi.org/10.1108/14754391011022244.

Offstein, E.; Morwick, J.; Koskinen, L. Making Telework Work: Leading People and Leveraging Technology for Competitive Advantage. *Strateg. HR Rev.* **2010,** *9* (2), 32–37. http://dx.doi.org/10.1108/14754391011022244.

Pearce, J. Successful Corporate Telecommuting with Technology Considerations for Late Adopters. *Organ. Dyn.* **2009,** *38* (1), 16–25. DOI: http://dx.doi.org/10.1016/j.orgdyn.2008.10.002.

Pein, C. The Moral Fabric of the Office: Panopticon Discourse and Schedule Flexibilities. *Res. Sociol. Organ.* **1991,** *8* (1), 241–268. http://faculty.babson.edu/krollag/org_site/org_theory/barley_articles/perin_panopt.html.

Perez, M. P.; Sanchez, A. M.; Carnicer, M. O. D. L. Benefits and Barriers of Telework: Perception Differences of Human Resources Managers According to Company's Operations Strategy. *Technovation* **2002,** *22* (12), 775–783. https://doi.org/10.1016/S0166-4972(01)00069-4.

Sang, K.; Gyi, D.; Haslam, C. Musculoskeletal Symptoms in Pharmaceutical Sales Representatives. *Occup. Med.* **2010,** *60,* 108–114. https://doi.org/10.1093/occmed/kqp145.

Sardeshmukh, S.; Sharma, D; Golden, T. D. Impact of Telework on Exhaustion and Job Engagement: A Job Demands and Job Resources Model. *New Technol. Work Employ.* **2012,** *27* (2), 193–207. DOI: 10.1111/j.1468-005X.2012.00284.x.

Schmidt, E.; Rosenberg, J. *How Google Works*; Grand Central Publishing: New York, NY, 2014.

Shamir, B.; Salomon, I. Work-at-home and the Quality of Working Life. *Acad. Manag. Rev.* **1985,** *10* (3), 455–464. http://www.jstor.org/stable/258127.

Shaw, J.; Sharma, B. A Survey Paper on Voice over Internet Protocol (VoIP). *Int. J. Comput. Appl.* **2016,** *139* (2). http://www.ijcaonline.org/research/volume139/number2/shaw-2016-ijca-909112.pdf.

Smith, C. Email Statistics. *Digital Company Stats,* 2017. http://expandedramblings.com/index.php/email-statistics/.

Sousa, K.; Oz, E. *Management Information Systems,* 7th ed.; Cengage Learning: Singapore, 2015.

Stephanie, A. S.; Alyssa, P.; Margaret J. P. Communication and Teleworking: A Study of Communication Channel Satisfaction, Personality, and Job Satisfaction for Teleworking Employees. *Int. J. Bus. Commun.* **2015,** *25* (1). DOI: 10.1177/2329488415589101.

Stone, D. L.; Deadrick, D. L.; Lukaszewski, K. M.; Johnson, R. The Influence of Technology on the Future of Human Resource Management. *Hum. Resour. Manag. Rev.* **2015,** *25* (2), 216–231. http://dx.doi.org/10.1016/j.hrmr.2015.01.002.

Stone, D. L.; Lukaszewski, K. M. An Expanded Model of the Factors Affecting the Acceptance and Effectiveness of Electronic Human Resource Management Systems. *Hum. Resour. Manag. Rev.* **2009,** *19* (2), 134–143. https://doi.org/10.1016/j.hrmr.2008.11.003.

Suhardi, A. R. Renewal of Performance Management System in Family Company. *Soc. Behav. Sci.* **2015,** *211,* 448–454. http://www.sciencedirect.com/science/article/pii/S1877042815053999.

Tavares, A. I. *Telework and Health Effects Review and a Research Framework Proposal*; 2015. https://mpra.ub.uni-muenchen.de/71648/1/MPRA_paper_71648.pdf.

Time. *Telecommuting: What Marissa Mayer Got Right—And Wrong*; 2013. http://time.com/money/2791618/telecommuting-what-marissa-mayer-got-right-and-wrong/.

Tomaskovic-Devey, D.; Risman, B. J. Telecommuting Innovation and Organisation: A Contingency Theory of Labor Process Change. *Soc. Sci. Q.* **1993,** *74* (2), 367–385. https://www.researchgate.net/profile/Donald_Tomaskovic-Devey/publication/247706006_Telecommuting_Innovation_and_Organization_A_Contingency_Theory_of_Labor_Process_Change/links/5729070908aef5d48d2c8fe4/Telecommuting-Innovation-and-Organization-A-Contingency-Theory-of-Labor-Process-Change.pdf.

Tsai, M.; Chuang, S. An Integrated Process Model of Communication Satisfaction and Organisational Outcomes. *Soc. Behav. Personal.* **2009,** *37* (6). DOI: https://doi.org/10.2224/sbp.2009.37.6.825.

Turetken, O.; Jain, A.; Quesenberry, B.; Ngwenyama, O. An Empirical Investigation of the Impact of Individual and Work Characteristics on Telecommuting Success. *IEEE Trans. Prof. Commun.* **2011,** *54,* 56–67. DOI: 10.1109/TPC.2010.2041387.

U.S. Census Bureau. *Census Bureau Report Shows Steady Increase in Home-based Workers Since 1999*; 2012. https://www.census.gov/newsroom/releases/archives/employment_occupations/cb12-188.html.

Vaishnav, C. *Voice over Internet Protocol (VoIP): The Dynamics of Technology and Regulation*; 2006. https://dspace.mit.edu/bitstream/handle/1721.1/34533/70958271-MIT. pdf?sequence=2.

Veerraju, R. P. S. P.; Rao, S. A. Benefits of Video Conferencing. *Int. J. Hybrid Comput. Intell.* **2011,** *12* (4). http://serialsjournals.com/serialjournalmanager/pdf/1332236849.pdf.

Watson, M.; Belanger, F. Media Repertoires: Dealing with the Multiplicity of Media Choices. *MIS Q.* **2007,** *31* (2), 267–293. http://www.jstor.org/stable/25148791.

Weise, E.; Swartz, J. As Yahoo! Ends Telecommuting, Other Says it has Benefits. *USA Today*, 2013. https://www.usatoday.com/story/money/business/2013/02/25/working-at-home-popular/1946575/.

West, M. D.; Anderson, J. Telework and Employees with Disabilities: Accommodation and Funding Options. *J. Vocat. Rehabilit.* **2005,** *23* (2), 115–122. http://kter.org/employment-research/telework-and-employees-disabilities-accommodations-and-funding-options.

Wright, P.; Oldford, A. Telecommuting and Employee Effectiveness: Career and Managerial Issues. *Int. J. Career Manag.* **1993,** *5* (1), 230–240. http://dx.doi.org/10.1108/09556219310024751.

Ye, L. R. Telecommuting: Implementation for Success. *Int. J. Bus. Soc. Sci.* **2012,** *15* (3), 20–29. http://ijbssnet.com/journals/Vol_3_No_15_August_2012/4.pdf.

# CHAPTER 13

# ORGANIZATION CONTINUOUS IMPROVEMENT PROGRAMS: MIS TECHNOLOGICAL ADVANCEMENT

## ABSTRACT

Continuous improvement programs (CIPs) came as a result of new challenges and changes that are facing organizations in the era of digital ecosystem. These changes include technological advances unprecedented rate, customers' behavior as well as ethical and environmental awareness of the community issues. CIPs have become an important tool for these organizations, it also described as a systematic process actively seeking new improved ways of performing tasks. CIPs are being used by organizations all over the world in order to increase their profits in the short-term whilst maintaining these gains in the long-term and also making the organization more competitive by making the business more efficient and effective. The introduction of management information system (MIS) facilitate the processing of data into a useful informative report where managers can use for further decision-making purposes such as sales analysis, and product performing. Effective decision-making is difficult as the business environment is constantly changing; therefore in order to ease the decision-making process, the decision support system (DSS) has been introduced. DSS allows users to analyze the massive data collection and provide better solutions to the problem and most importantly enable better decision making. This chapter explore the usage of CIPs within an organization includes the use of CIPs in matters relating to human resource management, customer relations management as well as product development. The scope of this report has been narrowed to these three fields as it was determined that these fields have been the greatest affected by the adoption of CIPs by organizations. Additionally, an explanation of the background including the characteristics of CIPs has also been provided.

## 13.1 INTRODUCTION

Continuous improvement programs (CIPs) came about as a result of new challenges and changes that are facing organizations in this globalized world. These changes include technological advances at an unprecedented rate, customers that are more informed and more demanding than before as well as issues regarding ethical and environmental awareness of the community. Because of this, organizations can no longer rest on their laurels, instead, they need to constantly strive to improve and not remain stagnant for long. CIPs have become an important tool for these organizations. CIPs can be defined as "both a philosophy of change that involves seeking opportunities for improvement in all work processes and a method for implementing change that is characterized by company-wide involvement and incremental improvements of existing processes" (Jørgensen and Hyland, 2014). It can also be described as a systematic process actively seeking new improved ways of performing tasks (Anand et al., 2009). CIPs are being used by organizations all over the world in order to increase their profits in the short-term whilst maintaining these gains in the long-term and also making the organization more competitive by making the business more efficient and effective (Aartsengel and Kurtoglu, 2013). As a result of the need to reduce costs as well as become more efficient and effective, it has led to the rise in importance of information systems, which are now deemed critical for the success of any organization as it can significantly improve productivity.

Information systems can be simply explained as the methods and ways in which organizations as well as people, in general, utilizes computers and information technologies to collect, process, store, use and distribute information (Agarwal and Lucas 2005; Laudon and Laudon, 2012). In the past, the information system was commonly used for electronic data processing (EDP), where it was used for recording, classifying, manipulating and summarizing data in a computer. EDP is also known as transaction processing systems (TPS) (Chow, n.d.). As technology advances, the introduction of management information system (MIS) facilitate the processing of data into a useful informative report where managers can use it for further decision-making purposes such as sales analysis, and product performing (Chow, n.d.). Effective decision-making is difficult as the business environment is constantly changing; therefore in order to ease the decision-making process, the decision support system (DSS) has been introduced. DSS allows users to analyze the massive data collection and provide better solutions to the problem and most importantly enable better decision-making (Chow, n.d.). In today's challenging businesses, the work tasks have become more complex than before, thus the implementation of CIP on an information system is essential to assure organizations are achieving their goals and objectives. Furthermore, the development of artificial intelligence (AI) into the business information systems prevents human errors and eases the progress of handling complex tasks. The introduction of expert systems (ES) and knowledge management systems (KMSs) enhance the capabilities of information systems. ES provides advice in specific areas whereas KMS supports the creation, organization, and distribution of knowledge within the organization (Chow, n.d.). By improving the information system, the decision may be made more accurately and quickly with fewer errors which eventually boost its effectiveness (Chow, n.d.). Hence, this will enhance business performance and enable the business to face upcoming challenges.

The purpose of this report is to explore the usage of CIPs within an organization. This includes the use of CIPs in matters relating to human resource management, customer relations management as well as product development. The scope of this report has been narrowed to these three fields as it was determined that these fields have been the greatest affected by the adoption of CIPs by organizations. Additionally, to help the readers gain a better understanding of CIPs, an explanation of the background including the characteristics of CIPs have also been provided.

## 13.2   LITERATURE REVIEW

### 13.2.1   BACKGROUND OF CIP

In today's modern and globalized world, organizations around the world are faced with tackling issues not encountered in the past. Digitization has broken down barriers and commercial distances between suppliers and consumers around the world making it more convenient and at times more financially viable to do business with parties located around the world instead of doing so locally (Mora, 2014). An example of this would be doing business through e-Commerce websites such as eBay.com where consumers are free to choose from vendors from around the world in order to obtain goods. Because of this paradigm shift, prices for goods and services are now being controlled by the market and/or consumer as they now have access to countless vendors around the globe who may offer better prices or have better quality products (Mora, 2014). As a result of this, organizations around the world have had to adjust their business operations in order to survive and remain competitive, and one of the ways this is being carried out is through the practice of CIPs as organizations no longer compete on processes but the ability to continually improve processes so as to reduce the operations costs in order to increase or maintain the profit (Anand et al., 2009). When talking about costs, it is not only limited to the sum of materials required to manufacture a particular product and salaries, but also hidden costs that may affect an organization's bottom line such as scrap or wasted material leftover from the manufacturing processes as well as administrative requirements such as pen and paper (Mora, 2014). All of these costs can be considered as waste, and it is the aim of CIPs to eliminate this waste as much as possible (Mora, 2014). Organizations also have to put up with ever more demanding customers who insist on better products, higher quality and shorter delivery times, amongst other things (Mora, 2014).

In order to tackle all the challenges that are facing organizations in this modern age, organizations around the world have started to rely on CIPs to reach their goals, as it allows organizations to become flexible and operate without having a fixed and clear strategy (Mora, 2014). This is because CIP is a philosophy of change that involves seizing opportunities for improvements in all work processes and is also a method for implementing changes which are characterized by organization-wide involvement and incremental improvements of existing processes (Jørgensen and Hyland, 2007). The implementation of CIP is a development process in itself as organizations would generally begin CIPs with small and sporadic improvements before gradually increasing in magnitude and scale as member of the organization becomes more competent with key CIP behaviors and eventually CIP may become a part of daily routine and culture for organizations (Jørgensen and Hyland, 2007).

### 13.2.2  SUCCESSFUL IMPLEMENTATION OF CIP

For the successful long-term implementation of CIP, a number of criteria must be achieved as stated by Jørgensen and Hyland (2007) which are:

- A clear strategic framework as it must be in line with an organization's mission and vision
- Needs to be managed strategically both in the short run and in the long run, with goals in place at different milestones as well as being properly communicated throughout the organization
- An underlying supportive culture where the importance and value of CIP is understood by all within the organization

- An enabling infrastructure where the organization adopts an organization structure which encourages efficient two-way communication and allows for the decision to be made at different administrative levels
- Needs to be managed as a process with the adoption of learning or problem-solving processes
- The availability and access for members of the organization to a set of common problem-solving tools

While the implementation of CIP may seem simple, it is anything but as unsuccessful implementation of CIP can be attributed to a number of causes, with the most common being an organization's failure to properly align the CIP with its mission and vision, a lack of effective organizational mechanism such as leadership and knowledge base as well as a lack of participation and involvement from members of the organization (Jørgensen and Hyland, 2007).

The success of CIP is also depended on a number of quality assurance and quality management tools in their improvement activities and decision-making processes. Today there is a huge numbers of tools available to organizations; however, it is important to choose which tools suit an organization best as the right tools would greatly benefit an organization, the wrong tools could jeopardize the organisations future, hence it is important which tools are to be used at which point in time (Soković et al., 2009). Tools are generally considered as a means of accomplishing change and out of the numerous tools available, this paper will focus on the seven fundamental quality tools which are the basis for all the other tools, the seven fundamental quality tools being flow chart, Pareto diagram, check sheet, control chart, histogram, scatter plot and cause-and-effect diagram (Soković et al., 2009). These simple yet effective tools of improvement are widely used as "graphical problem-solving methods" and as a general management tool which can be used in process identification and/or process analysis, with an illustration (Fig. 13.1) of which of the two process can the each of the seven fundamental quality tools perform (Soković et al., 2009). The successful application of the quality tools as mentioned before, the presence of an implemented quality management system is an advantage as it can act as a starting point for an organization's management striving for continuous efficiency improvement over the long-term and customer satisfaction (Soković et al., 2009).

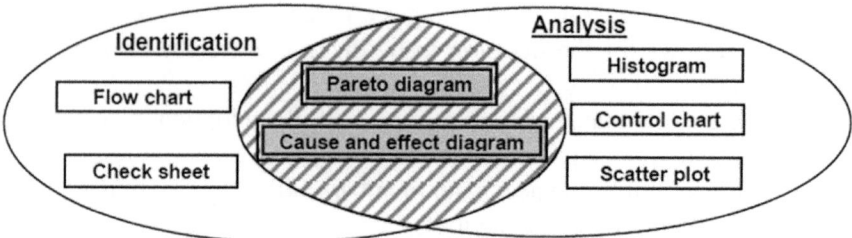

**FIGURE 13.1**   The simple improvement tool; graphical problem-solving methods.

CIP as a principle could not have been realized without quality tools which are presented through four groups of activities of Deming's quality cycle or plan–do–check–act or plan–do–check–adjust (PDCA)-cycle as shown in Figure 13.2 (Soković et al., 2009). The PDCA-cycle is an integral part of the management process and is designed to be used as a dynamic model as one cycle represent one complete step of improvement, which is why it is being used to coordinate CIP efforts (Soković et al., 2009). The PDCA-cycle emphasizes and demonstrates the fact that CIP must be carried out with careful planning in a never-ending continuous cycle in order to achieve innovation in a number of business objectives such as safety, quality, and morale (Soković et al., 2009).

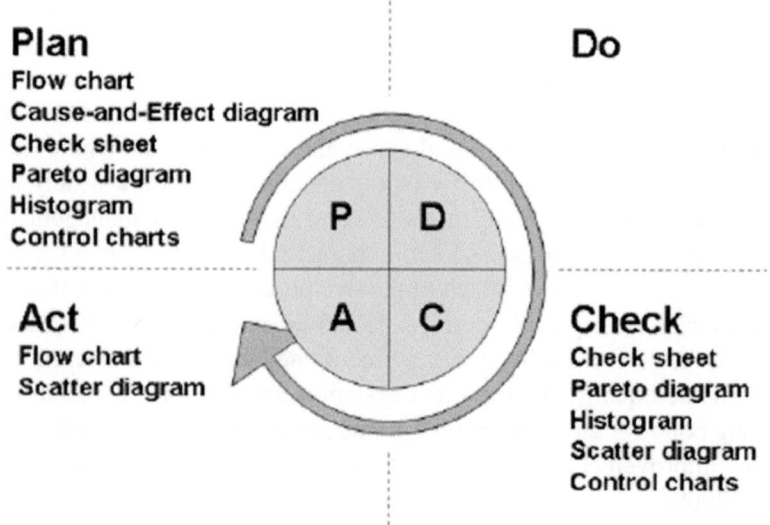

**FIGURE 13.2**   Deming's quality cycle; PDCA-cycle.

The PDCA-cycle consists of four stages, namely:

- Plan: At this stage, opportunities from which improvements can be made are identified and decisions to be made on what changes can be made to benefit from those opportunities,
- Implementation: Implement the changes that were decided upon in the previous stage,
- Check: Check control and management processes and products to ensure they reach the standards as set previously, and
- Act: React to the changes made or repeat the PDCA-cycle again to maintain continuous improvement (Soković et al., 2009).

### 13.2.3   CIP IN HUMAN RESOURCE MANAGEMENT

In today's highly competitive business environment, close interactions between customers and organizations are essential to provide fully personalized products and faster delivery of goods or services. In order to improve and maximize effectiveness and efficiency, organizations need to have a proper monitoring and controlling system which can coordinate all aspects of planning in the processes of an organization. Hence, it is necessary to have a standard software package which caters to the requirements of the individual organization, and most importantly be compatible with the current system which would help organizations to gain a competitive advantage so it can build a foundation to serve its customers better. In addition, technological advancements will also increase the demand for computer-related workers. It is evident that continuous improvements are needed on the long-term well-being of the organization in order not to be eliminated from its competitors.

According to Moon (2007), the purpose of enterprise resource planning (ERP) system is to integrate the MIS, both internally and externally across an organization. In layman's terms, ERP consists of one database, one application, and one user interface for the management to obtain information from every function which enables managers to plan, monitor and control the organizations activity. Furthermore, ERP system enhances manufacturer ability to accurately schedule production, fully utilize capacity, reduce inventory and meet promised shipping dates (Chartered Accountants India, n.d.). A large number of organizations use information system for ERP. It is crucial for the management of an organization to know how to use and apply any relevant information system at their workplace as the world is rapidly changing and improving whereby information systems actively supports core business. Boddy et al. (2008) stated that ERP enables the management to have access to current operating information which allows them to integrate customer and financial information, standardize manufacturing processes and reduce inventory, improve information for management decisions across site and enable online connections with suppliers' and customers' systems with internal information processing.

Prouty (2011) stated that ERP is also used for managing human resource activities which includes time attendance, time-off management, talent management, and other customized functions. They are either manually performed or part of a dedicated human resource system. Hence, ERP is able to influence the operation of human resource as it includes people, forms, policies, procedures, and data. When ERP is said to be linked with a human resource information system (HRIS) subsystem, therefore create a competitive advantage for that particular organization (Nagendra and Deshpande, 2013). Muriithi et al. (2014) explained innovations are crucial to tackle the challenges and determine the success of an organization. As technology advances, the use of information technology (IT) enables the development of a computer-based HRIS (Muriithi et al., 2014). Such innovation allows HRIS to not only perform simple functions but more complex ones. Nonetheless, the human resource system has to be fast, flexible and in line with trends as it can be used to determine the success of an organization (Dusmanescu and Martinovic, 2011). Likewise, it is fundamental to use information and communication technology to keep an organization up to date and to ensure continuous improvement. In addition Dusmanescu and Martinovic (2011) stated that the common terms for describing these systems include electronic human resource management (e-HRM), HRIS, and human resource management systems (HRMS), all of which would be further discussed later.

According to Dusmanescu and Martinovic (2011), HRIS primarily consists of collecting, storing and preparing data for reports, simplifying and accelerating the processes and controlling the available data, reducing labor costs for HR departments, and providing timely and diverse information to the management of the company, based on which it is possible to make quality strategic decisions related to human capital. The number of organizations who implemented HRIS to manage their daily human resources operations is gradually increasing. Muriithi et al. (2014) mentioned that in order for HRIS to have a positive impact on the overall growth of an organization, HRIS must satisfy the needs of the organization itself as well as its users. Effective human resource management takes place when the organization frequently updates information on the company's employees in the labor market where it provides a competitive advantage. Bulmash (n.d.) mentioned that various researches have indicated that companies who use technology to manage their HR functions experience advantage compare to those that do not. Most importantly, the Human resource department is able to focus on HR activities where managers can make effective HR related decisions. The HRIS includes systems and processes that connect the function of Human resource management and IT. The evolution of IT has made it easier and faster to obtain information and ease communication between employees by developing a proper HRIS system. Chakrabortya and Mansorb (2013) mentioned that people and information are the two important factors affecting the overall performances of a business. Therefore, the management of people and information are required in order to achieve greater success in the near future.

According to Chakrabortya and Mansorb (2013), organizations can perform human resource processes on a daily basis which benefits the organization itself by improving the efficiency and effectiveness and provides self-service human resource such as computer-based training and online recruitments. Furthermore, the automation of tasks and processes may lower the organization's HR costs (administration costs and process costs) and minimize the use of other resources such as financial, material and human (Chakrabortya and Mansorb, 2013). Boddy et al. (2008) also see that automating processes may cut cost which results in lower expenses to the organization. Based on a case study by Shiri (2012), it has been proven that HR cost is reduced after HRIS has been implemented and the efficiency of HR is able to produce more effective outcomes than that can be done in paper. Chakrabortya and Mansorb (2013) further explained that HRIS ease the data management for employees and line-managers where it has web applications to transfer data. Thus, the accuracy of data is ensured whereby employees can easily insert and update data on their own, hence save time and cost and largely reduce the probability of human error. According to Benfatto (2010), some other benefits of HRIS include effective human resource decision making, speed up transaction processing, reduce information errors and improve the tracking and control of human resource actions. Therefore, increase overall decision making efficiency for the management accordingly. HRIS enables easy access to obtain any relevant information of the employees in the company as managers can access anytime at one click. Traditionally, information was kept by manual processes such as handwritten and in paper forms. Thus with HRIS, it will minimize human errors as all the data is in the system. In addition, HRIS also benefits the employees where employees may access to data anytime when required like changing employee's personal data. According to Chakrabortya and Mansorb (2013), in order to further develop their skills and knowledge, they are also encouraged to attend internal training courses online. This will then encourage and improve employees' decision making where they can obtain information from the system with minimal efforts. Thus, it is obvious that HRIS is one of the most important MISs, which plays an important role in human resource administration functions of an organization where it can identify trends, evaluate and manage costs and for comparison purposes (Nagendra and Deshpande, 2013). There are two types of HRIS classified based on their usage known as "unsophisticated" and "sophisticated". Things like payroll and benefits administration as well as employee absence records keeping electronically are listed as "unsophisticated". Sophisticated usage includes the use of IS in the recruitment and selection of employees, training and development, HR planning and performance appraisal. A study led by Nagendra and Deshpande (2013) explained that the relationship between increased usage of HRIS results in an increase in both effectiveness and efficiency of the organization. With higher effectiveness and greater efficiency, the organization is able to ensure continuous growth and capable to perform well. Thus, recruitment of employees, planning functions and training are the main contributions of HRIS.

HRIS plays an important role in the successful implementation of corporate strategy. Other authors see it as a strategic organizational resource. The company should analyze the objectives of their organization and develop a strategy at the corporate level and then coordinate with Human resource strategy or any other strategies. According to Nagendra and Deshpande (2013), Human Resource strategy integrates an organization's culture, organization, people and systems which support the organization to achieve its strategic goals. Human resource strategy is core related to its corporate strategy as it will determine the company's effectiveness and success. HRIS plays a leading role in computerized HR Systems. Aside from the corporate business plan, Human resource strategy plans should also synchronize with organizational information systems strategic plan to ensure continuous growth.

The introduction of a human resource information system or HRIS, to the human resource department, enables them to go from HRM to e-HRM which will enhance efficient HR services (Maier et al., 2012). According to Chuttu (n.d.), Technology Acceptance Model studies how an individual accepts information system and uses it to influence the system and their behavior toward using it. Chuttu added that individual motivation can be explained by three factors that are perceived ease of use, perceived usefulness and attitude toward using the system. The impact of human resource employees' perception of information system helps us to better understand the consequences of HRIS. When the organisation introduced a new information system in their department, every individual has a different attitude toward the acceptance of it. Some may have positive attitudes and vice versa. According to Maier et al. (2012), assuming that there is a relationship between attitudes toward a newly implemented IS and work-related outcomes such as job satisfaction and turnover intention, which can lead to better understanding of the perceptions of an IS and attitude toward using IS. If the employee knows how to use the system, he/she will perceive it as useful and have a positive attitude toward it. Shiri (2012) mentioned that HRIS automates administrative processes, support strategic decision making which will then enhances efficiency. Moreover, HRIS is also used for employee management (Nagendra and Deshpande, 2013), knowledge development, career growth and development and managers can access the information they need to legally and ethically (Shiri, 2012).

According to Maier et al. (2012), the introduction of e-HRM will have an impact on both employees' job satisfaction and turnover intention. If an employee's job satisfaction declines, the employee may be inclined to resign from their post (Cunningham, 2006). He added an individual will have a higher chance of turnover intentions if he/she is unable to commit to his/her job and unable to cope with change especially if she/he evaluates the change as a negative one which will then affect job satisfaction. Maier et al. (2012) stated that HRIS is the change-inducing object that influences that individual's job satisfaction and turnover intention during and after system implementation. During the implementation stage of HRIS, level of job satisfaction could be affected by the difficulty of the use of HRIS and poor communication regarding the reasons for the change may give a negative impression of HRIS. Introducing a new system may require employees to change work habits which will likely to create additional work and more stress consequently create a low level of satisfaction and jeopardized improvement and growth. Furthermore, job satisfaction is a major contributing factor to turnover intentions. An employee who experiences poor job satisfaction at work or finds it difficult to fit in the new changes role will quit his/her job. The objective of CIP is vague when the employee turnover rate is high as the organization could not fully utilize the system on their employees.

## 13.2.4   CIP IN CUSTOMER RELATION MANAGEMENT

In order for businesses to achieve competitive advantage, installing IT is one of the elements that help businesses to strive for it. Thus, businesses need to have a good right mix of updated IT, data management and employees' initiative to achieve such a goal. Bahrami et al. (2012) stated that investing in IT leads to businesses gaining profit and productivity. Furthermore, cost reduction is interrelated with IT. As a result, new product development time, manufacturing time and a number of employees can be minimized. Furthermore, it helps businesses to expand its activity area by creating good chemistry with its customers and suppliers (Bahrami et al., 2012). Generally, IT plays a key role in businesses. Beck et al. (2005) emphasized that installing IT in business generate advantages which include communication improvements, coordination and interaction with customers and supplies and customer relationship management (CRM).

Swift in Ngai (2005) defined CRM as creating a mutual relationship with its customers by developing a one-to-one relationship with these customers who affect customer retention, acquisition, loyalty, and profitability. CRM strategy focuses on creating and retaining profitable customers where it has evolved around three aspects include customer profitability, customer acquisition and customer retention as obtaining new customer has a higher cost compared to retaining the old ones (Swift in Ngai, 2005).

In today's advanced technological era, CRM has constantly been affecting the organization's marketing strategy in products and services. Such technological applications are computer and mobile phone. Business has to improve their technology continuously in order to keep up with the constant changes in the dynamic environment. Viljoen et al. (2005) mentioned that there are several technology applications which are used in the development of CRM. The three main technology applications of CRM are Operational CRM, Analytical CRM and Collaborative CRM (Viljoen et al., 2005).

Operational CRM refers to the service provided by the organization to handle customer's inquiries. This includes a customer-facing application such as sales force automation, enterprise marketing automation, and customer service. On top of that, customer call centers have been known the main component in the CRM system. As a result, customer's information will be recorded through all kinds of interactions which then allow the organization to gather customers' data (Viljoen et al., 2005).

Xu and Walton (2005) has also mentioned that the collected data via operational CRM will be analyzed completely to provide a clearer view of the customer. With this, the organization is able to examine their behavioral patterns and make significant adjustments to the businesses to deliver suitable marketing and promotional strategies to their customers. Moreover, according to Xu and Walton (2005), the processes that used to obtain data are capturing, storage, processing, interpretation and reporting of customer data that stored in data warehouses.

The combination of new and traditional communication technologies enables organization to have more effective interaction with customers. Therefore, collaborative CRM involved all members of organization, that is, suppliers and other partners to have a better response to the needs of customers. Furthermore, it also provides a point of interaction between the customers and the channel (Xu and Walton, 2005).

Apart from that, businesses have to constantly improve in both IT and information system. Customers have become more socialized these days where they would like to talk about their ideas, purchasing experiences and reviews and ratings. So in order for businesses to fit in the needs of customers, the newest CRM application software is E-CRM which is relates to electronic businesses. This application helps businesses to improve their interaction with customers more effectively. Xu and Walton (2005) defined e-CRM whereby organization that has the ability to contact as much of their customers and channels as possible through internet and intranet. In today's online businesses, it provides presales and post-sales services to their customers in order to develop a better relationship with them. Such service is collecting their customers' data while providing frequent contact with their customers hence this closed up the relationship between both parties. In addition, it also acts as a tool that creating trust between both parties and making their customers accepts the production of products and easily attract new customers and retain them. Furthermore, social CRM is known as a business strategy that is supported by technology platforms and social characteristics which engage its customers in a collaborative conversation in order to provide benefits for both parties in a business. Social has improved CRM by introducing a new element that is dealing social customers with conversation instead of just dealing with data. The emerging technology through social media has improved CRM by introducing a new element that is dealing social customers with conversation each other, such as; Facebook, Twitter, Instagram, YouTube, and LinkedIn. For instance, Southwest Airlines is a large brand on Twitter that has the ability to build a relationship with their customers by listening to their customers whom they have conversations with. In contrast, the old tradition of CRM is where businesses only can collect data and information and do not communicate with any social media.

Vilijoen et al. (2005) mentioned that improving the development of CRM technology is highly dependent on the level of IT used in building a customer relationship. There are four phases of development in CRM, namely:

- Non-IT-assisted CRM method, for example, the manual recording system
- IT-assisted CRM, for example, call centers and databases
- IT-automated CRM, for example, E-commerce
- Interactive CRM (i-CRM), for example, analytical CRM

The development of CRM technologies is identified in the following plan. According to Harvey (2003), predictive CRM is the idea of knowing your customers by analyzing complex data about them. On top of that, it is known as the subdivision of data mining that allows organizations to examine deeper or obtain new information from the existing one. Examining through the customer's behaviors allows businesses to obtain such information; this can be done by following the steps below:

- With predictive CRM customer's action can be tracked and determined whether the customers will withdraw their subscription plan, this is also known as Customer's churn rate. In general, whether customers will stop their subscription services from an organization (Xu and Walton, 2005, p 965).
- Potential customers can be identified and deliver the most premium customer services to them.
- Select the most effective sales channels to reach their customers. Blumstein in Harney (2003) emphasized that knowing which channels and by combining them together to deliver the best relationship to their customers.
- It can help businesses to choose the best targeted promotional materials to increase sales and improve customer services.

Hence, predictive CRM helps the organization to determine its customer behavior, which increases its effectiveness of the strategy. According to Shearer (2004) research showed that a Japanese-based computer and software reseller has increased its profit by 200% by using predictive CRM.

Following with i-CRM, which is known as the instant communication with customers in the website where the data are analyzed. Moreover, it helps customers to keep in touch with the organization. Therefore, i-CRM is used to enhancing a long-term relationship between both parties and improving both customer care and website productivity (Viljoen et al., 2005). On top of that, Trepper (2002) mentioned, i-CRM is also installed in e-commerce where customers' queries and difficulties can be resolved online instantly. Hence, this increases customers' satisfaction and trust.

In order to provide fully personalized services for customers, organization use i-CRM to gain a deeper understanding of the individual customers. For example, a customer service representative may discover specific customer preferences and needs through the live interactive web chat session and with this discovered information will then be integrated with other web CRM information and therefore adjust the website accordingly to the customer needs. After all, if the needs are accurately targeted, then the website contents can be adjusted accordingly to improve customer's satisfaction, thus increases customer's loyalty and sales. According to Trepper (2002) mentioned, the knowledge that obtained from i-CRM will then be combined with another web CRM knowledge (which convert the data into a valuable knowledge), hence it improves sales and customer relationships.

In the 1980s, CRM was a one-way communication where businesses send email communication to their customers but had no ways to track the responses from their customers. In contrast where today's CRM strategy works in two-way communication. The interactive technologies tools that used in CRM to provide interactive communications are:

- Collaborative chat: a live collaborative chat allows customers to press on a live customer service icon where a chat box will be popped out for customers to question on the products or shipping details. Customers details will be stored if he/she has ever shopped on the specific web pages with the registered account and therefore customer services representative will have addressed them by their name (Vilijoen et al., 2005).
- Self-help: Vilijoen et al. (2005) mentioned that traditional FAQs have evolved to today's intelligent FAQs, virtual agents or bots that can chat with customers, answer their questions and providing a suggestion as real as a live person. Technological advancements have enabled AI to be equipped with learning algorithms that further improves the rate of accuracy of responses experienced with each consumer. Furthermore, inquiries that are put on hold are redirected to a live agent. As a result, with the fast and accurate responses to customer's queries, it increases the customer's satisfaction and customer retention.
- Telephony: Aside from the capability of CTI to be integrated as web pages, a new feature is designed that allows consumers to input their

contact number where they can be reached for questions and other services. The requested call can also be transferred to a specialized CSR that possess related skills and knowledge on the products and services as advertised on the specified web page. This in turn, increases the probability of acquiring new customers as well as higher sales (Vilijoen et al., 2005).

### 13.2.5   CIP IN NEW PRODUCT DEVELOPMENT

For a business organization to gain a competitive edge over its competitors, the business must be the first to have its new and innovative product to be ready to be launched into the market. Today's business organizations often implement rapid prototyping to efficiently and effectively produce their product prototypes (Ahmad et al., 2015). Three-dimensional (3D) models created by the rapid prototyping process can help product development teams to have a better understanding of the design of their prototype in contrast with mere conventional drawings. The 3D model can be tested to check and analyze its performance. By rapid prototyping, business organizations have can find out the flaws of the design of the prototype and take necessary corrective actions at an early stage of the new product development process in order to meet expectations. Hence, this can largely help business organizations to save time and resources.

### 13.2.6   CIP IN ADMINISTRATION

In the past, business transactions were recorded manually by a clerk. This means that the process of recording business transactions was not only labor intensive but also has a higher risk of errors and loss of data if the records are misplaced (Mahar, 2003). In contrast, many of today's businesses adopt a TPS to allow for efficient and effective recording of business transactions (Mahar, 2003). TPS consists of many subsystems such as a payroll system, accounts receivable system, and accounts payable system (Mahar, 2003). TPS enables a business organization to eliminate the time it takes to process huge volumes of information and transactions. This is because TPS automates the process of recording transactions. This means that transactions are processed instantly. TPS processes and manages various documents such as a company's invoices, customer orders, and payments. For example, if the debt of accounts receivable is due soon, the TPS will issue a statement to remind payment (Mahar, 2003). With the ease of use of TPS, fewer employees are needed in the administrative department (Mahar, 2003). This can help companies cut down the cost of hiring many clerks.

### 13.2.7   CIP AND KNOWLEDGE MANAGEMENT

As cited by Assegaff et al. (2013), in today's competitive business environment, business organizations need to attain new information, utilize and manage it well in order to gain a competitive edge. Knowledge management is vital for a business's success. The main purpose of knowledge management is to encourage a learning environment in an organization whereby the organization can maintain and make full use of the knowledge and ideas of its employees. To fully utilize the intellectual property brought in by employees, organizations have encouraged interaction and knowledge sharing amongst employees. With the advancements in technology, people can now interact and communicate with each other in a virtual way within a virtual community. Organizations create virtual communities to facilitate communication, knowledge sharing and information exchange on specific work-related problems across departments within the organization. Effective knowledge management can bring substantial benefits to an organization such as new opportunities for the business, improved business performance, innovative thinking, and improved customer satisfaction.

According to Assegaff et al. (2013), it is recommended for organizations to adopt a KMS to stay competitive in the market. KMS plays an important role in the knowledge management activities of an organization. Barber et al. (2006) argued that KMS can facilitate continuous improvement by providing a platform that gathers meaningful information. As cited by Assegaff et al. (2013), a KMS has two main functions. Firstly, a KMS allows effective employee communication, collaboration and knowledge sharing through networks. Secondly, a KMS functions to manage, store and retrieve information to allow better decision making by managers. Without a proper KMS, useful information and knowledge are not maintained and stored in the organization's database. That knowledge will surely be lost when the employees no longer work for the organization. In addition, a KMS is able to connect a user to the right personnel instantly (Assegaff et al., 2013).

### 13.2.8   CIP AND TOTAL QUALITY MANAGEMENT

Quality is a key indicator for analyzing the competitiveness of a business (Addae-Korankye, 2013). As cited by Addae-Korankye (2013), many organizations have adopted total quality management (TQM) to achieve a higher competitive position. TQM is a practice that aims to produce quality end-products that meet the expectations of the customer. Sheikholeslam and Emamian (2016) argue that TQM and CRM should work together to satisfy customers in order to contribute toward continuous improvement. TQM monitors and controls the quality at every phase of the production process (Addae-Korankye, 2013). Under TQM, the quality of products are improved and maintained and as a result, it can lead to reduced wastage. Furthermore, TQM can also help the business improve its customer loyalty, lower production costs, perform better financially and gain positive recognition from consumers (Addae-Korankye, 2013). According to Addae-Korankye (2013), measures must be taken to ensure the integration of the practice of TQM at all levels of the organization is done smoothly in order to achieve better outcomes.

## 13.3   DISCUSSION

HRIS contains various information so that managers can make use of the data to manage HR such as employees' job performance, recruitment of new employees, and other aspects of human resources. Therefore choosing the right tool or system is important for a company to achieve a competitive advantage. Aforementioned in the literature review, it is clear that HRIS is a system which an organization used to acquire, store, manipulate, analyze and retrieve information. By implementing such a system enable to maintain management control, allowing a much more efficient and cost-effective HR workflow. Normalini et al. (2012) commented that with faster information processing, HRIS improves administrative efficiency as there are less complicated procedures compared to past times. Since most of the HRIS software is based online, employees can access anytime and any day they want as long as it is convenient for them provided that they have internet access. Furthermore, more accurate information can be obtained when the data is analyzed and manipulated by the system itself. Using data or information in the system can ease the procedures of generating a report. Thus, higher management can make use of these data to review its weakness and most importantly to do a comparison with other competitors in the market. Ultimately, HRIS reduce costs such as administration costs like papers, printers, and ink. Besides, an automated system can minimize human errors as previously data is entered manually. Thus, it leads to a higher productivity of the HR.

HRIS ease the recruiting process, a company can now accept and filter online applications. This will save time for managers to review and sort out the applicants' information first before proceeding to the next step that is face to face interviews. Consequently, speed up the hiring process. Besides, HRIS recruitment systems also make sure the company is on track with the planned budget to ensure no other extra expenses incurred when recruiting. Furthermore, suggested that HRIS can be used for internal promotions as well. According to Rietsema (2017), adopting HRIS in recruiting employees will encourage their empowerment. This will help the managers to ensure only the right person gets the job as managers can observe the performance of an individual when employees themselves have control over their job. Trust is created when employees can make decisions on their own such as beneficiary information or any other personal information. In addition, HRIS can help make employees more autonomous (Rietsema, 2017), as they are able to access certain information which does not require managers' approval. An employee may proceed to their task right away which results in greater efficiency. In addition to that, more time is available for employees to focus on other tasks assigned. Perform automation of tasks and processes may lower organization's HR costs which have been discussing on the literature review Chakrabortya and Mansorb (2013). Moreover, employees are able to work on training courses when an organization introduces the Learning Management System. Rietsema (2017) mentioned that HRIS fosters communication between peers. This will create an open door policy and aid employee's engagement as they are able to voice out their opinions and provide feedback which will create a sense of belonging. With this sense of belonging some employees may feel motivated as they feel useful. This will enhance employee satisfaction which will then create deep loyalties. Thus, increasing the level of productivity. Rietsema (2017) also revealed that HRIS lowers risks associated with empowerment. Feedbacks from the customers enable managers to monitor employees' performance to make sure they are not abusing the empowerment. With the help of HRIS, managers now can easily view employees' productivity and reward to those who performed well.

Necessary training for employees is required to ensure continuous improvement and performing effectively. Unfortunately, HRIS solutions cannot provide direct employee training but it can be a valuable aid such as training aids or portals to train employees. When employees are exposed to company information, they can learn more about an organization added HRIS solutions help to track employees' training progress by giving online quizzes or tests as friendly competitions will encourage employees to complete training faster with a better score. Thus, it is essential to establish a clear sense of goal and motives which gives motivation to them. According to employee training is closely interrelated to the employee's job performance and goal achievement. Like, managers will make use of the system that keeps a record of related details such as customer compliments or complaints, to monitor their job performances. With close monitoring, employees will perform at their best to hit their achievement milestone. Employee's personal information like skills, training and qualification are kept confidentially in the system. With HRIS, information can be extracted easily where succession planning is much easier. In simple words, HRIS can tell if an individual has met the requirement of that particular position. Sometimes, employees may be too busy and unable to attend training or some even avoid training intentionally which results in inconsistencies training. Therefore, a proper implementation of HRIS with online access to training portals will ensure employee involvement and to make sure they are able to train at their convenience and alert managers those who are absent from training.

Regardless of all the benefits how HRIS can bring to the organization, the adoption of HRIS is not easy as said. It can be frustrating if the system is down and takes a longer time to process tasks. According to (Rietsema, 2016), apparently human is the main factor why HRIS projects failed. The adoption of a new system is not accepted by everyone in the company as some employees are loyal to the old system and refuse to change. Technical knowledge is one of the biggest challenges (Rietsema, 2016). Employees may have insufficient technical knowledge of the system which results in the process of adopting a system more difficult and takes some time. Usually, before officially adopt a new system to the company, they will provide training to the employees to ensure they are aware of it and know how to operate the system. Some employees may require one to one training which results in a longer time of adoption. There could be some technical challenges when integrating the new software with existing system and software which lead to delay of the HRIS project. Issues may arise during and after the transition period and this will have a negative impact on the employees itself. It will discourage employees from using the system, which will then again affect their job performance and productivity. Furthermore, outdated hardware and poor internet connection may worsen the situation (Rietsema, 2016). For example, when the internet is down, employees are not able to perform their task as everything is online. Poor internet connection will affect their job performance and indirectly affecting their level of productivity. Sometimes, data may be entered incorrectly which will result in analyze wrong data.

In order to have a positive impact on the employees or the company itself, the implementation of HRIS has to be effective and do not hinder company performances. (Rietsema, 2016) suggested to actively promote the new system first before the real adoption of HRIS begins. This will create a physiological effect whereby employees have positive image of the system and they will be excited to know and learn more about the system. This will also reduce the possibility of employees who refused to give up the old system. Extra time and effort are needed to train certain employees to make sure they adapt to the new system well. Furthermore, the company has to make sure there are people available with both technical and HR knowledge to fix any bugs occurred during the transition period. A slow system frustrates the users and reduces employee satisfaction and decreasing their productivity. Therefore adopting a user-friendly hardware and system will ease employees in using the system. Stable internet connection is essential to support the system. Hence, boost employee's productivity. Managers should also perform housekeeping: system maintenance (Rietsema, 2016) to make the sure system run smoothly and is still meeting the company's objectives. The company can determine any changes that are needed at this stage. Annual maintenance is required to evaluate the effectiveness of the system and also determine the need for improvements. Companies should keep themselves up to date by attending relevant meetings and conferences regarding the HRIS they adopt. Thus, with all these recommendations, the continuous growth of the company in meeting their goals and objectives is ascertained.

As mentioned in the literature review, CRM strategy could enhance a long-term relationship with customers as CRM focuses on obtaining the knowledge about customers and analyze it to improve the interaction between both business and customers (Jelonek, 2015). Hence, this creates a mutual relationship with them. In order for businesses to shift where they had a stand in before, they have to continuously improve by installing such an information system and IT. Therefore, linking back to the literature review mentioned above, operational CRM, Analytical CRM, and Collaborative CRM are the three main technology applications that have been used in the development of CRM.

Operational CRM often was known as front-office which focuses on automation and enhancement of business process through customer-facing such as sales automation, marketing automation and service automation (Jelonek, 2015). Sales-force automation is an application that business uses to automate its tasks such as sales processing, tracking customer interaction, inventory control, and analyzing sales forecasts (MSG Experts, n.d.). According to MSG Expert (n.d.), marketing automation is a software application that businesses used to automate and measure its marketing workflows and task. This, in turn, will increase operational efficiency and revenue. Service automation- Service staff and manager of the organization used service automation as a supporting system to achieve theirs objectively. The domains of service automation are contact centers, call centers and help desk. By implementing this application, it enhanced productivity where the number of customers is organized and schedule and the time taken to provide services are reduced therefore this increasing productivity (Jelonek, 2015).

Analytical CRM is the process of analyzing the data collected through operation CRM and later examines the customer's behavior and implements the most suitable strategies for them. According to MSG Expert (n.d.), analytical CRM is a platform that uses the analytical application to predict scales and optimize customer relations. Therefore, implementing analytical CRM could result in several benefits which ensure continuous growth of the organization. Like, retaining potential customers and attract new customers through the use of advanced analysis. In addition, it can also help in meeting the customer's needs and improving the relationship for both new and old customers efficiently. Hence, this builds up the customer's satisfaction and enhances business long-term growth.

Collaborative CRM is handled with the synchronization of the communication between customers and channels. Moreover, different departments from an organization such as sales, finance, and services are involved in the collaborative CRM where the customer's information are shared between each other in order to have a better understanding of the customers. Businesses who have implemented collaborative CRM could end up receive several advantages (MSG Expert, n.d.) which includes:

1) Valuable communication between customers and channels.
2) Online collaborative helps to cut down service cost of customers.
3) Understands the customer's detail during interaction and provide them with the best solution.

Product differentiation and product development contribute to the continuous improvement of a business. With rapid prototyping, a company can shorten the time it takes to develop new products (Ahmad et al., 2015). For example, a toy company wants to add a new airplane product into its product portfolio. The company can use computer-aided design (CAD) software to quickly produce an appropriate design of the airplane. Once the design of the airplane is complete, the product development team can use a special machine to produce a three dimensional model of the airplane within a short period of time. Then, the model of the airplane can be examined and tested by the product development team to monitor the take-off, flight, and landing of the airplane model as well as to check for errors. In addition, the prototype can also be used by the marketing department to conduct focus group tests to verify that the prototype actually meets the expectations of focus groups. If the prototype of the airplane proves to be satisfactory, the Computer Aided Manufacturing software can be implemented to use the electronic drawings and product specifications of the airplane in order for machines to manufacture the parts of the airplane and assemble the airplane. As a result, the integration of rapid prototyping, CAD and CAM can lead to continuous improvement in terms efficiency and effectiveness of the design and manufacturing processes of a new product.

On the other hand, if the prototype was found to have some flaws in its design or did not meet customer expectations, the airplane will have to be redesigned again using CAD. Even so, rapid prototyping can still help companies save time and resources because it helps companies discover errors in the product earlier, enabling the product development team more time to redesign the product.

Furthermore, by enabling new products to be developed quickly, rapid prototyping can also allow a company to exploit the demands of its target market before its competitors do. The product will be viewed by consumers as innovative since it is the first of its kind in the market. This can result in the company gaining a large market share and even becoming the market leader. Being at the right place and at the right time with a new and desired product allows a business to grasp the opportunity to exploit the market and improves its organization's profitability. This shows that by reducing the time to market, rapid prototyping can contribute to continuous improvement in terms of profitability.

TPS can enable a company to continually keep its business transactions updated through real-time transaction processing (Mahar, 2003). With current information, managers can make better decisions. Hence, TPS can improve organizational performance as well as support continuous improvement. TPS can automate the process of sending messages to debtors for reminding payment whenever payment is due (Mahar, 2003). This will greatly help the business improve its cash flow as well as its financial capability.

In addition, TPS can also increase the efficiency of inventory management (Mahar, 2003). Whenever inventory of a product drops to it reorder level, the TPS will automatically order new stocks from a supplier to maintain proper inventory levels. Due to the automated processing performed by TPS, fewer staff members are needed in the company's administrative department and management can reduce the cost of hiring many clerks (Mahar, 2003). This means that more resources are available to be allocated to more important business activities. By strengthening the financial capabilities of the business and enabling better decision-making by managers, TPS can provide an opportunity for continuous improvement.

Knowledge management is one of the most important activities in a successful business. KMS can be integrated into a company's business functions for the company to achieve continuous improvement (Barber et al., 2006). By adopting a KMS, a company can improve its overall business performance, facilitate innovative thinking and improve customer satisfaction. In addition, KMS enables employees and managers to have easy access to a large pool of useful and informative resources (Assegaff et al., 2013). This allows management and staff members to make better decisions when undertaking important tasks and eventually, leads to continuous improvement in terms of better decision-making in a timely manner.

By storing the knowledge of key employees into the organization's database, KMS can retain useful information and knowledge of those employees within the organization (Assegaff et al., 2013). Therefore, even if managers or key employees leave the organization, their knowledge and expertise will not disappear from the organization but remain with the organization. The knowledge and information can easily be retrieved by the KMS. This would lessen the negative impacts of employee turnover.

Furthermore, by connecting employees together, KMS can allow employees to communicate online and brainstorm ideas together to solve specific work-related problems. Besides that, KMS can gather and store data about the knowledge and expertise of knowledgeable employees in the company's database (Assegaff et al., 2013). This useful data can be easily accessed by new employees later on in the future to find solutions to a similar problem. The employees need not to spend so much of their time trying to find answers. Therefore, KMS can reduce the time it takes for new employees to tackle a past problem and hence, improves employee productivity. As a result, KMS can lead to the continuous improvement of an organization's business activities.

CIP and TQM

A successful business must ensure that the quality of its products is consistent in order to keep its customers satisfied. TQM ensures that only quality products that comply with customer expectations are allowed to leave the production floor and to be marketed to potential clients. TQM leads to continuous improvement in the reputation for quality of a business because TQM works to ensure that quality is always maintained and consistent (Addae-Korankye, 2013).

TQM can help a company saves costs as it prevents the need to withdraw a batch of faulty product items from the market (Addae-Korankye, 2013). Besides, TQM helps to maintain the reputation of a company for its products. For example, a newly bought electric kettle is being used and is left unattended. A few minutes later, it catches fire due to a defect in its thermostat. This incident would hurt the reputation of the company producing the electric kettle. On top of that, the company would lose the trust of its loyal customers. Sales of the electric kettle would plummet as consumers begin to associate the product or brand with a fire incident and switch to a different brand. Eventually, it leads to a fall in market share. In addition, the company would incur the cost of having to withdraw the batch of possibly faulty electric kettles from the market. The company would also need to take measures to regain the confidence of consumers in its product. All of these facts made it seem worthwhile for companies, especially for those in the electronic industry, to adopt a TQM system to prevent these unprecedented problems from happening.

According to statistics, companies that have achieved quality awards seem to have been performing better than those without one (Addae-Korankye, 2013). By winning quality awards and quality certificates, a manufacturer or business is able to gain international recognition for quality management. Hence, the business is in a better position to reassure its clients that its product is of good quality and instill confidence in its clients that they have made the right choice. This means that TQM together with quality awards can help a business improve its customer loyalty.

To achieve continuous improvement, TQM should focus on satisfying customer demands (Sheikholeslam and Emamian, 2016). This can be achieved by integrating TQM and CRM. A business should continually gather information about the expectations and demands of its consumers through CRM activities. This information can be used to set a benchmark for the quality of a product that the business needs to achieve or even exceed through TQM activities. Through the integration of CRM and TQM, a business can produce high-quality products that meet and exceed customers' expectation and satisfy its customers. Because finding out customer demands through CRM activity is a continuous process, the benchmark for quality will also gradually increase. Therefore, TQM enables continuous improvement in terms of quality and improved customer satisfaction.

Addae-Korankye (2008) highlighted that how TQM is integrated into the organization just as important as implementing it. The practice of TQM can radically change the business processes, especially the manufacturing process in the organization. Like all management practices, TQM needs to be integrated and communicated throughout all levels of the organization. Therefore, Information Systems can play an important role in the successful implementation of TQM by allowing effective communication between lower-level employees and the upper management.

## 13.4   CONCLUSION

In conclusion, it is important to choose which tools benefits an organization best as the right tools would greatly benefit an organization. Continuous growth of an organization is to make sure they are not eliminated by its competitors. With the help of HRIS, an organization can easily improve administrative efficiency, enhance employee communications, analyze and manipulate data with greater information accuracy, and improve overall HR productivity. Reduced labor cost appeared to be one of the many advantages. Automation of tasks and processes may lower the organization's HR costs and minimize the use of other resources such as financial, material and human. Effective HRIS generates employees' productivity and enhances decision-making process. HRIS do have some drawbacks, however with proper monitoring and planning problem could be solved. Thus, selecting the right HRIS software can benefit the organization and enhance continuous growth. In addition, the development of IT results in an improvement in the relationship between the business and customers is many ways. Collecting and analyzing of customer's purchasing patterns and behavior enable the organization to distinguish specific products and services that are preferred by customers. As a result, businesses could easily target the market by providing the right products or services they desire. Thus, increases profit as well as customer's satisfaction and retention. While there are several factors contributing to the success of IS, CRM achieves the organization's communication and information needs through the combination of marketing, services and IT. In order to successfully carry out this strategy, the organization has to constantly improve in building a relationship with customers. However, implementing CRM is considered as a huge investment in IT where not all sized businesses are affordable, but profitable output is foresaw. Apart from that, businesses must ensure that they have chosen the right CRM strategy that fits in their organization and being applied properly in order to gain profitable outcome and sustain a long-term growth. With that, organizations are capable to perform well in the long term. The adoption of KMS can improve an organization's business performance as well as its ability to make better decisions through effective knowledge management. However, before adopting a KMS, managers should take into consideration the cost of KMS as it can be very expensive. Only if the cost of implementing a KMS can be justified by the benefits that come with it, should an organization consider adopting a KMS? In order for TQM to support continuous improvement within an organization, it needs to be customer oriented. Hence, by constantly finding out what customers want through CRM activities and produce quality products that meet customer expectations, TQM enables a business to continuously improve customer satisfaction.

## KEYWORDS

- **continuous improvement programs**
- **technological advances**
- **wide involvement**
- **incremental improvements**
- **transaction processing systems**
- **artificial intelligence**

## REFERENCES

Aartsengel, A. V.; Kurtoglu, S. *A Guide to Continuous Improvement Transformation Concepts, Processes, Implementation*; Springer: Berlin, Heidelberg, 2013. DOI: 10.1007/978-3-642-35904-0_2.

Addae-Korankye, A. Total Quality Management (TQM): A Source of Competitive Advantage: A Comparative Study of Manufacturing and Service Firms in Ghana. *Int. J. Asian Soc. Sci.* **2013,** *3* (6), 1293–1305.

Ahmad, A.; Darmoul, S.; Ameen, W.; Abidi, M. H.; Al-Ahmari, A. M. Rapid Prototyping for Assembly Training and Validation. *IFAC-PapersOnLine* **2015,** *48* (3), 412–417.

Anand, G.; Ward, P. T.; Tatikonda, M. V.; Schilling, D. A. Dynamic Capabilities Through Continuous Improvement Infrastructure. *J. Operat. Manag.* **2009,** *27* (6), 444–461.

Assegaff, S.; Hussin, A. R. C.; Dahlan, M. H. Knowledge Management System as Enabler for Knowledge Management Practices in Virtual Communities. *Int. J. Comput. Sci. Issues* **2013,** *10*, (1), 685–688.

Bahrami, M.; Ghorbani, M.; Arabzad, M. S. Information Technology (IT) as An Improvement Tool for Customer Relationship Management (CRM). *Proc. Soc. Behav. Sci.* **2012,** *41*, 59–64.

Barber, K. D.; Eduardo, J. M. H.; Keane, J. P. Process-based Knowledge Management System for Continuous Improvement. *Int. J. Qual. Reliab. Manag.* **2006,** *23* (8), 1002–1018.

Beck, R.; Wingand, T.; Konig, W. The Diffusion and Efficient Use of Electronic Commerce Among Small and Medium-sized Enterprises: An International Three-industry Survey. *Electron. Mark.* **2005,** *15* (1), 38–52.

Benfatto, M. *Human Resource Information Systems and the Performance of the Human Resource Function*; 2010. www.eprints.luiss.it/653/1/benfatto-20100224.pdf.

Chakraborty, A. R.; Mansor, N. N. Adoption of Human Resource Information System: A Theoretical Analysis. *Proc. Soc. Behav. Sci.* **2013,** *75,* 473–478.

Chow, D. Evolution of Information Systems; n.d. http://www.hkiaat.org/images/uploads/articles/AAT_Paper8_Oct09.pdf.

Chuttur, M. Overview of the Technology Acceptance Model: Origins, Developments and Future Directions; *All Sprouts Content* 2009, Paper 290; n.d. 14.161.30.37:9989/viewfile/test/pdfs/TAMReview.pdf.

Cunningham, G. The Relationships among Commitment to Change, Coping with Change, and Turnover Intentions. *Eur. J. Work Organ. Psychol.* **2006,** *15,* 29–45. http://dx.doi.org/10.1080/13594320500418766.

Dusmanescu, D.; Aleksandra, B. M. *The Role of Information Systems in Human Resource Management. Research Monograph on the Role of Labour Markets and Human Capital in the Unstable Environment*; 2011. https://mpra.ub.uni-muenchen.de/35286/ (accessed Aug 20, 2017).

Jelonek, D. *The Evolution of Customer Relationship Management System*; 2015. http://www.inase.org/library/2015/zakynthos/bypaper/COMPUTERS/COMPUTERS-01.pdf.

Jørgensen, F.; Hyland, P. Human Resource Development's Contribution to Continuous Improvement. Paper Presented at Academy of Human Resource Development (AHRD Conference), Indianapolis, Indiana, United States, 2007.

Mahar, F. Role of Information Technology in Transaction Processing System. *Inform. Technol. J.* **2003,** *2* (2), 128–134.

Mora, J. N. Continuous Improvement Strategy. *Eur. Sci. J.* **2014,** *10* (34), 117–126.

MSG Expert. *Strategic CRM: Maintaining Long Term Relationship with Customers*; n.d. http://www.managementstudyguide.com/collaborative-crm.htm (accessed Aug 25, 2017).

Muriithi, J. G.; Gachunga, H.; Mburugu, C. K. Effects of Human Resource Information Systems on Human Resource Management Practices and Firm Performance in Listed Commercial Banks at Nairobi Securities Exchange. *Eur. J. Bus. Manag.* **2014,** *6* (29), 47–55.

Nagendra, A.; Deshpande, M. Human Resource Information Systems (HRIS) in HR Planning and Development in Mid to Large Sized Organizations. *Proc. Soc. Behav. Sci.* **2014,** *133,* 61–67.

Ngai, E. Customer Relationship Management Research. *Planning* **2005,** *23* (6), 582–605.

Normalini, Md. K.; Ramayah, T.; Kurnia, S. Antecedents and Outcomes of Human Resource Information System (HRIS) Use. *Int. J. Prod. Perform. Manag.* **2012,** *61* (6), 603–623.

Rietsema, D. *Supporting Job Training with HRIS*; Jan, 2016a. https://www.hrispayrollsoftware.com/supporting-job-training-with-hris/.

Rietsema, D. *HRIS Issues that may be Undermining Your Productivity*; Oct, 2016b. https://www.hrispayrollsoftware.com/hris-issues-may-undermining-productivity/.

Rietsema, D. *6 Ways HRIS Helps with Employee Empowerment*, 2017a. https://www.hrispayrollsoftware.com/6-ways-hris-helps-employee-empowerment/.

Rietsema, D. *Using HRIS for Recruitment*, 2017b. https://www.hrispayrollsoftware.com/hris-and-recruitment/.

Rietsema, D. *Challenges to HRIS User Adoption*, 2017c. https://www.hrispayrollsoftware. com/challenges-hris-user-adoption/.

Shearer, C. *One-to-one CRM with Predictive Analytics*; 2004. http://www.crmbuyer.com/ story/36059.htm (accessed Aug 25, 2017).

Sheikholeslam, M. N.; Emamian, S. TQM and Customer Satisfaction Towards Business Excellence. *Int. J. Learn. Manag. Sys.* **2016**, *4*, 35–42.

Shiri, S. *Effectiveness of Human Resource Information System on HR Functions of the Organization: A Cross Sectional Study*; 2012. https://www.researchgate.net/file.PostFileLoader. html?id...assetKey...

Soković, M.; Jovanović, J.; Krivokapić, Z.; Vujović, A. Basic Quality Tools in Continuous Improvement Process. *J. Mech. Eng.* **2009**, *55*. http://lab.fs.uni-lj.si/lazak/assets/qc-tools-sokovic_zl-1-.pdf (accessed Aug 21, 2017).

Sritharakumar, S. *Human Resources Information System (HRIS)-enabled Human Resource Management (HRM) Performance: A Business Process Management (BPM) Perspective*; 2015. usir.salford.ac.uk/38034/1/PhD%20Thesis_S.%20Sritharakumar_Final.pdf.

The Institute of Chartered Accountants India. n.d. www.icaiknowledgegateway.org/.../ chapter-7-an-overview-of-enterprise-resource-planni.

Viljoen, M.; Bennett, J. A.; Berndt, A. D.; Zyl, C. *The Use of Technology in Customer Relationship Management (CRM)*; 2005. https://www.actacommercii.co.za/index.php/acta/ article/download/75/75.

Xu, M.; Walton, J. Gaining Customer Knowledge through Analytical CRM. *Ind. Manag. Data Sys.* **2005**, *105* (7), 955–971.

# CHAPTER 14

# CONTINUOUS IMPROVEMENT PROGRAMS: THE DIGITAL TECHNOLOGY ERA OF HUMAN RESOURCE MANAGEMENT

## ABSTRACT

Organizations face new opportunities and challenges as the use of technology in organizations to increase their productivity through supporting an IT emerging technology in the era of digital ecosystem, when management information tool to support organizations called by management information system (MIS) is applied. Most organizations are fundamentally supported by MIS to be able to handle various information and business processes smoothly. Here, continuous improvement (CI) is a process that involves strategies and goals to improve operation and production processes efficiently and effectively that eventually satisfy customer needs. CI allows people to share their knowledge and experience of learning systematically. Some of the vital elements of CI are all of the people in the organization must be involved from planning until evaluation, it should be a step-by-step approach to achieve goals, and it should be a cycle where the process continues to maintain their quality and productivity. In other hand, Six Sigma also have the concept of focusing on customer satisfaction and eliminating waste to improve efficiency systematically but both practices have a different cycle to achieve their goals. Six Sigma consists of five-step process minimize defects; defining the target customers and their problems, measure its performance, analyze the causes of having performance failure, improve the poor performance, and in control to maintain good performance to have DMAIC as their consistent improvements.

## 14.1 INTRODUCTION

### 14.1.1 THE EMERGENCE OF MANAGEMENT INFORMATION SYSTEM AND TECHNOLOGY

Globalization and technology are considered as essential drivers that have enhanced the importance of innovation and competition through information systems (ISs) (Achilov, 2016).

Due to globalization, organizations face new opportunities and challenges each day, and it is their duty to keep up with the changes and understand them to maintain or become more competitive (Mora, 2014). As the use of technology in organizations is increasing, it becomes a management information tool to support organizations, which is also called management information system (MIS). Most organizations are fundamentally supported by this system to be able to handle various information as to guide management for the purpose of running the business smoothly, either internally or externally (Flett, 2011).

### 14.1.2 THE IMPORTANCE OF HUMAN RESOURCE INFORMATION SYSTEM

Due to globalization, most organizations have started using IS in different departments, especially the human resource management (HRM) department. Traditionally, HRM is labor intensive, but in the modern era, HRM has changed to technology-intensive, that uses a computer-based system called human resource information system (HRIS) to improve administrative paper processing (Karikari et al., 2015). With the implementation of HRIS, this reinforces the strategic use of the human resource (HR) of a business, from manpower planning to budgeting and turnover analysis. This can eventually help the business or organization to function more smoothly (Karikari et al., 2015).

#### 14.1.2.1 OBJECTIVES OF HRIS

The purpose of HRIS is to produce a competitive advantage by providing more efficient information in order to develop effective decision making. It is very useful for HR as it can keep, recover, update, and analyze data easily in the organization. This system is continuously used to achieve some objectives such as tracking information about the employee's qualifications and their job, to have up-to-date information at low cost, and to have security and individual privacy (Karikari et al., 2015).

### 14.1.2.2   NEED OF HRIS

Organizations are usually using HRIS after the implementation of enterprise resource planning (ERP) and customer relationship management (CRM) to improve decision making (Doral and Martinovic, 2011). The reasons of using HRIS in an organization are as follows: it gives information about the operations to departments; it improves quality of time and decision making; and it produces different types of reports to enhance administrative departments (Karikari et al., 2015).

Each of the stakeholder, HR professionals, management personnel, and employees in the organization has various interest of using HRIS. First, HR professionals' main reason for using HRIS is owning a single database that contains all information about the employees. Second, management has several reasons for using HRIS such as the effectiveness and efficiency of overall decision making, clear business mission, and vision as well as transparency. Third, employees have the advantage of using HRIS because of their involvement where they are able to access data and make decisions (Doral and Martinovic, 2011).

### 14.1.3   THE IMPORTANCE OF CONTINUOUS IMPROVEMENT PROGRAM

#### 14.1.3.1   DEFINITION OF CONTINUOUS IMPROVEMENT

Continuous improvement (CI) is a process that involves strategies and goals to improve operation and production processes efficiently and effectively that eventually satisfy customer needs. The purpose of CI is to allow people to share their knowledge and experience of learnings in a systematic way (Bhuiyan and Baghel, 2005). Some of the vital elements of CI are all of the people in the organization must be involved from planning until evaluation, it should be a step-by-step approach to achieve goals, and it should be a cycle where the process continues to maintain their quality and productivity (Pineda and Madrigal, 2013).

### 14.1.3.2 HISTORY OF CONTINUOUS IMPROVEMENT PROGRAM

There are various programs that organizations' implemented due to global-ization. In the 1800s, incentive programs were a well-known strategy to motivate and satisfy their employees. In 1900s, the emergence of technology enabled management to analyze any problems that they encountered using IS. With continuous improvement program (CIP) and IS, organizations continue to improve and evolve, especially the manufacturing sector, where they produce large quantities systematically, not manually. In 2000s Kaizen, Lean Manufacturing and Six Sigma were introduced. These strategies focus on minimizing waste and improving efficiency. In addition to this, there are also some famous recent CIP such as total quality management (TQM), balance scorecard and many more.

### 14.1.3.3 THE NEED TO IMPLEMENT CIP

The implementation of CIP is important to increase skills and to share knowl-edge internally and externally such as online discussions, presentations, and many other strategies. With CIP, individuals can learn from previous mistakes and get the opportunity to improve themselves, that is, by sharing of knowledge and experience to enhance the value to learnings of individ-uals and their members. CIP can also increase opportunities by continuously having a step-by-step improvement (Ambler, 2015).

### 14.1.3.4 BENEFITS OF HAVING CIP

Though CI is hard to sustain, there are numerous benefits of imple-menting CI, such as, it creates a synergy between employees and manage-ment, resulting in enhanced transparency and development of mutual trust, employees' motivation and satisfaction because they have a sense of ownership by involving in CI that makes them a flexible and committed workforce, gaining and retaining satisfied customers because of improved innovative products/services, and it can reduced overstock inventory and time to produce products/services through lean production. Therefore, all of these benefits can increase the competitive advantage of the organization (Business Case Studies, 2017).

CI can help employee's morale by altering the mindset of the employee to neglect the idea of perfection. By having this mindset, employees can freely express themselves without the fear of failing and show the full extent of their capabilities with the backing of the company or organization on their side. Although CI does not condone mistakes as an excuse, the concept of CI should be used to broaden the mindset of employees by seeing mistakes as an opportunity to improve rather than seeing it as a failure (Langer, 2014).

Most common CI programs and the influence of ISs in their processes are used in many organizations. Some of the CI strategies and programs will be further discussed in this essay.

## 14.2   LITERATURE REVIEW

### 14.2.1   CI STRATEGIES

#### 14.2.1.1   KAIZEN STRATEGY ("THE JAPANESE WAY")

Most organizations are strongly influenced by a Japanese concept called Kaizen. It is a strategy of an ongoing improvement that focuses on improving every function in the organization that involves all of the employees to proactively work together in order to achieve long-term efficiency and effectiveness (G, 2015). By ongoing improvements means it is a cycle where it is usually referred to four steps; Plan, Do, Check, and Act (PDCA). This is when organization's practitioners; Plan to define problem that should be addressed and collect the appropriate data to solve the problem; Do to run and carry out the solution; Check to evaluate and compare both before and after results; and Act to share results about the changes and get feedback from others to solve the problem in the new cycle. In other words, the Kaizen approach is a problem-solving process (Watanabe, 2011).

Kaizen needs top management especially HR to provide training for employees such as job rotation, coaching, and on-the-job training so that employees will be involved in any decision making and committed as well as responsible to solve any problems and provide solutions for it in order to improve their organization. The recognition and empowerment of employees can change individuals to become better. It says that the major reason for the success of Kaizen is the national culture in Japan but in fact, the organization's strategies played the essential role in Kaizen's success (Watanabe, 2011). For example, in the case of motor manufacturing joint venture have a teamwork environment where both organization train and learn from each other about practices that they implemented respectively to become productive in the organization (Watanabe, 2011).

Thus, Kaizen is one of the best systematic approaches to implement in order to boost the employee's confidence to be involved in any events, and have a sense of learning and direction to achieve performance and quality improvement as well as organization's objectives (Van Aken et al., 2010).

## 14.2.1.2  LEAN MANUFACTURING PRACTICE (LEAN PRODUCTION)

Kaizen is a concept to build and involve all of the employees in the implementation of CI but in lean manufacturing, it is natural that both management and employees are involved. Lean manufacturing is a CI practice that focuses on eliminating waste to systematically enhance their performance and ensure sustainable success (Elbadawi et al., 2010). It involves Six Sigma, TQM, just in time (JIT), quality circles, and many more (Jasti and Kodali, 2016). Like Kaizen, lean also have a cycle throughout the process. To eliminate waste, there are five-step process to consider; identify the customers' expectations and satisfaction, identify and eliminate any steps taken on each part of the product that does not add value, create an organized flow of information of product to customers, customers are able to pull value to improve the product, and continue the process until all wastage is eliminated (Elbadawi et al., 2010).

Waste can have an impact on production (Elbadawi et al., 2010). Such waste, like overproduction, means the number of inventories exceeds the expectation with the addition of undetected fault thus, waste of time in waiting as well as transportation. Waiting means delivery is not on time and transport means unnecessary transportation within the operation. Another waste is over processing that consists of no value. This is possibly because of a lack of communication between management and employees. Apart from that, defect waste is when customers' expectation and satisfaction are not met due to management take action differently. There is also other waste that the organization should consider before begin with any CI strategies.

In the United States, most of the manufacturing industries are becoming more toward customer-oriented that makes lean manufacturing is widely known. Environmental Protection Agency (EPA) created and introduced toolkit for organizations to use and adopt lean manufacturing practice. There are a variety of cases and reports about the benefits of using this practice. Most of the organizations find lean eliminates wastage and prevents pollution, and at the same time, it can produce high-quality products with low cost that meets customer's expectation and satisfaction (Layfield, 2013).

## 14.2.1.3  SIX SIGMA

Like lean, Six Sigma also have the concept of focusing on customer satisfaction and eliminating waste to improve efficiency systematically but both practices have a different cycle to achieve their goals (Munson, 2012).

Six Sigma has the vision to produce products or services in an optimal standard. To achieve the optimal standard, the operation must not produce defects or deficiency for more than 3.4 per million. It consists of five-step process minimize defects; defining the target customers and their problems, measure its performance, analyze the causes of having performance failure, improve the poor performance, and in control to maintain good performance in order to have a CI (DMAIC , Define, Measure, Analyze, Improve and Control) (Munson, 2012). It is the top management duty to regularly check the progress of the operation and gives support for anything that they need such as technical, financial and political support. Apart from that, it is also their duty to analyze and evaluate results, develop any appropriate systems and continuously modify action plans to have consistent improvements (Vella et al., 2009).

## 14.2.1.4  EMPLOYEE INVOLVEMENT

Employee involvement is a process where organizations utilize and encourage employee participation aid further in achieving individuals and organizational goals. The relationship between employee involvement and organization's performance is significant as it enables the organization to look a little deeper on how the organization function as a whole and pinpoint the problems and make improvements that can be instrumental to the success of the organization (Sofijanova, 2013). In order for employee involvement to succeed, managers should abandon the command control system. Managers should act as facilitators and granting the employees more freedom. This allows the employee to feel more liberated and have the sense of working with a superior rather than working for a superior. Building teamwork out of employee involvement is a great way to minimize organizational hierarchy and increase the organization's performance (Sofijanova, 2013).

## 14.2.2   LIST OF PROGRAMS WITH THE HELP OF IS

### 14.2.2.1   TOTAL QUALITY MANAGEMENT

A tremendous number of organizations are using and adapting TQM as their strategy. This is because quality management is important to survive in the highly competitive business world (Ganapavarapu and Prathigadapa, 2015). TQM is defined differently over the years by different theorists. Some theorists define it as satisfying customers by involving employees, while others define it as an advanced tool to manage organizations and guidance for CI on quality that aims for a long-term success (Patyal and Maddulety, 2015). In general, TQM has three principles; customer-oriented by designing and delivering products or services to achieve customers' expectation and satisfaction; CI by maintaining technical processes; and teamwork between management and employees (IC et al., 2014).

### 14.2.2.1.1   TQM PROGRAM

TQM program (TQMP) is a series of stages with necessary strategic actions to improve the organization's competitive advantage continuously. It concentrates on cost reduction, enhancing high-quality products and services for customer satisfaction, and to develop employees' full participation or involvement (Gul et al., 2012). It is the top management responsibility to regularly check the progress of the operation and gives support for the organization's needs such as technical, financial and political. Apart from that, it is also their duty to analyze and evaluate results, develop any appropriate systems and continuously modify action plans to have a consistent improvement (Vella et al., 2009). With the success of TQMP, it can improve the organization's overall performance efficiently and effectively.

There are a number of programs that consists TQM process but the main program must divide into subprograms with a specified period of time to finish the operation. Organizations must measure their performance to know their progress in terms of objective and checkpoints and timeframe of the specific subprogram (Gul et al., 2012). Organizations with TQMP usually use the Kaizen strategy of PDCA method (Watanabe, 2011) and also the Six Sigma strategy of five-process DMAIC to minimize defects to have a successful TQMP (Jehangiri, 2017), with the addition of help and support of management's functions such as finance for equipments and HR for hiring and training employees (Gul et al., 2012).

### 14.2.2.1.2   The Role of HR in TQMP

HR professionals play a vital role in the implementation of TQMP. The HRD department has close alliance with TQM International Organization for Standardization (ISO) to have consistent operations, professional culture, and employee morale, improve product quality and meet customer's requirements (Gul et al., 2012). There are 3 stages that they need to undergo; preparing, implementing and sustaining. In the preparation stage, the HR department assists the top management in planning and drafting the strategy and an appropriate system, hiring HRD professionals either an internal or an external consultant to guide and achieve objectives, have a proper internal and external communication, make a decision with top management to proceed with the implementation of TQM. In the implementing stage, HR department and HR professionals work together to organize workshops and create awareness within the organization about the importance of quality. In the sustaining stage, employees are given training to learn in order to have more knowledge and develop the necessary skills so that employees feel more comfortable with managing change and develop consistent performance. With the three stages, it will achieve the objectives of TQMP (Bisk Education, 2017).

### 14.2.2.1.3    The Usefulness of IS in TQMP

Technology also plays an important role that consists of two parts; current and emerging technologies. Current technologies are a database management system, data processing, networks and other technologies that the organization currently uses. Emerging technologies are computer-aided engineering (CAD), internet, groupware, multimedia and many more (Open Learning World, 2011). When implementing TQMP, IS is essential to ensure high quality of information and maintain software cohesion with an appropriate design architecture and technology. IS generates quality information to become high level in giving accurate and reliable information, flexible in changing needs such as quick response and process time, and reduce cost in maintenance (Siam et al., 2012. At the same time, IS must be aligned with the organization's strategies to achieve TQMP's objectives as well as the organization's mission and vision (Siam et al., 2012).

An example of an industry that use IS strategy to have a successful TQMP is the aviation industry. Aviation industry is known for excellent quality of service because they achieve maximal effectiveness such as installing IS for maintaining quality and following international standard operations (Qasim and Zafar, 2016). They have their IS strategy to increase overall operation's performance; defects, speeds, design, and costs but before that, they usually use Force Field Analysis to understand IS strategies that they used.

One of the IS strategy that they are using is AvPro Software which is the powerful software tool for quality management such as easy database application and operation maintain independently, and also other stand-alone software like scheduling of employees and flight, maintenance of materials, asset management, and many others. Addition to that, a demo version for this software is also available for trial (Qasim and Zafar, 2016). Another IS strategy that they are using is radio frequency identification (RFID) application to assure the quality and effectiveness of administrations. RFID can be used to track the location of the travelers in air terminals if they are not around so that the planes leave on time. Apart from that, RFID is also used for security inspection and discharging after checking (Qasim and Zafar, 2016).

The role of IS in the aviation industry with TQMP is an improvement in customer satisfaction, management, maintenance, and communication. With IS, they can ensure that the customers can access information conveniently, management can save cost, flexible to respond and exchange information, and having an efficient communication. Moreover, IS also has the capability to identify issues such as maintenance, training and communication issues, as well as to have better quality control. This industry has a CI due to having proper management and IS strategies (Qasim and Zafar, 2016).

### 14.2.2.2 LEADERSHIP AND EDUCATOR PREPARATION PROGRAM (TO ENABLE A BETTER ONLINE EDUCATION EXPERIENCE)

The use of data is becoming an outstanding strategy for innovation and improvement in the field of education across countries. The division of data collection often delays the ability of researchers, consultants, and policymakers to access and analyze the data regularly made in an organization or educational institution (Carlos and Stéphan, 2016).

Thus, the term MIS appeared from the main element that supports management of an organization or educational institution to capture, process, store, and recover relevant, latest and demand-driven data and information for management functions as well as daily activities inquiries. It is essential to introduce the benefits of using computers in a working environment as it helps the educational institution in decision making and communication via the Internet or wireless local area network (WLAN). In a collaborative or shared environment in the field of education, the frequent users in exchange and sharing of vital information in managing their duties are the managers, policymakers, educators, and students. It also helps them in making decision and valuable statistics in recruitment, admissions, examinations, student enrolment, course schedules and exam results as it provides search engines for the relevant information from the Website (Aldarbesti and Saxena, 2014).

### 14.2.2.2.1  The CSLA in Educator Preparation Program

In this era of results-oriented in which all educators are valued by their impact on student attainment, Educator Preparation Programs (EPPs) are created to provide data on the outcomes of both student and program, and also CI evidence. Student Learning Collaboration Analysis (CSLA) is an approach to develop critical research on student learning. Some universities used collaborative analysis in their EPPs to announce the practice and make any improvement during an accreditation cycle. Hence, the CSLA is successful when administrators and faculty are strongly committed; protocols and processes are purposefully designed; loyalty to the CSLA technique; and a concentration on professional growth. Through CSLA, programs are able to produce significant data and meet the requisite for accreditation. In order for educator preparation providers to promote the importance of evidence to generate the preparation programs effectively is by developing techniques, tools, and process that achieve multiple purposes (Colby et al., 2016).

The CSLA is a platform where teachers analyze and assess students' work to support and guide students' learning academically. The CSLA framework provides guidance for educator groups to analyze the potential of student responses whether they meet the objectives and standards, and to identify the suggestions for improvement. Information in student learning such as multiple-choice-question tests, classroom assessments, written essays, videotapes, posters, or student-peer assessments, reveals from any evidence or data that the educators collect, which also known as student work. Thus, educators analyze the student work by their point of view and create any immediate assumptions to improve student learning which formed a supportive environment (Colby et al., 2016).

### 14.2.2.2.2  CRM Concept in e-Education (Student Relationship Management)

In this 21st century, the advancement of new technologies had brought the pace of change in an organization and for this reason, the way people live, work, and perform worldwide is the significant effect of it. Indeed, new and emerging technologies had replaced most of the traditional process of learning and teaching, and also the management of education. In educational establishment, IT had a major impact in terms of collecting data and information, communication around the world, and any relevant education systems. As a consequence, the essential of IT in education because most data and information are accessible via Internet; one of the main subject as students are required to become IT literacy; and help in reducing cost (Knezek and Christensen, 2008).

CRM is an organization's strategy to identify, attract, develop, and maintain a successful relationship with customer continuously with the intention to retain the increased number of profitable customers. The CRM concept is applicable to electronic learning (e-learning) field by defining the goal, strategies, adaptation, and implementation of CRM. The CRM strategy allows the student to interact with any educational institutions and also provides an educational institution with a clear picture of each individual and their relevant activities. Hence, with the aim to improve the current electronic education (e-education) system by emerging methods for managing relationships with students—called student relationship management (SRM). This concept is suitable for most educational institutions because the demands and desires of students have to be met that makes them belong to the service industry. Besides, an educational system and its subsystems (student, educator, and communication method) are called independent teaching and learning.

The demand of students with technology skills that increases simultaneously that causes the integration of CRM into e-learning extensive and challenging process. By definition, the term SRM is a business relationship systematic care between the students and university that inquires service quality. Therefore, student satisfaction and mutual trust are the evidence for CI once they graduated. In addition, SRM becomes a new vision in the education system, in which student comes out for learning opportunities and become the central subject of the teaching process. For example, teaching methods and ways of communication are customized by the students themselves (Marko Vulic et al., 2014).

### 14.2.2.2.3    Link Between E-Learning and HRM

E-learning is a concept of the training process that involves the use of information and communication technologies (ICT). This means the content of the training is presented using IT such as text, video and voice transference in order to create two-way communication between educators and students. Thus, users can either communicate directly through IT like video conference, virtual classes and chat; or indirectly through electronic mail (e-mail), assignments and councils. On the other hand, HR exists to attract, develop and maintain the skills and qualification of the employees with the intention to achieve an organization's goals (Sodagar et al., 2013).

In addition, HRM has inspired in generating e-Learning and sharing knowledge from individual to operational level and they also used this concept for the communication between the employees and the organization as a continuous change to meet vital needs in the market. Hence, there are five basic steps to implement an effective e-Learning as a change in an organization; analyze the employees' attitude as their performance can affect the development of organization; modify e-learning content to meet the employees' capabilities; let the leaders start first so that the training program will be effective; strengthen the practice of new skills with a follow up session; and evaluation of the e-Learning training program for future improvement (DeSmet et al., 2017).

Nowadays, the role of education has acknowledged as a major factor for human development where it also helps in developing socio-economic and technological aspects. In order to achieve educational objectives and goals, administrators and policymakers of educational institutions get information from sound ISs. This is because most ISs can be found in all developed countries that have education enterprise connection whether within (in individuals and institutions such as libraries and information centers) or outside countries (such as telecommunication and information networks). Apart from this, comprehensive educational IS should contain the needs for information to process efficiently to help in designing the education policies and management and also to provide valid information for educational decision-making in the future (Shafique and Mahmood, 2012).

### 14.2.2.2.4 The Usefulness of IS in Education

An individual with IT literacy is not only has a job opportunity in many sectors due to increased demand for IT professional but also, an advantage for them to perform work effectively by using productivity tools either offline or via Internet and WLAN within or outside countries. For example, purchasing the e-book for a particular module; using cloud computing to do, share and store data and information for education purposes; and get access to an electronic library (e-library) to find references. In educator's' perspective, they make use of ISs to do some research that functions as continuous process improvement in managing the students' learning skills and developing vital teaching methods. In students' perspective, the knowledge they gain either from an educational institution or individual skill and talent in IT will give a significant impact on their performance and commitment with their studies (Ellis and Castle, 2010).

Besides, most students are already exposed to ICT since primary school or at home. In tertiary and university levels, students are required to fill in the online application form and attached some documents before they get accepted in that educational institution for further studies. Hence, this online processing system is one way of education system improvements to reduce cost and for time management. What is more, it is unavoidable that students will encounter increase number of information for personal growth and empowerment in terms of learning and making decision effectively as a preparation for information-based environment and rapid change education system.

Moreover, many schools and educational institutions across the world used a grading system in education to measure the capability and performance of a student which based on points only. There are several types of grading systems from given a grade A to either a 'pass' or 'fail' only, according to the standard of grading systems they incorporated with (Ming-Lang et al., 2011).

Therefore, the National Education System for the 21st Century (SPN21) is introduced in which they emphasized on the importance of ICT in education. The MOE is providing wireless technology in the classroom and also change traditional library in almost every schools and higher education into the technology-based environment. Thus, they are using bar-code readers and search engines to find online reference and materials via WLAN and Internet connection. In terms of education administration, the computer system is the main management tool and automatically changes the operation of the organization. Also, ICT helps the communication between educators, policy makers, and students more effective by using e-mail, a relevant education website and e-Learning (Siti Norainna, 2016).

### 14.2.2.3   ERP PROGRAM

The entire planning and management of HRs can be integrated into one computer system called ERP. The ERP is a computer system that integrates internal and external administrative functions of HRs, finance, accounting and operations into one database (Jackson, 2010). ERP's are excellent systems for employees to calculate HR costs, provide real-time data access, file, and track employee information, administrative recordkeeping, contractual agreements with unions, and business functions of the organization (Tomal and Schilling, 2013).

### 14.2.2.3.1 The Importance and Benefits of ERP

Installing an ERP system has various benefits. Some of the benefits are business integration, adaptability, improved analysis and planning capabilities, use of most recent technology, enhanced efficiency, information integration for better decision making, quicker response time to customer questions, improved corporate image, enhanced customer goodwill and satisfaction (Parthasarthy and Leon, 2014). The benefits which can be quantified or measured are called tangible benefits. Tangible benefits of ERP are it can reduce lead time, manifold increase in business, improved inventory system, and reduced cycle time. Whereas, those which cannot be seen or tasted or measured are called intangible benefits. ERP's intangible benefits are customer satisfaction, reduced cost of quality, improved vendor performance, better information accuracy, fast communication, helps in decision making (Bagad, 2008).

ERP systems are valuable because they allow firms to precisely evaluate and tightly coordinate production capabilities and to cultivate responsive relationships with customers based on reliable and precise information. Additionally, firms can coordinate with suppliers to look after the entire supply chain more efficiently and effortlessly through links between ERP systems (Bendoly and Jacobs, 2005). Moreover, ERP systems as implementations are highly non-substitutable (Bendoly and Jacobs, 2005).

The connections encoded in ERP provide a roadmap for enhancing the structural elements of a firm's social capital. ERP data flows and network connections present a valuable opportunity to enhance a firm's configuration of impersonal links between people and units. For example, people and units that rely on ERP data have inherent interdependencies. If these interdependencies are made visible and if people are rewarded for facilitating effective coordination across parts of the system, then an enterprise-wide view of the firm can be developed (Bendoly and Jacobs, 2005).

### 14.2.2.3.2　The Implementation of ERP Program

The implementation of the ERP program is essential to achieve users' expectation and satisfaction such as the flow of communication within the organization; delivery, and up-to-date. With ERP system, the organization's operation, as well as the performance of the individuals in the organization, will be improved efficiently and effectively, thus achieving their strategies and objectives (Dezdar and Sulaiman, 2011).

### 14.2.2.3.3　Example of Successful Organizations Using ERP Program

The first example is the banking industry, where they have various common activities such as lending services, deposits, and money transmissions. Most banks have huge branch networks in which many of them spread across the globe. They need to maintain strict secrecy about customer affairs and follow stringent regulations. ERP is the most used system in the banking sector. It is used mainly by highest-level banks, which means only those banks who can afford the expense of implementing such a high profile software system. ERP software applications help synchronize and integrate widespread financial processes, HRM, and support services. It also practically covers all functional areas that define an effective banking system (Almeida et al., 2010).

The second example is Fuze Energy Drinks, where they usually faced some challenges such as the numbers of stocks are rapidly increasing and they need support from business functions because they have major finance problems. Thus, they implement ERP program to have a cost-effective strategy and minimize difficulties in terms of maintenance of their computers. They introduce ERP Sage Accpac which is a solution system that has particular contents that match the individual needs. By having this system, Fuze ensures that they have control time over stocks that are coming in and out the organization, the information of the stocks such as the expiry date, and making a decision within the organization (Wong et al., 2015).

The third example is LG Electronics (LG), like Fuze, they faced some challenges such as the cost of maintenance is extremely high, limited resources available, decision-making is not efficient, employees are not committed and responsible because they are lacking training and learning, lack of transparency and every operation processes is manual (Seth, 2014). Because of these challenges, they implement ERP performance management program solution thus, they are now minimizing the cost of maintenance, enhance employees' productivity due to they are now learning and training, transparent in terms of recruiting employees, and overall improve performance (Seth, 2014).

### 14.2.2.4 EMPLOYEE INVOLVEMENT PROGRAM

Nowadays business environment has drastically changed with the growing number of turbulence and unpredictability. Organizations have to seek new strategies to adapt to these changes to stay competitive. This turmoil brings about the importance of employee involvement where a skilled and talented employee can contribute with ideas and knowledge to solve problems thus utilizing the full potential of the employees and in addition to build a cross-functional relationship in the workforce (Sofijanova, 2013).

Employee involvement can be significant to the competitiveness of a company that wants to edge out its competitors. Gallop Organization, a company that made several researches on employee involvement in 7939 business units found out that significant changes in performance in a variety of area that includes decreased employee turnover, increase productivity and customer satisfaction were associated with employee involvement (Konrad, 2006). An effective way for an organization to conceive the notion of teamwork and equality in the workplace is to encourage the practice of employee involvement. This practice allows the employee to give creative input to the company and break free from the rigid job description and hierarchy that confines them from giving their own idea or perception on the matter in hand.

Employee involvement cannot be achieved overnight. It takes time, effort and expertise. This is partly due to the transition of culture and policy within the organization from a rigid top-down hierarchy to one that takes employees to make particular decisions. Some examples of employee involvement programs like think tank, monthly training program, safety committee program, idea campaign, peer picking program and participation management. There are several factors that can contribute to employee willingness to be involved such as incentives where it is in terms of monetary or recognition. Also, they received the support and encouragement from the superiors. Managers play a crucial role in the effectiveness of the employee involvement program. Furthermore, employees need to be given the jurisdiction to partake in making a significant decision.

### 14.2.2.4.1    The Integration of IS Helps Intensify Employee Involvement

The integration of IS will enhance the effectiveness of the employee involvement programs that the company is adopting. In the modern era, technology is vital for a business to stay ahead of its competition. The organization needs to embrace the technologies that are made available to them. The introduction of IS system can make it easier for managers and employee alike to use the given data to make a better decision. There is numerous platform for the company to utilize IS in their employee involvement programs. Here are some examples:

*Online Community*: Online community offers many cost-effective alternatives for staff and management alike to communicate with each other in a virtual sense. Managers or even the HR team can post newsletters, chat rooms, memos and upcoming event announcement happening within the company. In addition, developing an online group collaboration projects can be more than beneficial for the company where an employee can have the chance to communicate and interact with other peers and exchange valuable information in the context of work. The possibility does not stop there. An employee can also speak out any concern they have and give ideas.

*Discussion Forums*: Management can create online forums on the organization's network for the employees to communicate with each other and share about work-related thoughts and ideas. The online forums are only accessible for the employees and should have an authorized mediator to monitor all of the conversation happening inside the forum. Managers can also create a topic of discussion that applies to work projects where the employees can give their insight on the matter and help bring about solutions. Other members can also post valuable insight or information related to the company, for example, a change in the company's policy. This can be a one-stop domain for an employee to perform better with use of the information given.

*Team-based Online Employee Training*: Employees are getting used to the old-fashioned traditional training provided by the company. In conclusion, companies find it difficult to engage the employee in job training. Companies now opted for online training as a solution. Recent research has shown that the addition of a team-based online training increase employee's willingness to learn. In a team, individuals are more like to push themselves to seeing that their peers are doing the same as well. It is very effective rather than a solo training session. In team-based training, team members can share different ideas and eventually come up with the best possible solution. Moreover, this process can help them learn how to work in a team (Heinrichs, 2008).

Companies adopting Employee involvement with the help of IS:

*Xerox*: The well-known document solutions company launched a management-training program called *Stepping Up to Management*. The program sent participated employees on an online quest that simulated real work environments. Employees that face problems will be helped by management to monitor how well the employees work through the conundrum. The program also provides leader boards to show the progress of each participant making the training program competitive and fun. The best participant will be rewarded accordingly (Kuo, 2015).

*Hyatt Hotels and Resorts*: Hyatt is a multinational company with multicultural personnel. Hyatt introduces the Hyatt Thrive platform to tackle cultural barriers. It mirrors the social networking sites accessible for Hyatt's employees to socialize between peers nationwide. The employees can also post photos, ask questions, join forums in the platform in the context of their own respective franchise. One of the main aims for the introduction of this platform is to build long lasting camaraderie between peers and shows the unity conjured by virtual communication (Anon, 2017).

### 14.2.3   THE ROLE OF HR

#### 14.2.3.1   FUNCTION OF HR

Some theorist believed that the function of HR is to look at the organization at the standpoint of line managers, and not just be the HR specialists. Organization performance relies on HR's strategic plans and objectives (Hamid et al., 2017).

There is an eight-step model of HR strategy to have competitive advantage improve continuously. Through a review of literature of management innovations, they able to create a model which aligns the strategic role of HR with business processors to achieve improvement in quality, cost, speed, and services, in other words, to provide a competitive advantage in a changing scenario (Kaylani and Sahoo, 2011).

*Step 1: HR as a Source of Competitive Advantage*

There is greater recognition now that high capability and commitment of an organization are obtained from distinct organizational culture, highly developed employee skills, management skills, and processes. Today, HR is highly involved in activities that create committed, competent and customer oriented employees such as flexible pay structure, restructuring strategy, training systems, and flexible time management.

*Step 2: Environment Scanning Process that Assists HR in the Management, Development, and Design of HR Practices*

This analyses the organization's internal and external environment, looking at its strengths and weaknesses, and its opportunities and threats. The internal environment looks at the social, political, technological and economic aspects; whereas the external environment is about balancing structure, the culture of the organization and the HR systems and processes.

*Step 3: It Is All About the Vision*

The vision of HR provides long-term direction toward developing abilities and commitment of the workforces to make the organization competitive. HR vision is the transformation of beliefs to goals, culture to strategy, dreams to reality.

*Step 4: The HR Assessment of Improvements Needed for Flaws and Skills that Need to be Upgraded*

This gives better affiliation of HR with other resources as well as become a benchmark of the maturity level of the HR function in the organization.

*Step 5: Strategic Planning is Essential to Help an Organization Achieve its Goals by Having Efficient Workforce and Linking that with the Strategic Business Planning*

HR makes things happen and hence should be a focal point within the process of defining business strategy. While formulating SBP of the organization it is essential to diagnose, define and consolidate all the HR issues and integrate the results into the main HR strategy.

*Step 6: Defined Objectives to Fulfill HR Vision*

Objectives provide sustainable direction to HR function and help in the formulation of HR strategy. HR seeks to achieve objective using its own function and roles.

*Step 7: Accomplishment of Objectives by Having an Action Plan*

Focus on how objectives can be achieved, its methods, what needs to be done, by whom and when. Action planning increases the confidence of managers about the achievability of an objective and elucidates contributions and responsibility of each manager.

*Step 8: HR Evaluation Strategies of the Short and Long-Term Business Plans of an Industry*

It provides inputs required to assess all aspects of competitiveness and assign the HR score for the industry on a number of dimensions. Its main objective is to align the HR function with business goals or to create a business-driven HR function. HR uses various methods such as workshops, observations, and interviews. It has to be business-driven and comprehensive. HR audit has a tremendous impact in business in areas of strategic planning, top management styles, improvement in HR systems, culture and TQM interventions. Quality provides an HR audit is vital in identifying the reasons for success or failure and in developing a definite plan of future action.

HR looks into maximizing quality and minimizing costs (Mulang, 2017) and thus they help in planning and providing a strategy to managers and employees as guidance. With the help of HRIS, it can produce competitive advantage and maintain in managing their human capital to achieve the organizational strategies and vision.

### 14.2.3.2   HR AS A CHANGE AGENT

During the implementation of CI, the involvement of HR is essential. With HR's assistance, CI programs or initiatives in various fields can continuously enhance organizational and individual CI capability which eventually improves quality and productivity (Hamid et al., 2017). HR should be flexible to any needs of improvements to be able to cater to customers' needs as well as to work productively with various stakeholders to ensure proper HR process. To ensure future sustainability, HR should implement CI programs in order to maintain their quality and productivity, as well as to grow the strong partnership with various stakeholders in the organization which include senior managers and any those involved in the process of management of HR (Hamid et al., 2017).

For example, traditional organizations have a tendency of having a hierarchical organizational structure where participation is often lacking but most HR emphasize the importance of employee participation in the organization. Participation of employees is highly important to maintain an organizational climate of ownership, trust, and fellowship (Jorgensen et al., 2007).

### 14.3   RESULT AND DISCUSSION

#### 14.3.1   EFFECTS OF IMPLEMENTING CIP

##### 14.3.1.1   EFFECTS ON ORGANIZATION

###### 14.3.1.1.1   Organization's Performance

Implementing CIP can have an immediate increase in performance and productivity. These improvements were found to be as a result of technological developments and market competition. Organizations not only focus on giving high-quality products or services but also efficiency. By using ERP system, operations can get rid of unnecessary actions in order to minimize time and effort to finish up the products or services. Therefore, faster and less waste means organization is competent to serve more customers; either new or existing customers (Patyal and Maddulety, 2015). In addition to that, proactive employees can help boost up the productivity of a company. When employees are given the freedom they become more efficient in doing their work and also more self-sufficient without minimal help from the managers or superiors. This allows the managers or any other top management staff to cater to individual duties rather than micromanage the employees (Sofijanova, 2013). With this, the organization can improve their performance.

### 14.3.1.1.2   Management Involvement

Top management must be involved to do a proper plan and support operations within the organization. For example, an organization with Six Sigma practices has a huge responsibility where they need top management especially HR management to delegate and give training using IS in order to have knowledge, skills and eventually transformed the employees to adapt to Six Sigma culture (Hamid et al., 2017). Based on the literature review, the involvement of HR is vital to act as a mediator between management and employees and not only that, HR is more than just to manage its HR because HR portrays the role of change agent. With proper communication from HR, the process of CIP will be done more smoothly.

### 14.3.1.2   EFFECTS ON EMPLOYEES

### 14.3.1.2.1   Training and Learning

When implementing CIP, employees training and learning in IS is essential component of employees management to develop knowledge and skills in order to understand and have the ability to overcome challenges and provide solutions in their current situation and also in the future efficiently and effectively (Patyal and Maddulety, 2015). With training and learning in IS, employees become open-minded and eager to explore opportunities more depth for their own development as well as for the benefits of the organization. An outstanding organization has a well systematic and innovative training and learning methods and trainer professionals so that they will get employees with advanced knowledge and skills that eventually meet customers' needs (Patyal and Maddulety, 2015).

For example, training and learning in IS is a significant factor in Six Sigma practices. Employees must be given chances to grow and develop themselves through training and learning with customized needs of IS that is useful for individuals to adapt current techniques before implementing Six Sigma practices (Thomas et al., 2008). In terms of education, customized and interactive training and learning can make e-learning programs effective because each employee have different learning needs to be applied in their daily job (Mittal, 2008). Another example is most organizations that adopting TQM practices have employees that are self-motivated for personal development and they have an extraordinary workshop with learning environments by creating variety of IS training tools and techniques such as encourage them to use IS to process and analyze data, identifying issues or errors and as a sense of communication (Gul et al., 2012).

Therefore after training and learning, employees are more deeply understand the intention of having CI to achieve organization's strategies and objectives (Patyal and Maddulety, 2015).

### 14.3.1.2.2  Employees' Involvement

Employees can enhance their commitment, motivation, and satisfaction by giving them empowerment to give ideas and make decisions according to their assigned tasks. With this, it can develop teamwork thus, lead to excellent quality products and improve the organization's performance. Employee involvement can strengthen the relationship between employees and managers alike due to the fact that the employees are given more freedom and independence. It is a win/win situation for both parties. Hence, a workplace with more self-governance is more likely to minimize the dependency of employees on their respective managers or superiors (Sofijanova, 2013).

The top management must communicate openly and clearly with their employees frequently so that the employees can ask and respond to any situation by using the current trend, IS. Such IS that can be used in employees' involvement are online training, online community, and discussion forums. For example, when the organization going to implement TQM practices, top management should be consistent and clear when giving instructions to employees such as which IS to use and ways or stages to achieve goals and incentives for whoever produce quality work. Every employee must be involved in order to create respect and harmony, and apart from that, recognition and rewards are all that matters for employees.

### 14.3.1.3   EFFECTS ON CUSTOMERS

#### 14.3.1.3.1   Customer Satisfaction

The sense of communication is essential when dealing with customers because this is to know the quality expectation and satisfaction of the customers and at the same time, customers must give feedback by giving appraisal through IS to know if the organization met their requirements (Patyal and Maddulety, 2015). Organizations that implement CIP usually use the voice of customers data to detect problems or errors and customer's feedback. The success of CIP is when there are various customers who give feedback and compliment the organization and a huge number of loyal customers (Patyal and Maddulety, 2015). In other words, this can generate a snowball effect afterward that with great satisfaction achieved; the company certainly gained the trust of the customers hence customer loyalty.

### 14.3.2   CHALLENGES OF CIP

From the literature review above, IS supports most business processes when it is integrating with the supply chain of other businesses (Goncalves and Spateiro, 2008). However, many organizations faced some challenges during the implementation of CIP and how ISs can help them. An effective CIP is not limited to practicing process-improvement and problem-solving techniques. The program usually starts with enthusiasm progress, but often fails toward the end. Hence, one of the reasons is employees become demotivated and likely to practice their old habits. CIP is viewed as a competitive differentiator (Hammad, 2012). The most common critical factors were divided into seven problems:

#### 14.3.2.1   PROBLEM 1: OBJECTIVES

At first, an organization needs to clearly identify what are the objectives for the program and consider the possible consequences of it before the implementation phase is underway. According to some research, it is proof that the wrong objectives will lead to a failure. Thus, this will not only affect the organization itself but also the employees' attitude toward the program from the beginning. When the employees are less motivated and have a negative expectation with the program, this can results them in misunderstand, distrust and insecurity in which can cause the program to fail (McLean and Antony, 2014).

### 14.3.2.2   PROBLEM 2: ORGANIZATIONAL CULTURE AND ENVIRONMENT

Culture and environment of an organization must be considered because this can show their reactions in accepting the program and start practicing the changes. Some organizations may predict their abilities to cope with the new culture, but the main influence here is the employees themselves as when they find that the organization's culture is a contrast with their current ways, they will become unsupportive and likely to disobey. In addition, external environment such as the right time to present the program must be considered as well (McLean and Antony, 2014).

### 14.3.2.3   PROBLEM 3: THE MANAGEMENT LEADERSHIP

When there is no support and commitment from the management and leadership, the program is likely to fail for the reason that change program needs high levels of involvement from top management to the employees. Hence, they must demonstrate their leadership skills at all levels of management and set the change program to become one of their priorities in developing the organization (McLean and Antony, 2014). Lack of proper management caused the CIP ineffective.

### 14.3.2.4   PROBLEM 4: IMPLEMENTATION APPROACH

Next, the implementation approach is essential, which it should not be poorly implemented, select and perform as it can also cause the failure of the change program. Tools and techniques of implementation need to be integrated with current practices and have a roadmap from the start (McLean and Antony, 2014).

## 14.3.2.5   PROBLEM 5: TRAINING

It is HR duty to provide training for the employees to adapt with new environment or idea in the organization. Furthermore, with the implementation of CIP, the training program should have an accurate standard and application. Trainers will have the opportunity to manage themselves in learning into practice. Unsatisfied employees and a high number of employee turnover could be an issue especially in terms of financial capacity in hiring and training new trainees (McLean and Antony, 2014).

## 14.3.2.6   PROBLEM 6: EMPLOYEE INVOLVEMENT LEVELS

The employee's' desired contribution could be impacted from some issues such as role conflict, time allocation, and participation levels. Consequently, the new CIP will need more number of employee involvement and empowerment in order to success and top management should consider how practical the new changes to employees are in the organization. This is because some employees will feel insecure about the change program especially when ICT are involved as they face difficulties to change efforts (McLean and Antony, 2014).

## 14.3.2.7   PROBLEM 7: FEEDBACK AND RESULTS

Then, the evaluation phase takes place. Failure to get feedback and suggestion from the implementation of CIP could also lead to failure in management. Indeed, inaccurate or poor review and communication are the major issues in order for the organization to improve their operations. A follow-up feedback on any change program provides a picture of the future plan in terms of financial, structure and content of the training program (McLean and Antony, 2014).

Apart from the seven problems that were listed above, several technologies and methods are often lead to IS implementation and development failures. Goncalves and Spateiro (2008) also stated that failures in developing IS are due to the organizational culture and environment in which it changes the ways the employees communicate and work together. This is because the employees have their own preference and find the traditional practice is efficient and quick as they already used to it for many years (Goncalves and Spateiro, 2008).

## 14.4 CONCLUSION

In conclusion, most of the organization implements CI strategies to enhance their competitive advantage and to have a sustainable long-term success. Such strategies like Kaizen, lean production, Six Sigma, and employees' involvement are common strategies that organization used in their programs such as TQM, leadership and educator program, ERP program and employees involvement program. Management, employees, and customers have their own important or essential role to get quality expectation and satisfaction internally and externally. There are varieties of benefits of CIP but there are some challenges that the organizations will encounter during the implementation of CIP. This is when management, employees, and customers work together to minimize risks as well as to improve their performance.

## KEYWORDS

- DMAIC
- Six Sigma
- continuous improvement
- digital technology era
- digital ecosystem
- total quality management

## REFERENCES

Allerin. *10 Most Popular ERP Systems Used in the Banking Industry*; 2017. https://www.allerin.com/blog/10-most-popular-erp-systems-used-in-the-banking-industry-2 (accessed Aug 28, 2017).

Achilov, N. Evaluation of Innovation Policy in Kazakhstan in the Period of Globalization Trends. *Bus. Eco. J.* **2016,** *7* (4). https://doi.org/10.4172/2151-6219.1000251.

Aldarbesti, H.; Saxena, J. P. Management Information System for Education. *J. Res. Method Edu.* **2014,** *4* (1), 36–44.

Almeida, T.; Teixeira, L.; Ferreira, C. In *Enterprise Resource Planning system in a Multinational Enterprise: Users' Attitude Post Implementation*, International Conference on Enterprise Information Systems, 2010; pp 264–273. https://link.springer.com/chapter/10.1007/978-3-642-16419-4_27 (accessed Aug 28, 2017).

Ambler, S. *Why do We Need a Continuous Improvement Program?* 2015. http://www.disciplinedagiledelivery.com/why-continuous-improvement/ (accessed Aug 22, 2017).

Bagad, V. *Management Information Systems*; 2008. https://books.google.com.bn/books?id=7fCgRL1-gGAC&pg=RA1-PA9&dq=benefits+of+ERP&hl=en&sa=X&ved=0ahUKE

wjS_4D85vnVAhVFi7wKHUQiCjQQ6AEIMjAC#v=onepage&q=benefits%20of%20 ERP&f=false (accessed Aug 28, 2017).

Barsh, J. *Innovation Management: A Conversation with Gary Hamel and Lowell Bryan*; 2007. https://socialknowledge.files.wordpress.com/2007/10/hamel-future-of-management. pdf (accessed Aug 28, 2017).

Bendoly, E.; Jacobs, F. *Strategic ERP Extension and Use*; Stanford Business Books: California, 2005.

Bhuiyan, N.; Baghel, A. An Overview of Continuous Improvement: from the Past to the Present. *Manag. Decis.* **2005,** *43* (5), 761–771. http://www.emeraldinsight.com/doi/pdfplu s/10.1108/00251740510597761 (accessed Aug 18, 2017).

Bisk Education. *Total Quality Management in Human Resources*; 2017. https://www. villanovau.com/resources/hr/implementing-tqm-in-human-resources/#.WZ6cgOkRXIU (accessed Aug 24, 2017).

Bosschaa, P.; Coetzeea, R.; Terblanchea, P.; Gazendam, A.; Isaac, S. Smart Factory: The Challenges of Open and Low-cost ICT in the Small Manufacturing Industry. *South Afr. J. Sci.* **2006,** *102*, 335–338.

Business Case Studies. *Continuous Improvement as a Business Strategy: A Corus Case Study*; 2017. http://businesscasestudies.co.uk/corus/continuous-improvement-as-a-business-strategy/the-benefits-of-continuous-improvement.html (accessed Aug 23, 2017)

Chapman, R.; Hyland, P. Complexity and Learning Behaviors in Product Innovation. *Technovation* **2004,** *24* (7), 553–561.

Colby, S.; Lambert, M.; McGee, J. Utilizing Collaborative Analysis of Student Learning in Educator Preparation Programs for Continuous Improvement. **2016,** *6* (4), https://doi. org/10.1177/2158244016673131.

Cooper, L. P. A Research Agenda to Reduce Risk in New Product Development Through Knowledge Management: A Practitioner Perspective. *J. Eng. Technol. Manag.* **2003,** *20* (1–2), 117–140.

DeSmet, A., McGurk, M.; Schwartz, E. *Getting More from Your Training Programs*; 2017. http://www.mckinsey.com/business-functions/organization/our-insights/getting-more-from-your-training-programs (accessed Aug 22, 2017).

Dezdar, S.; Sulaiman, A. Successful Enterprise Resource Planning Implementation: Taxonomy of Critical Factors. *Ind. Manag. Data Sys.* **2009,** *109* (8), 1037–1352.

Dezdar, S.; Sulaiman, A. The Influence of Organizational Factors on Successful ERP Implementation. *Manag. Decis.* **2009,** *49* (6), 911–926.

Dorel, D.; Martinovic, A. B. *The Role of Information Systems in Human Resource Management* 2011. https://mpra.ub.uni-muenchen.de/35286/ (accessed Aug 20, 2017).

Elbadawi, I.; McWilliams, D. L.; Tetteh, E. G. Enhancing Lean Manufacturing Learning Experience Through Hands-on Simulation. *Simul. Gaming* **2010,** *41* (4), 537–552.

Ellis, C.; Castle, K. Teacher Research as Continuous Process Improvement. *Quality Assur. Edu.* **2010,** *18* (4), 271–285.

Shafique, F.; Mahmood, K. Scanning the Information Infrastructure of Pakistan: A Step Towards the Development of a National Educational Information System. *Library Rev.* **2012,** *61* (7), 511–525.

Flett, A. Information Management Possible? Why is Information Management So Difficult? *Bus. Inform. Rev.* **2011,** *28* (2), 92–100.

Ganapavarapu, L. K.; Prathigadapa, S. Study on Total Quality Management for Competitive Advantage in International Business. *Arab. J. Bus. Manag. Rev.* **2015,** *5* (3). https://doi. org/10.4172/2223-5833.1000124.

Gonçalves, N. P.; Sapateiro, C. M. Aspects for Information Systems Implementation; Challenges and Impacts. A Higher Education Institution Experience. *Polytech. Stud. Rev.* **2008,** *4* (9), 1–17.

González-Sancho, C.; Vincent-Lancrin, S. *Transforming Education by Using a New Generation of Information Systems*; 2016. http://journals.sagepub.com.ezproxy.ubd.edu.bn/doi/full/10.1177/1478210316649287 (accessed Aug 26, 2017).

Gul, A.; Jafery, S. A. S.; Rafiq, J.; Naeem, H. Improving Employees Performance through Total Quality Management. *Int. J. Econ. Manag. Sci.* **2012,** *1* (8), 19–24.

Hamid, M.; Maheen, S.; Cheem, A.; Yaseen, R. Impact of Human Resource Management on Organizational Performance. *J. Account Mark.* **2017,** *6* (1). https://doi.org/10.4172/2168-9601.1000213.

Hammad, M. *Why do Most Continuous Improvement Programs Fail?* 2012 https://pixelballads.wordpress.com/2012/02/01/why-do-most-continuous-improvement-programs-fail/ (accessed Aug 24, 2017).

Heinrichs, W. L.; Youngblood, P., Harter P. M. Simulation for Team Training and Assessment: Case Studies of Online Training with Virtual Worlds. *World J. Surg.* **2008,** *32* (2), 161–170.

Hyatt. *Corporate Responsibility Scorecard*; **2016.** https://thrive.hyatt.com/content/dam/Minisites/hyattthrive/reports/CR-Scorecard-2015.pdf (accessed Aug 27, 2017).

Nwakanma, I. C.; Ubani, E. C.; Asiegbu, B. C.; Ngene, S. C. Total Quality Management: A Critical Component for Effective Delivery of Manufacturing Project. *Telecommun. Sys. Manag.* **2014,** *3* (2). https://doi.org/10.4172/2167-0919.1000116.

Jackson, L. A. Enterprise Resource Planning Systems: Revolutionizing Lodging Human Resources Management. *Worldw. Hospitality Tour. Themes* **2010,** *2* (1), 20–29.

Jalu, G. Achievement of Quality, Productivity for Market through Kaizen Implementation in Ethiopia. *Arab. J. Bus. Manag. Rev.* **2015,** *6,* (1). https://doi.org/10.4172/2223-5833.1000170.

Jasti, N. V. K.; Kodali, R. Development of a Framework for Lean Production System: An Integrative Approach. *J. Eng. Manuf.* **2016,** *230* (1), 136–156.

Jehangiri, R. Identification of Critical Success Factors for Total Quality Management Implementation in Organizations. *Int. J. Econ. Manag. Sci.* **2017,** *6* (3). https://doi.org/10.4172/2162-6359.1000420.

Jorgensen, F., Laugen, B. T., Boer, H. Human Resource Management for Continuous Improvement. *Creativ. Innov. Manag.* **2007,** *16* (4), 363–375.

Cardona Mora, J. N. Continuous Improvement Strategy. *Eur. Sci. J.* **2014,** *10* (34), 117–126.

Kalyani, M.; Sahoo, M. P. Human Resource Strategy: A Tool of Managing Change for Organizational Excellence. *Int. J. Bus. Manag.* **2011,** *6* (8). http://doi.org/10.5539/ijbm.v6n8p280.

Kaplan, G.; Lopez, M.; McGinnis, J. *Transforming Health Care Scheduling and Access: Getting to Now*; 2015. https://www.ncbi.nlm.nih.gov/books/NBK316143/ (accessed Aug 28, 2017).

Karikari, A. F.; Boateng, P. A.; Ocansey, E. O. N. D. The Role of Human Resource Information System in the Process of Manpower Activities. *Am. J. Ind. Bus. Manag.* **2015,** *5*, 424–431.

Knezek, G.; Christensen, R. *The importance of Information Technology Attitudes and Competencies in Primary and Secondary Education*; 2008. https://link.springer.com/chapter/10.1007/978-0-387-73315-9_19 (accessed Aug 27, 2017).

Konrad, M. A. *Engaging Employees Through High-involvement Work Practices;* **2006.** https://iveybusinessjournal.com/publication/engaging-employees-through-high-involvement-work-practices/ (accessed Aug 27, 2017).

Kuo, I. *When Xerox Gamifies Employee Training, Everybody Wins, Gamification.co, Enterprise;* 2015. http://www.gamification.co/2015/06/18/when-xerox-gamifies-employee-training-everybody-wins/ (accessed Aug 28, 2017).

Langer, J. *3 Big Benefits of Continuous Improvement;* 2014. https://jessicalanger.com/blog/2016/1/14/3-big-benefits-of-continuous-improvement (accessed Aug 24, 2017).

Layfield, K. *Analysis of Lean Practices as a Continuous Improvement Program in the Manufacturing Industry;* 2013. http://wvuscholar.wvu.edu/reports/Layfield_Ky.pdf (accessed Aug 23, 2017).

Leon, A. *Enterprise Resource Planning;* McGraw-Hill Education: New Delhi, India, 2014. https://books.google.com.bn/books?id=s5YiBAAAQBAJ&printsec=frontcover&source=gbs_ge_summary_r&cad=0 (accessed Aug 28, 2017).

McLean, R.; Antony, J. Why Continuous Improvement Initiatives Fail in Manufacturing Environments? A Systematic Review of the Evidence. *Int. J. Prod. Perform. Manag.* **2014,** *63* (3), 370–376.

Tseng, M.-L.; Lin, R.-J.; Chen, H.-P.; Evaluating the Effectiveness of e-Learning System in Uncertainty. *Ind. Manag. Data Sys.* **2011,** *111* (6), 869–889.

Mitta, M. Evaluating Perceptions on Effectiveness of e-Learning Programs in Indian Banks: Identifying Areas for Improvement. *Develop. Learn. Organ.* **2008,** *22* (2), 12–14.

Mjema, E. A. M.; Victor, M. A. M.; Mwinuka, M. S. M. Analysis of Roles of IT on Quality Management. *TQM Magazine,* 2005; Vol. 17 (4), pp 364–375. http://dx.doi.org/10.1108/09544780510603206.

Mulang, A. International Human Resource Management. *J. Socialomics* **2017,** *6* (3). https://www.omicsonline.org/open-access/international-human-resource-management-2167-0358-1000203.php?aid=92179 (accessed Aug 20, 2017).

Munson, C. L. Six Sigma: The Time to Act is Now. *Ind. Eng. Manag.* **2012,** *1* (1). https://www.omicsonline.org/open-access/six-sigma-the-time-to-act-is-now-2169-0316.1000e105.php?aid=6118 (accessed Aug 23, 2017).

Nurmilaakso, J. M. ICT Solutions and Labor Productivity: Evidence from Firm-level Data. *Electron. Comm. Res.* **2009,** *9* (3), 173–18.

Open Learning World.*Com;* 2011. *TQM of Information Systems.* http://www.openlearningworld.com/books/Fundamentals%20of%20MIS/TECHNOLOGY%20OF%20INFORMATION%20SYSTEMS/TQM%20of%20Information%20Systems.html (accessed Aug 24, 2017).

Paradice, D. *Emerging Systems Approaches in Information Technologies: Concepts, Theories, and Applications: Concepts, Theories, and Applications;* Information Science Reference: Hershey, PA, 2009. https://books.google.com.bn/books?id=9IYUQmGCs4EC&printsec=frontcover&source=gbs_ge_summary_r&cad=0#v=onepage&q&f=false (accessed Aug 28, 2017).

Parthasarthy, S. *Enterprise Resource Planning: A Managerial & Technical Perspective;* New Age International Pvt. Ltd. Publishers, 2007.

Patyal, V. S.; Maddulety, K. Interrelationship Between Total Quality Management and Six Sigma: A Review. *Global Bus. Rev.* **2015,** *16* (6), 1025–1060.

Pineda, H. J. Q.; Madrigal, J. Sustaining Continuous Improvement: A Longitudinal and Regional Study. *Int. J. Eng. Bus. Manag.* **2013,** *5* (43), 1–13.

Qutaishat, F.; Khattab, S.; Abu Zaid, M.; Al-Manasra, E. The Effect of ERP Successful Implementation on Employees' Productivity, Service Quality and Innovation: An Empirical

Study in Telecommunication Sector in Jordan. *Int. J. Bus. Manag.* **2012,** *7* (19), http://doi. org/10.5539/ijbm.v7n19p45.

Qasim, S.; Zafar, A. Information System Strategy for Total Quality Management (TQM) in Aviation Industry. *Int. J. Comput. Appl.* **2016,** *135* (3), 37–42.

Seth, S. *LG as a Case Study of a Successful Enterprise Resource Planning System. Investo-pedia* **2014.** http://www.investopedia.com/articles/investing/111214/lg-case-study-successful-enterprise-resource-planning-system.asp (accessed Aug 28, 2017).

Siam, A. Z.; Alkhateeb, K.; Al-Waqqad, S. The Role of Information Systems in Implementing Total Quality Management. *Am. J. Appl. Sci.* **2012,** *9* (5), 666–672.

Siti Norainna, H. Impact of ICT in Brunei Darussalam System of Education. *J. Akad. Penga-jian Brunei* **2016,** 66–75.

Sodagar, H.; Afrasyabi, R.; Asgarzade, H. E-learning Role in Promoting Human Resource Personnel's Training in the Environment Protection Department of North Khorasan. *Int. J. Acad. Res. Bus. Soc. Sci.* **2013,** *3* (1), 550–564.

Sofijanova, E. Employee Involvement and Organizational Performance: Evidence from the Manufacturing Sector in Republic of Macedonia. *Trakia J. Sci.* **2013,** *11* (1), 31–36.

Soltani, E.; Azadegan, A.; Liao, Y.; Phillips, P. Quality Performance in a Global Supply Chain: Finding Out the Weak Link. *Int. J. Prod. Res.* **2011,** *49* (1), 269–293.

Stratman, J. K.; Roth, A. V. Enterprise Resource Planning (ERP) Competence Constructs: Two-stage Multi-item Scale Development and Validation. *Decis. Sci.* **2002,** *33* (4), 601–628.

Sumner, M. *Enterprise Resource Planning*; Pearson Education Inc.: New Jersey, 2005.

Taher Qutaishat, F.; Ahmed Khattab, S.; Abu Zaid, M. K. S.; Al-Manasra, E. A. The Effect of ERP Successful Implementation on Employees' Productivity, Service Quality and Innova-tion: An Empirical Study in Telecommunication Sector in Jordan. *Int. J. Bus. Manag.* **2012,** *7* (19), https://doi.org/ 10.5539/ijbm.v7n19p45.

Tomal, D.; Schilling, C. *Managing Human Resources and Collective Bargaining*; 2013; pp 34. https://books.google.com.bn/books?id=WQmsIo1QD7oC&printsec=frontcover#v=on epage&q&f=false (accessed Aug 28, 2017).

Van Aken, E. M.; Farris, J. A.; Glover, W. J.; Letens, G. A Framework for Designing, Managing, and Improving Kaizen Event Programs. *Int. J. Prod. Perform. Manag.* **2010,** *59* (7), 641–667.

Vella, R.; Chattopadhyay, S.; Mo, J. P. T. Six Sigma Driven Enterprise Model Transformation. *Int. J. Eng. Bus. Manag.* **2009,** *1* (1), 1–8.

Vulic, M.; Labus, A.; Despotovic-Zrakic, M. *Implementation of CRM Concept in e-Educa-tion,* **2014.** https://link.springer.com/chapter/10.1057/9781137402226_18 (accessed Aug 23, 2017).

Wong, W.-P.; Veneziano, V.; Mahmud, I. Usability of Enterprise Resource Planning Software Systems: an Evaluative Analysis of the Use of SAP in the Textile Industry in Bangladesh. *Inform. Develop.* **2015,** *32* (4), 1027–1041.

Watanabe, R. M. Getting ready for Kaizen: Organizational and Knowledge Management Enablers. *VINE* **2011,** *41* (4), 428–448.

# INDEX